‘Doing ethnography in the construction industry is not easy – especially in project-based settings. It takes practice, time and patience and often meets with resistance from the organisations and people being studied. This important book brings together the experiences, insights and lessons from a wide range of researchers who have employed ethnography in a built environment context. It will no doubt be seen as an essential authoritative resource for those contemplating using this approach in this and many other related fields.’

**Distinguished Professor Martin Loosemore**
*School of Built Environment, University of Technology Sydney, Australia*

# EMBRACING ETHNOGRAPHY

This book calls for those interested in robust construction research to embrace ethnography – in all its forms, including rapid ethnographies, ethnographic-action research, autoethnography, as well as longer-term ethnographies.

The diversification of ethnographic approaches, as well as ethnographers, will lead to rich insights that can advance the industry theoretically and practically. We share experiences, key considerations and recommendations from leading construction ethnographic researchers from around the world to provide discussion, reflection and understanding into doing ethnography in the construction industry.

This book is aimed at academics, students, consultants, editors, reviewers, policymakers, funders and others interested in robust research in the construction industry and built environment but will also be useful for those undertaking research within organisations in other industries.

**David Oswald** is an Associate Professor in the School of Property, Construction and Project Management at RMIT University, Australia. He has undertaken award-winning ethnographic research in the fields of construction management and the built environment, including the '2020 Best Paper Award' in a leading journal in the field, the *Journal of Construction, Engineering and Management*.

**Léon olde Scholtenhuis** is an Associate Professor in the Department of Construction Management and Engineering at the University of Twente, the Netherlands. Léon's lab conducts practice-based research, focusing on the intricacies of planning and construction operations and how technological innovations enter this context. Léon publishes ethnographic action research work in technology and construction management outlets such as the *Journal of Construction Engineering and Management* and the ASCE Monograph Series.

# EMBRACING ETHNOGRAPHY

## Doing Contextualised Construction Research

*Edited by David Oswald and Léon olde Scholtenhuis*

Routledge
Taylor & Francis Group
LONDON AND NEW YORK

Designed cover image: © Daniël Stronks

First published 2024
by Routledge
4 Park Square, Milton Park, Abingdon, Oxon OX14 4RN

and by Routledge
605 Third Avenue, New York, NY 10158

*Routledge is an imprint of the Taylor & Francis Group, an informa business*



*British Library Cataloguing-in-Publication Data*
A catalogue record for this book is available from the British Library

*Library of Congress Cataloging-in-Publication Data*
Names: Oswald, David, editor. | Olde Scholtenhuis, Léon, editor.
Title: Embracing ethnography : doing contextualised construction research /
edited by David Oswald and Léon olde Scholtenhuis.
Description: Abingdon, Oxon ; New York, NY : Routledge, 2024. | Includes
bibliographical references and index.
Identifiers: LCCN 2024000225 (print) | LCCN 2024000226 (ebook) | ISBN
9781032459929 (hardback) | ISBN 9781032455716 (paperback) | ISBN
9781003379584 (ebook)
Classification: LCC HD9715.A2 E4195 2024 (print) | LCC HD9715.A2 (ebook)
| DDC 624.068--dc23/eng/20240327
LC record available at https://lccn.loc.gov/2024000225
LC ebook record available at https://lccn.loc.gov/2024000226

ISBN: 978-1-032-45992-9 (hbk)
ISBN: 978-1-032-45571-6 (pbk)
ISBN: 978-1-003-37958-4 (ebk)

DOI: 10.1201/9781003379584

Typeset in Times New Roman
by KnowledgeWorks Global Ltd.

# CONTENTS

# CONTRIBUTORS

**Richard Brett** is a Senior Lecturer in Construction Management at Anglia Ruskin University, Chelmsford, UK. Ten years spent working on construction sites informs his research interest in construction workers and especially the everydayness of building activity.

**Leon Crascall** is a Senior Lecturer and Course Leader for Architecture at the University of Suffolk, Ipswich, UK. His career has been an unorthodox journey through the construction industry and architectural education, in which he continues to antagonise the relationship between educational theory and practice relating to built environment education.

**Carrie Sturts Dossick** is a P.D. Koon Professor of Construction Management in the College of Built Environments at the University of Washington, Seattle, USA. She has studied emerging technologies in the built environment for over two decades and uses qualitative social science methods to investigate how people work through innovation with information technologies.

**Natalie Galea** is a Senior Research Fellow in the Faculty of Architecture, Building and Planning at The University of Melbourne, Australia. She is an Adjunct Senior Lecturer at the Australian Human Rights Institute, UNSW Sydney. Her research focuses on human rights and gender justice in the construction sector and in elite sport.

**Henning Grosse** is a Professor for Civil Engineering at IU International University Campus Hamburg, Germany, and Postgraduate Supervisor in the Gloucestershire Business School, at the University of Gloucester, UK. He has 30 years of construction experience, obtained a PhD for an autoethnography in construction management, runs his own business, and teaches and supervises students in project management and qualitative research.

**Timo Hartmann** is a Professor at the Technische Universität Berlin, Germany. He proposed ethnographic action research as a method for developing decision support systems in engineering contexts that account for specifics of local work cultures. Timo has also explored engineering work using ethnographic interview and observation techniques in settings as wide as early stage conceptual engineering design, use of simulation tools in detailed engineering efforts or support of construction site coordination. His work on ethnography is published in journals, such as *Advanced Engineering Informatics*, *Automation in Construction* or the *Journal of Construction Engineering and Management*.

**Jan Hayes** is a Professor at the School of Property, Construction and Project Management at RMIT University, Melbourne, Australia. She is a sociologist with 35 years of experience in safety and risk management. Her research interests also include professionalism, expertise and decision-making.

**Andrea Yunyan Jia** is a Research Associate at Hong Kong Polytechnic University. Before joining PolyU, she worked as a Senior Lecturer at the Faculty of Architecture, Building and Planning at University of Melbourne, and held academic positions at National University of Singapore, Curtin University Australia, University of Hong Kong and Sichuan University in China, where she did ethnographic studies in many different projects and organisations, including a construction firm in the Netherlands.

**Helen Lingard** is an RMIT Distinguished Professor at the School of Property, Construction and Project Management, RMIT University, Australia, with more than 25 years of experience in undertaking applied research in work health, safety and well-being in the construction industry. Helen's research is typically undertaken in partnership with private and public sector organisations, industry associations and unions. Her research work using ethnographic approaches has won multiple awards.

**Emmanuel Manu** is an Associate Professor in Quantity Surveying and Project Management at the School of Architecture, Design, and the Built Environment, Nottingham Trent University, UK. Emmanuel was a Principal Investigator on the joint CITB and industry-funded Innovation Driven Procurement project. He is the lead author of the book *Secondary Research Methods in the Built Environment.*

**Evelijn Martinius** works as a doctoral ethnographer at the Department of Organization Studies at the Vrije Universiteit. Her research examines the management of underground public infrastructures in urban regions in the Netherlands. She writes about imagination, routines and grass-roots pioneering solutions to overcome issues with bounded manageability in societal transitions.

**Sarah Maslen** is a Professor of Sociology at University of Canberra, Australia. She is working principally in science and technology studies (STS) with a program of work spanning engineering, medicine, and digital life. She has previously published extensively on professional practice, addressing decision-making processes, skill acquisition, and professional ethics.

**Léon olde Scholtenhuis** is an Associate Professor in the Department of Civil Engineering and Management at the University of Twente, Enschede, the Netherlands. He leads the ReDUCE programme, a long-term university-industry collaboration that develops technology to reduce damages during excavations in the urban underground. His lab combines ethnography with implementation of digital technologies in the field.

**Joan Evelyn Ongodia** is a Research Assistant and Doctoral Candidate at the Technische Universität Berlin, Berlin, Germany. Joan has benefited from construction ethnography throughout her decade-long experience in industry involving large infrastructure civil and geotechnical engineering projects.

**Laura Osburn** is a Senior Research Scientist in the Department of Construction Management at the University of Washington, Seattle, USA. She researches how teams in the building industry work, organise, communicate and integrate technology. Laura used ethnographic methods to study collaborative work within the contexts of smart building cybersecurity and sustainable design.

**David Oswald** is an Associate Professor and Deputy Program Manager in the School of Property, Construction and Project Management at RMIT University, Australia. His award-winning ethnographic research has focused on health, safety and well-being within the UK and Australia.

**Sarah Pink** is a Professor and Director of the Emerging Technologies Research Lab at Monash University. Sarah is a design and futures anthropologist and documentary filmmaker, known for her methodological innovation, interdisciplinary and engaged practice and her current focus on technology, environmental, human and other species futures. Sarah has collaborated to develop ethnographic research in the construction industry in the United Kingdom and Australia for over 15 years, and her publications in this field include the co-edited (with Dylan Tutt and Andrew Dainty) Ethnographic Research in the Construction Industry (2013). Sarah's most recent books include her sole-authored *Emerging Technologies/Life at the Edge of the Future* (2023) and co-authored *Design Ethnography* (2022).

**Kevin Porter**, Independent Consultant, Queensland, Australia, has over 20 years of experience as a social anthropologist helping clients make sense of the cultural worlds in which they operate. He has conducted ethnographic research on three continents and across several diverse industries, including the military, the health industry, non-governmental organisations, manufacturing and construction industries.

**Ani Raiden** is an Associate Professor in Responsible Management at Nottingham Business School, Nottingham Trent University, UK. Ani was a Co-Investigator on the joint CITB and industry-funded Innovation Driven Procurement project. She is the lead author of Social Value in Construction and Social Value in Practice.

**Fred Sherratt** is the Associate Director of Research for the Construction Safety Research Alliance at the University of Colorado, Boulder, USA. Her award-winning work brings new and enlightening methodologies to the construction safety space, grounded in the social sciences, including the use of ethnographic research in the field.

**Simon Smith** is a Reader in Construction and Project Management at the School of Engineering, University of Edinburgh, UK. He spent his early career in industry and has nearly 30 years of experience teaching and researching construction issues. Simon supervised ethnographic studies including gender issues, safety and language management in construction.

**Madison Snider** is a Researcher at the Siegel Family Endowment, New York, USA. She is a qualitative social scientist, and her research takes place at the intersection between science and technology and critical infrastructure studies. Madison used ethnographic and participatory methods to study multi-sector collaboration in sustainable building, technology adoption and technology policy implementation.

**Sarah Swider** is an Associate Professor at the College of Liberal Arts and Sciences, Wayne State University. Her ethnographic research when living and working with migrant workers led to the book *Building China: Informal Work and the New Precariat* This was awarded the 2016 Book Award, Sociology of Development Section, American Sociological Association.

**Marc van den Berg** is an Assistant Professor at the Department of Civil Engineering and Management at the University of Twente, Enschede, the Netherlands. He develops circular urban mining futures through rethinking the existing built environment as an urban mine and uncovering the industry practices that make the future circular.

**Leentje Volker** is a full Professor in the Department of Civil Engineering and Management at the University of Twente, Enschede, the Netherlands. She is an expert in public-private collaborations, innovation and change in the construction industry, contributing to the societal challenges of today by performing qualitative empirical research and educating civil engineers in these areas.

**Hans Voordijk** is an Associate Professor in the Department of Civil Engineering and Management at the University of Twente, Enschede, the Netherlands. His research interests include philosophy of technology in civil engineering and supply chain management and ICT in construction. He focuses on post-phenomenological and ethnographic approaches in digital technologies, circular construction and design science research.

# FOREWORD

In an age where research productivity has become a valued measure, it is encouraging to read a book that calls for researchers to slow down and embrace ethnography in order to dwell in the throes of everyday life. As the editors rightly point out, ethnographic research enables us to search deeper, ask unexplored or under-explored questions and reveal the 'hidden gems' that allow us to better understand what really goes on in the practices of constructing the built environment. Over 20 years ago, as I was embarking on my own journey as a PhD student, I was indeed drawn to the power of ethnographic research. An early inspiration came from reading LeMasters' (1975) tavern approach and his attempt to grasp the culture of building workers, their attitudes to family and work life and their perspectives and suspicions of the 'other' (e.g. the managerial classes, women, ethnic minorities). LeMasters spent five years immersed in the life of a tavern where building workers socialised and, apart from writing up his observations, he also learnt to play pool and eventually became part of the winning team in the tavern's champion's league.

Enriched by LeMasters' (1975) study, I ventured into a much more modest ethnographic case study of the construction of a car park building at an airport in Scotland. Though I have been no stranger to building sites – my father was a general contractor and I had spent vacation periods working on sites – the experience of spending two days a week for an entire year on a Scottish site created feelings of both the familiar and the unfamiliar (Chan & Kaka, 2007). On the one hand, the pouring of concrete in-situ to create the building's structure was largely a similar technical process, both in equatorial Singapore where I grew up and in Scotland where the cold and dreary conditions presented a striking contrast. On the other hand, I remember being baffled by the construction workers having conversations about the placement of 'Mars Bars' before the ready-mixed concrete arrived on site. What I thought were those chocolaty snacks that I consumed as a teenager turned out to be concrete spacers the size of 'Mars Bars' to ensure an adequate gap between the reinforcement steel and the edges of the concrete slab.

It is this mixture of familiarity and strangeness that lies at the heart of an ethnographic approach. In *Embracing Ethnography*, the editors have assembled an eclectic range of reflective chapters in which authors elaborated on the complexities of how they navigated through the etic (outsider) and emic (insider) challenges of doing ethnography. These reflections covered issues of gender (i.e. women working in what is still a male-dominated sector), race and ethnicity (i.e. western and global north researchers engaging in eastern and global south contexts), and class (i.e. academic researchers coming into contact with industry practitioners and managers confronting workers). The golden thread that shines through every chapter is how important it is for ethnographic researchers to take the time to familiarise themselves with and to dwell in the context of their study, to suspend judgements of predefined categories (e.g. western, eastern, gender) and to relate to the 'other' in more critical ways. As Hans Voordijk in Chapter 7 stresses, it is in the richness of the multiplicity of meanings in everyday life that drives ethnographers to do what they do, that is, to provide rich, critical descriptions of being in the world.

In *Embracing Ethnography*, the editors acknowledge how ethnographic research has its origins in anthropology, which early in the nineteenth century aimed to study and understand unfamiliar tribes in a period of social and cultural change. In the post-WWII period, the ethnographic approach was gradually used to study industrialisation in and its effects on society. The seminal work of Trist and Bamforth (1951) offers one such example. By immersing themselves continuously in the coal-mining sector for two years, they brought to the fore the power of sociotechnical systems. Trist and Bamforth observed how the introduction of the new, more efficient longwall technique served to alter the ways mining workers related to one another. These new interactions were often associated with unintended and not-always-positive consequences for their collective sense of camaraderie and safety.

It is therefore reassuring to see how *Embracing Ethnography* covers not just the agendas of the managerial class, but also the well-being and safety of workers, as well as burning issues such as collaboration and innovation for the sustainability of people and planet. In some sense, the editors have curated a series of chapters that go beyond the usual suspects of the client, architect and contractor and that transcend typical agendas of time and cost performance (see Chan, 2023). As Jan Hayes and Sarah Maslen assert in Chapter 6, ethnographic research that seeks a deeper understanding of individual frontline workers, rather than to simply blame those at the coalface for the ills of industry, provides a much firmer basis for developing meaningful improvements in the sector.

Creating and sustaining the built environment is one of the most complex endeavours we undertake, with significant repercussions for conditioning the ways we live, work and recreate. *Embracing Ethnography* provides us with such a glimpse of the complexities that go beyond the study of an 'unfamiliar tribe'. In this compilation, the editors have taken care to feature studies that consider the multiple spaces where work happens, from collaboration across different organisations (Chapter 13), to observations in a multi-site context (Chapter 12), to the use of ethnography in the digital environment (Chapters 15 and 16), as well as the role of the researcher in autoethnographic expositions (Chapters 17 and 21). In assembling this varied collection of ethnographic studies in the built environment, the editors and the authors stress the need to learn by observing and to learn through intervening, the so-called ethnoventionist approach (van Marrewijk et al., 2010).

In many respects, the crafting of the built environment has a long pedigree of learning-by-watching and learning-by-doing through the master-apprentice relationship. As Richard Brett (Chapter 10) acknowledged, his attempt to undertake a short-term, more rapid form of ethnography was helped in no small part by his already long and varied apprenticeship in the building trades. Nowadays, of course, the apprenticeship route to knowledge production has given way to ever more professionalisation of (construction) managers through mass higher education. Renewed interest in ethnographic methods thus goes some way to encourage students of the built environment to connect in deeper and more meaningful ways with the contexts, objects and subjects of study. The growing movement of more engaged forms of scholarship (Van de Ven, 2007; Voordijk & Adriaanse, 2016; Van Marrewijk & Dessing, 2019) undoubtedly opens up our sensibility and sensitivity to the multiple truths, realities and subjectivities in organisational life, in order to produce knowledge on how things work (Watson, 2011).

*Embracing Ethnography* is a challenging undertaking, because there is often limited time in an ever-compressed world. Furthermore, it is impossible to be everywhere at the same time, through digital technologies can be helpful in enlarging our ability to sense and observe the environment. *Embracing Ethnography* is also demanding in terms of its writing. Ethnographic research requires space to provide rich descriptions and reflections. Books have thus been the traditional medium for conveying the twists and turns of ethnographic research. In the unfettered drive for research productivity and citational impacts, authoring books seems to run counter to modern academic life, at least in the field of construction management research. As part of the editorial team of Construction Management and Economics we have entertained the idea of permitting longer, 40,000-word articles, so that space can be created for ethnographic research to be published in the journal. Unfortunately, despite our willingness to be creative with our editorial requirements, no one took up this challenge. Therefore, I hope that *Embracing Ethnography* will cultivate new ways of doing and writing ethnographic research in construction management.

Paul W. Chan
Delft University of Technology, the Netherlands

## References

Chan, P., & Kaka, A. (2007). The impacts of workforce integration on productivity. In A. R. J. Dainty, S. Green, & B. Bagilhole (Eds.), *People and Culture in Construction: A Reader* (pp. 240–257). London: Spons.

Chan, P. W. (2023). Beyond the boundary and scale of the construction project. In A. Dainty, & R. Leiringer (Eds.), *Research Agenda for Construction Management Research* (pp. 117–139). Cheltanham: Edward Elgar.

LeMasters, E. E. (1975) *Blue-Collar Aristocrats: Life-Styles at a Working-Class Tavern*. Madison: The University of Wisconsin Press.

Trist, E. L., & Bamforth, K. W. (1951). Some social and psychological consequences of the Longwall method of coal-getting: An examination of the psychological situation and defences of a work group in relation to the social structure and technological content of the work system. *Human Relations*, 4(1), 3–38.

Van de Ven, A. H. (2007). *Engaged Scholarship: A Guide for Organizational and Social Research*. Oxford: Oxford University Press.

Van Marrewijk, A., & Dessing, N. (2019). Negotiating reciprocal relationships: Practices of engaged scholarship in project studies. *International Journal of Project Management*, 37(7), 884–895.

van Marrewijk, A., Veenswijk, M., & Clegg, S. (2010). Organizing reflexivity in designed change: The ethnoventionist approach. *Journal of Organizational Change Management*, 23(3), 212–229.

Voordijk, H., & Adriaanse, A. (2016). Engaged scholarship in construction management research: The adoption of information and communications technology in construction projects. *Construction Management and Economics*, 34(7–8), 536–551.

Watson, T. J. (2011). Ethnography, reality, and truth: The vital need for studies of 'how things work' in organizations and management. *Journal of Management Studies*, 48(1), 202–217.

**SECTION I**

# Introducing the construction context and culture

# 1

# INTRODUCTION

*David Oswald and Léon olde Scholtenhuis*

*RMIT University, Melbourne, Australia*

*University of Twente, Enschede, The Netherlands*

Ethnography dates to the 1800s in early anthropology, but it was not until the late 1900s when the first ethnographic studies were being performed within the construction industry. The excerpt below is from one of these initial ethnographers, Sykes (1969), who undertook participant observation and interviews during the construction of a hydroelectric plant in Scotland.

> Minor accidents occurred daily and were ignored, but a serious accident set off a contest between the injured individual and the employer. If the employer won, the man got very little; if the man won, by exaggerating his injuries and the employer's negligence, he secured excessive compensation. The men would describe with glee how particular individuals had "done the firm". For example, they told how a man recently killed had been drunk at the time, but they kept this information from the employer who had compensated the widow. The men told various stories about men who had simulated crippling injuries and in consequence received vast sums in compensation – sums far greater than any Scottish court would have paid for such injuries. When the writer was injured he was warmly congratulated on his luck, and advised on how to "play it up". Men who had never spoken to him before came up and said: "you've got a good one there"; "don't let them beat you down, get a good lawyer"; "watch them, don't take a thing until they settle"; "be careful, don't do a stroke of work until it's all settled". Discussion as to what the accident would be worth and what was the best way to handle it was general for some time…
>
> *(Sykes, 1969, p. 24)*

Sykes' writing demonstrates many hallmarks of ethnographic research. He was *close to the action*, was able to *observe* and spend *extended periods in the field*. He had *built rapport* with the workers. They trusted him with sensitive information that the employer did not know about, they had accepted him as one of them and provided him advice when he was injured. Even in this short *fieldnote*, he has provided many insights on topics that are still relevant within the industry today, including: the construction site being a dangerous place (see Tutt & Pink, 2019), alcohol misuse (see Flannery et al., 2021), the lack of reporting of (minor) accidents (see de Silva et al., 2018), the 'us' and 'them' between workers and the management (or the employer) (see Andersen et al., 2015), the exaggeration of accidents for fraudulent claims (see Oswald et al., 2018) and the researcher's own safety during fieldwork (see Morgan & Pink, 2018).

DOI: 10.1201/9781003379584-2

Sykes's ethnographic work was able to provide deep insight into group behaviour patterns when accidents occurred on the construction project.

For construction engineering and management researchers from positivist backgrounds, the above fieldnote may not initially appear to them as 'research'. Instead, they may be more familiar with common engineering methods, such as laboratory tests with technology or controlled testing of material, and other popular approaches such as questionnaire surveys that are often used to consult professionals about their perceived factors influencing construction performance. Yet, ethnography is a robust research methodology that has been used extensively in anthropology and sociology. It should be noted that there are, of course, other qualitative methods used within construction that researchers can consider, including case studies (Piekkari & Welch, 2018) and interviewing techniques, but the focus of this book is ethnography.

Ethnography is 'the study of social interactions, behaviors, and perceptions that occur within groups, teams, organisations, and communities' (Reeves et al., 2008, p. 337). It has become more common within construction management research within the last 15 years but is still a rare approach when compared to the traditional quantitative methods, rooted in positivism, that dominate the field.

## The slow growth of ethnography in construction

The shift towards greater use of ethnography (from the sporadic studies pre-millennium, such as the work of Sykes) was indirectly initiated following the great construction management debate of the mid-late 1990s (see Seymour & Rooke, 1995; Seymour et al., 1997) and then reinforced a decade later after the more direct calls for more construction ethnographies (see Laryea & Hughes, 2008; Bresnen, 2009; Styhre, 2009; Phelps & Horman, 2010; Pink et al., 2010). Since these direct calls, there has been a clear increase of ethnographic studies in the construction industry (Oswald & Dainty, 2020). While still a minor pursuit within construction, this proliferation was likely contributed to by the work of Pink et al. (2013), who published 'Ethnographic research in the construction industry'. The purpose of their book was to show how the research approach could 'generate fresh insights' and to 'provide an indication of the trajectories along which future work could travel' (p. 3).

Our book builds upon this work, by providing insight on how to undertake robust ethnographic research within the dynamic construction environment by presenting ethnographic experiences of leading ethnographic researchers in the field. Ethnographic research approaches have many benefits including being able to offer deep insights into construction practice. Yet, there has been little academic work on the process of doing ethnographic research within this setting. This book provides in-depth coverage on doing different forms of ethnographic approaches within the construction industry. It should be of great interest for anyone undertaking ethnographic research within the construction industry and built environment. This could be construction management scholars, as well as researchers from other fields and disciplines, such as organisational studies. As the field of construction management moves towards closer understanding of local phenomena, site dynamics and development of conceptual insights from rich thick descriptions and narratives, the community benefits from demonstrated evidence and a wide variety of cases in which the methodology is applied. With ethnographic research approaches gathering popularity within the field (Oswald & Dainty, 2020), there is a need for a book that focuses on doing ethnographic research.

The recent increase in popularity has also led to a diversification of ethnographic work. The 'classic' ethnographies, where the researcher spends extended period of time (often years) immersed in a setting, observing, participating and writing fieldnotes, are no longer the main approach to doing ethnography within the construction industry (Oswald & Dainty, 2020). Instead, various forms now exist including short-term rapid, online, digital, multi-sited, inter-organisational, autoethnography, ethnographic interviews and ethnographic-action research. We call to embrace all the variety of existing and emerging forms of ethnography.

## Embracing ethnography

This book demonstrates an overview of the possibilities that ethnography may bring to the researcher, practitioner, publisher, manager and policymaker. Altogether, the chapters provide insight through discussions and reflections of ethnographic researchers who share their experiences within the construction industry. The following paragraphs provide a brief overview of the sections of the books, as well as highlighting some central planks of ethnographic research.

### *Section I*

Within Section I, we position ethnography within *the context of the construction industry*. We discuss culture, which is central to ethnography given the anthropological roots. In Chapter 2: The culture of construction in Western society, Kevin Porter reflects on doing construction ethnographies in a Western society, and in Chapter 3: The culture of construction in Eastern society, Dr. Sarah Swider reflects on doing construction ethnographies in an Eastern context. In Chapter 4: Negotiating gender in construction ethnography, we turn the focus to the male-dominated nature of construction, with Dr. Natalie Galea discussing gender within the industry. Across the 22 chapters in this book, the majority of contributors are women. This is interesting, because in Pink et al.'s book (2013), they note that construction is 'one of the most gender-segregated sectors', that 'women who have done construction ethnography are few and far between' and 'it is therefore not surprising that the contributors to this volume are also predominately men' (p. 5).

Gender variation is particularly important because women undertaking construction ethnographies can provide research that is 'very revealing of the working cultures of construction sites for ethnographer and worker' (Pink, 2013, p. 5). The increase in female contributors (and other genders), in a male-dominated sector, will help to provide new avenues to knowledge, since the positioning of the ethnographic researcher within a natural setting leads to different data, as participants may 'open up' in various ways to different identities. This shows that positioning of the researcher is important with respect to the data that is collected and is thus one reason why being self-reflexive is an essential activity for the ethnographer.

### *Section II*

In Chapter 5: Ethnographic methods in construction, Dr. David Oswald provides a historical overview of ethnography in construction. This introduces us to Section II of this book, which focuses on *methodological considerations for ethnographic work*. Within this section, emphasis is placed on key considerations relevant to ethnographic work, such as research philosophy, research contributions, ethics and working with industry. It should be noted that more general methodology chapters (such as on research questions, research proposals or reviewing literature) that involve research design (that could be found in broader research methods textbooks) are not included to provide a central focus on doing ethnographic work robustly with construction and the built environment.

Working closely within an industry setting brings with it a consideration of ethics, since there is human participation. In Chapter 6: Doing ethical research in an industry context, Dr. Jan Hayes and Dr. Sarah Maslen discuss ethics when working closely with industry. A continuous focus on the implementation of ethics is important in a dynamic natural setting, such as the construction site, where unforeseen and unexpected situations can arise. Such unexpected ethnographic events and issues previously reported include a riot (Hamid & Tutt, 2019), sexual exploitation (Parry, 2014) and an accident to Sykes (1969) that ended his study. In Chapter 7: Philosophical reflections on ethnography in construction research, Dr. Hans Voordijk provides philosophical perspectives from ethnographic application – perspectives that are different from positivism. By approaching research from different research philosophies with the use of ethnography, there is potential for new knowledge to be produced, which otherwise would have been left untouched from positivist only research. In Chapter 8: How ethnographic methodologies contribute to construction practice, Dr. David Oswald and Dr. Helen Lingard provide examples of the types of knowledge that can be generated from ethnographic work, proposing that ethnographers should consider how to innovatively design opportunities for immersive encounters that can lead to rich data. In Chapter 9: Ethnography for Construction 5.0, Dr. Sarah Pink discusses the potential for various ethnographic approaches to produce knowledge on construction future, considering smart construction, robotics and Construction 5.0. This leads us into the various types of ethnographic research that have been performed within the construction industry, which are discussed and reflected upon within Section III.

### *Section III*

Within peer-reviewed academic journals, ethnographic approaches are discussed and justified within their respected methodology sections, but these are restricted in length. Thus, a more detailed reflection on *doing ethnographic fieldwork within the construction context* is less well documented in academic literature. Section III provides an opportunity to focus and reflect upon doing these different types of ethnography, when applied to a wide range of varied settings in which construction research can be conducted.

In Chapter 10: Doing short-term ethnography, Richard Brett reflects on the energy-intensive rapport building with participants in rapid timeframes, proposing video use to supplement traditional fieldnotes. In Chapter 11: Doing long-term ethnography, Dr. David Oswald, Dr. Fred Sherratt and Dr. Simon Smith discuss the challenges of 'fitting in' and 'finding a role', when many participants may not fully comprehend why ethnographic research can take significant periods of time. In Chapter 12: Doing multi-site ethnography, Dr. Andrea Yunyan Jia explains the different forms of multi-sited ethnography and encourages the ethnographer to show empathy by being interested in and sincere to the fields of study – to strategically build networks and rapport. In Chapter 13: Doing case-based inter-organisational ethnography, Evelijn Martinius and Prof. Leentje Volker propose three questions around places, people and project to provoke continuous ethnographic reflections. In Chapter 14: Doing ethnographic-action research, Dr. Léon olde Scholtenhuis shows how ethnographic-action research can enhance information system implementation in infrastructure projects, by use of deeper ethnographic encounters with practice, self-reflexivity and awareness of different role intensities. In Chapter 15: Doing ethnographic video interviewing, Joan Evelyn Ongodia and Prof. Timo Hartmann apply ethnographic video interviewing to uncover tacit knowledge that project teams have – about process workflows and project requirements – which could be re-used in the early design stages of construction projects. In Chapter 16: Doing online ethnography, Dr. Laura Osburn, Dr. Madison Snider and Dr. Carrie Sturts Dossick compare in-person and online ethnographies raising considerations around ethics and challenges with rapport building in virtual settings. In Chapter 17: Doing auto-ethnography, Dr. Leon Crascall and Dr. Fred Sherratt discuss positioning, methodological rigour and ethnographic writing within the context of built environment education.

### *Section IV*

The last section of this book, Section IV, provides *case examples of doing ethnography* across different settings in the construction industry and built environment. This section shows, besides extensive methodological detail, what possible outcomes of ethnographic work can deliver. It begins with Chapter 18: An ethnography of co-creating collaboration, innovation, and value, in which Dr. Ani Raiden and Dr. Emmanual Manu showcase an ethnographic-action research study at the design stage. Dr. David Oswald and Dr. Helen Lingard then present a multi-sited ethnography across the construction project stage in Chapter 19: An ethnography exploring health and safety communication in construction workgroups. Following on Chapter 20: An ethnographic account of circular demolition practices: realising reuse potentials, Dr. Marc van den Berg then focuses on the end-of-life stage, demonstrating how ethnographic approaches can move the discourse from problems and barriers to solutions. The final case example in Chapter 21: An autoethnography to reveal the bitter truth: I'm the problem is from the University setting within the built environment. This autoethnography by Dr. Henning Grosse reflects on how the research changed his behaviour with respect to conflict situations in meetings, and gender inequality.

These four sections on *Embracing Ethnography* show that the forms and applications of ethnography are increasing within the construction and built world. In the final Chapter 22: Conclusions, we call for all these forms, including auto-ethnography, ethnographic-action research, online ethnography, rapid and long-term ethnographies, to be embraced. This book discusses these various emerging forms of ethnographic approaches, displays exemplary cases of ethnographic research within the field, highlights the challenges that researchers have in dynamic construction settings and proposes ways to take towards robust research.

We hope you enjoy the book while passing through all the examples and reflections provided by the authors. As you go through, you will notice that some passages are quite 'unlike' other scientific articles in construction. Specifically, we encouraged the use of first-person narratives where appropriate. This follows the writing styles traditional to ethnographic conventions and can help increase the authenticity and clarity of the various encounters that the authors narrate throughout the book.

Good luck in your ethnographic adventures!

## References

Andersen, L. P., Karlsen, I. L., Kines, P., Joensson, T., & Nielsen, K. J. (2015). Social identity in the construction industry: Implications for safety perception and behaviour. *Construction Management and Economics*, 33(8), 640–652.

Bresnen, M. (2009). Living the dream? Understanding partnering as emergent practice. *Construction Management and Economics*, 27(10), 923–933.

De Silva, N., Rathnayake, U., & Kulasekera, K. M. U. B. (2018). Under-reporting of construction accidents in Sri Lanka. *Journal of Engineering, Design and Technology*, 16(6), 850–868.

Flannery, J., Ajayi, S. O., & Oyegoke, A. S. (2021). Alcohol and substance misuse in the construction industry. *International Journal of Occupational Safety and Ergonomics*, 27(2), 472–487.

Hamid, W., & Tutt, D. (2019). "Thrown away like a banana leaf": Precarity of labour and precarity of place for Tamil migrant construction workers in Singapore. *Construction Management and Economics*, 37(9), 513–536.

Laryea, S., & Hughes, W. (2008). How contractors price risk in bids: Theory and practice. *Construction Management and Economics*, 26(9), 911–924.

Morgan, J., & Pink, S. (2018). Researcher safety? Ethnography in the interdisciplinary world of audit cultures. *Cultural Studies↔ Critical Methodologies*, 18(6), 400–409.

Oswald, D., & Dainty, A. (2020). Ethnographic research in the construction industry: A critical review. *Journal of Construction Engineering and Management*, 146(10), 03120003.

Oswald, D., Sherratt, F., Smith, S., & Dainty, A. (2018). An exploration into the implications of the 'compensation culture' on construction safety. *Safety Science*, 109, 294–302.

Parry, J. (2014). Sex, bricks and mortar: Constructing class in a Central Indian Steel Town. *Modern Asian Studies*, 48(5), 1242–1275.

Phelps, A. F., & Horman, M. J. (2010). Ethnographic theory-building research in construction. *Journal of Construction Engineering and Management*, 136(1), 58–65.

Piekkari, R., & Welch, C. (2018). The case study in management research: Beyond the positivist legacy of Eisenhardt and Yin. In C. Cassell, A. L. Cunliffe, & G. Grandy (Eds.), *The SAGE Handbook of Qualitative Business and Management Research Methods* (pp. 345–359). Thousand Oaks, CA: Sage Publications.

Pink, S., Tutt, D., Dainty, A., & Gibb, A. (2010). Ethnographic methodologies for construction research: Knowing, practice and interventions. *Building Research and Information*, 38(6), 647–659.

Pink, S., Tutt, D., & Dainty, A. (2013) *Ethnographic Research in the Construction Industry*. Oxon: Routledge.

Reeves, S., Kuper, A., & Hodges, B. D. (2008). Qualitative research methodologies: Ethnography. *BMJ*, 337.

Seymour, D., & Rooke, J. (1995). The culture of the industry and the culture of research. *Construction Management and Economics*, 13(6), 511–523.

Seymour, D., Rooke, J., & Crook, J. (1997). The role of theory in construction management: A call for debate. *Construction Management and Economics*, 15(1), 117–19.

Styhre, A. (2009). Tacit knowledge in rock construction work: A study and a critique of the use of the term. *Construction Management and Economics*, 27(10), 995–1003.

Sykes, A. J. M. (1969). Navvies: Their work attitudes. *Sociology*, 3(1), 21–35.

Tutt, D., & Pink, S. (2019). Refiguring global construction challenges through ethnography. *Construction Management and Economics*, 37(9), 475–480.

# 2

# THE CULTURE OF CONSTRUCTION IN WESTERN SOCIETY

*Kevin Porter*

*Independent Consultant, Brisbane, Australia*

**Key takeaways**

- Culture is a complex phenomenon to understand within the construction industry, partly due to the sector having various mix of trades, professions, nationalities, ethnic backgrounds, and individual personalities, all conforming to differing contractor rules, practices, policies, and societal expectations.
- Ethnography can capture and provide the rich comprehension within this complex setting, to make the cultural changes needed for improving the construction industry.
- This chapter provides insight on construction site challenges within a Western society context, such as on contractor collaboration, workplace diversity, worksite pressures, and worker vulnerability, to show examples of the type of knowledge that can be gathered through ethnographic methods.

## Introduction

Culture is an important concept for us all to understand. Why? Because the world is imbued with culture: all humans, social groups, institutions, and organisations, as well as the markets. Culture determines how we behave as human beings, how business and work are undertaken, how customers select products, and how markets rise and fall. How we communicate, provide feedback, organise people into teams and hierarchies, decide on leadership models, develop, and maintain trust, and social relationships, make decisions, or organise time are all culturally determined. Understanding culture and related concepts is important to avoid making costly assumptions and to find realistic and long-lasting solutions to human-centric problems.

As the purpose of ethnographic research is to make sense of culture it should be no surprise that one of the most important concepts for the ethnographer to define is the concept of *culture.* I have defined culture, for the last 20 years, as, 'all taught and learned behaviour that is shared with other members of the group'. This includes, but is not limited to, shared beliefs, values, practices, narratives, symbols, norms, meanings, rules, histories, and language. Whether we are looking at national cultures, ethnic cultures, regional cultures, or organisational cultures, this way of understanding culture is the same. This does not mean that each individual member of the group perceives culture in exactly the same way. Individual factors such as age, gender, upbringing, social status, education,

DOI: 10.1201/9781003379584-3

intellect, and profession also play a role. Cultures are dynamic, they change and evolve over time, influenced by the individuals that make up the group, technological advances, intergroup relationships, and social trends.

To explore culture in depth, I decided to conduct ethnographic research on construction sites in Australia in 2019. This was after I read a report by Aurora Marketing (2019) for the Queensland Major Contractors Association (QMCA) and contributed to a discussion on procurement and tender processes in Queensland. The report, presented at a networking event, identified several issues affecting the industry. Two of the prime concerns were the rising costs of tendering and the 'lowest cost' approach by private and government customers, which were causing stress, anxiety, absenteeism for industry workers and problems recruiting and retaining qualified staff.

A two-day workshop, organised by QMCA and Construct Australia for senior engineers, designers, and team leaders, from a variety of professional backgrounds, further highlighted a number of areas of concern centring around cultural practices, such as written and verbal communication, collaboration, diversity, toxic behaviours, professional knowledge, the lack of transparency, a lack of shared values, unreal expectations, and the legal complexities that govern the industry. The information I gathered from these discussions raised many questions that I could not find satisfactory answers to in the literature or from industry professionals. The more questions I asked, the more curious I became about the culture within the construction industry.

Then, during the pandemic lockdowns, I participated in a number of online conferences and industry meetings organised by the Australian Constructors Association (ACA). During these conversations, it was argued by some participants that one of the contributing factors to the issues faced by the industry was caused by the conservative masculine culture and values within the industry, which it is rooted in and a lack of understanding when it comes to key social concepts, such as 'culture' or 'identity' and the plethora of concepts that are inextricably linked to them. So how does one go about addressing cultural problems? How do you determine what aspects of an organisational culture might be causing problems?

It is incredibly difficult for anyone to fix what they do not comprehend, and if they do not have the tools required in the shape of knowledge, social theories, models, and methodologies for collecting the data, then that task is nigh on impossible. Ethnographic fieldwork helps anthropologists and ethnographers collect and make sense of such information. An ethnography is a qualitative research methodology devised by social and cultural anthropologists in the early 20th century that aims to understand and describe the behaviours and social interactions of a particular group of people by gaining an *emic* (insider) perspective and contrasting it with an *etic* (outsider) perspective (Morris et al. 1999). It involves immersing oneself in a community, or setting, in order to gain a deeper insight and understanding of its members' practices and beliefs, through the use of *participant observation*, interview techniques, surveys, focus groups, and literature reviews (LeCompte & Schensul 2010). By employing ethnographic methods, researchers can gain valuable insights into the impact that policies, actions, and events have on the lives, motivations, and productivity of project teams and clients. This information can help business leaders make more informed decisions and find productive solutions to common and uncommon problems.

Conducting ethnographic research in any industry can be challenging. There may be problems gaining access and acceptance into the group or groups intended to be studied. There are considerations about how the presence of the ethnographer affects and influences the participants' behaviour. There are methodological facts relating to what social models and theories should be employed and what kind of data collection methods can be effectively used for the study according to locations and environments. There are also ethical and legal considerations related to confidentiality, informed consent, non-disclosure agreements, and representation.

This chapter aims to provide insight into doing ethnographic research on construction sites within a Western society. It begins by explaining the importance of the construction industry to society, before considering how it operates within the Western world. I then take a closer look at the culture on Australian construction sites, before presenting a case example that highlights findings from ethnographies on construction sites in Queensland.

## The importance of the construction industry

The construction industry plays a vital role in shaping the built environment and supporting the economic growth of nations around the globe. A report commissioned by the ACA in 2022 found the construction industry generated approximately 7% to Australia's GDP and employed around 1 in 10 of the country's working adults. Similar figures can be seen in other Western societies. The construction industry added approximately 6% to the GDP of the German economy according to the Federal Ministry of the Interior and Community (2022) and the German Federal Statistical Office (2023), 5.6% in the UK according to the Office for National Statistics (2023), and around 4% for the same period in the US according to the Bureau of Labor Statistics (2023).

Apart from this direct contribution to the economy, the industry also indirectly contributes a substantial amount to the national economy by providing society and its industries with the infrastructure required to create products and services. These include road and rail networks, bridges, ports and harbours, power plants, industrial complexes, schools, universities, hospitals, as well as commercial and residential properties. It should therefore come as no surprise that if the construction industry falters or struggles the knock-on effects will be felt widely throughout society.

Globally, the construction industry has struggled to attract and retain workers (McGrath-Champ et al. 2011; ABC 2023; Ding et al. 2023; Jones 2023) and there are growing challenges with procurement, tender processes, and innovation (Hovnanian et al. 2022; Beruguette 2023; Turk 2023). Some changes have been implemented, but according to Jon Davies, CEO of the ACA: 'they have been painfully slow, inadequate, or ineffective.' The industry is obviously not immune to the various problems, challenges, and crises that affect the global economy. The recent COVID pandemic along with the wars in the Middle East and the Ukraine have affected global supply chains, interrupted manufacturing and consumption practices, and increased energy prices and material costs, have added to the industry's problems and threatened its future. According to the Australian Constructors Association (2022), these problems are being felt globally. A report produced by German Management Consulting firm Roland Berger in October 2022 highlighted concerns about the consequences of economic uncertainty on the future of the industry in the DACH region (Germany, Austria, and Switzerland). Further, a recent article by Reuters (April 2023), titled 'Struggling UK economy buttressed by construction as other sectors flag', discussed how the current economic crisis in the UK has been lessened by stronger than expected results from the construction sector over the previous six-month period (Bruce 2023). Such articles and reports highlight just how important the construction industry is to the larger economy and why a healthy construction industry is so vital to a healthy economy and society.

## Defining construction culture in a Western setting

What do people mean when they talk about 'Western culture'? Is there even such a thing? When talking about the 'West' most people refer to the countries that make up Europe, especially the European Union and the UK, the US, Canada, Australia, New Zealand, and to a lesser extent Central and South American nations, depending on who you ask (Bonnett 2017). What do these nations have in common that allows them to be considered 'Western' cultures? What they primarily share is a number of social, political, and religious beliefs that set them apart from their Asian and African counterparts; these include the promotion of individual human rights and responsibilities, freedom of speech, and the freedom of religious and political affiliation without persecution. Many nations in the West separate political, religious, and legal institutions favour democratic political systems, universal primary and secondary education, and a host of other values, too many to list here. While these macro concepts might be shared amongst Western countries, each nation still maintains its own cultural flavour and uniqueness because their norms, rules, practices, and values have been shaped over hundreds, if not thousands of years by geography, climate, and the history of shared symbols and experiences.

When people talk about Western cultures sharing the same values it has to be made clear that what is being shared are basic ideas of certain concepts, but what constitutes a given concept may be different from group to group and from culture to culture. An example I like to use is borrowed from Christian Madsbjerg, co-founder of the Danish consulting firm ReD Associates and author of a number of bestselling books. In a recoded lecture given to faculty members at Princeton University (November 2017), Madsbjerg considers how the meanings of what appear to be shared concepts might differ greatly from one culture to another. Using the concept of 'freedom' he argues that in some parts of the US freedom may mean something that includes 'small government', 'low taxes', 'the right to shop 24/7', or 'the right to own a firearm', whereas in a country like Denmark freedom could mean 'large government', 'high taxes', 'strong social institutions such as health and pensions', and 'free or affordable education' (Madsbjerg 2017).

Identity is also a product of culture and is shaped by concepts such as the person and the self (Cohen 2002). It makes us different and unique from others and, simultaneously, makes us similar to the other members of the group we belong to (Byron 2002). Our identities are shaped by the cultures we originate from and grow up in (Demossier 2012: 5) and while many cultures around the world promote collective identities, (see Hofstede 1980), most Western cultures have been emphasising individualism over collectivism for the last few centuries. Individualism and collectivism represent two contrasting approaches to understanding human relations and organising societies. Individualism emphasises personal freedom, autonomy, self-interest, and the pursuit of individual goals, while collectivism highlights the common good, cooperation, and social responsibility (Triandis 2001). Both ideologies have their strengths and weaknesses, and the choice between them depends on cultural, historical, and societal factors. Cultures that promote individualism over collectivism can often have problems with communication and collaboration amongst the group members due to their pursuit of self-interests, which also affects trust and reciprocal relationships, whereas collective cultures often struggle with innovation because of the strict nature of their hierarchical structures, where individual members of the group are assigned specific roles and tasks that place collective goals above those of the individual (Triandis 2001). This is one reason why striking a balance between individualism and collectivism, especially in relation to industry, productivity, and the wider economy, is crucial to creating a society that respects individual rights while fostering social cohesion and shared prosperity.

However, as a former British colony, modern Australia has its roots in Western culture. English is the national language; the country operates a Westminster style government and shares a number of social values with other Western nations. Much has been written about Australian culture and identity in recent years, both in academia and the popular press. While (white) Australian culture has its roots in Anglo-Celtic culture it has developed its own distinct features over the last 230 years. Its citizens are described as 'friendly', 'liberal', 'egalitarian', 'relaxed', 'humorous', 'tolerant', and fond of sports and a drink (Crawford 2003; Toby 2013; Muir 2019; Salt 2019). Whilst most of these are cultural stereotypes that date from the foundations of Australia as a penal colony up to the 1950s, many still persist today. Despite the influx of immigrants from the Mediterranean, Asia, and the Middle East since the 1950s, this image of Australian cultural identity has remained largely unaffected in popular imaginations both at home and abroad.

Most national cultures and identities traditionally originate from masculine stereotypes, and Australia is no different. In the 1950s when non-indigenous Australians were predominantly of Anglo-Celtic descent, historians such as R.B. Ward described the Australian national identity as one partly based on the myth of the rugged bushman from the outback: a rough and ready, foul-mouthed improviser who likes a drink and a gamble, a man who believed in hard work and that everyone should get a fair go. The Australian male was also sceptical of authority and strangers but a loyal friend to his mates (Ward, 1956). More recently Callum Scott described Australian national identity as 'built upon hard masculine foundations, using the harsh nature of the bush to exemplify these character traits'. Today, these masculine images persist (Flood 2018; Winton 2018; Willis 2019). Of course, these do not represent all Australian men's ideas of Australian culture or identity, but they are still promoted as national ideals that influence behaviour, expectations, and judgement. Australians are also fiercely individualistic and are encouraged to be self-sufficient from a young age. Some of these stereotypes are on display in popular national sports, military culture, and identity, and they can still be found on construction sites in Queensland, Australia, even if they do not represent the average Australian male's ideals in the 21st century.

## Conducting ethnographic research on cultures in construction

For simplicity and clarity, the construction industry can be divided into two main cultural sectors, the 'corporate' and 'site' sectors. The work that is performed in each sector helps to shape and determine their culture; this is because culture is all taught and learned behaviour that is shared with other members of the group, and what you do, and how you do it, shapes who you are. Each sector comprises smaller cultural groups and units whose culture, in turn, is shaped by their individual members. The corporate culture consists of 'white-collar' workers, while the 'site' culture primarily comprises 'blue collar' workers. I have interviewed and spoken to a substantial number of corporate sector professionals regarding the culture within the industry over the years, but all of my ethnographic research has been conducted on 'worksite' culture, which is my focus here. The reason for this particular focus was due to the fact that I had previous experience working on construction sites in Europe and because many of the cultural problems within the construction industry in Queensland were being attributed to the actual worksite culture rather than the corporate culture. Therefore, I address my experiences in this sector by first describing the Australian context, and then zooming in on my fieldwork.

Although most of the discussions emanating from within the Australian construction industry talk about 'culture' in the singular, Australian construction sites are complex microcosms of diverse subcultures and identities that are variations and adaptations of Australia's non-indigenous national culture. They are hierarchical spaces where different roles and positions are assigned based on expertise, experience, and organisational structure. They have rituals and rules, in the shape of daily toolbox briefings, safety protocols, operational processes and dress codes, and like sites in many Western nations in the 21st century, they are multicultural, reflecting society and the national culture at large. Major sites tend to be run by a primary contractor that subcontracts work to specialised companies. Construction sites are dynamic spaces with lots of moving parts, where people always come and go. Multiple companies are usually onsite simultaneously, which can further shape the dynamics and the workflow onsite affecting the overall culture, depending on the levels of communication and collaboration between companies and teams. The industry also brings together individuals from various nationalities, ethnic and cultural backgrounds, age groups, professions, and skill sets. It is not uncommon to see Australian construction workers rub shoulders with workers from around the world. Over the years I have worked alongside fellow Brits, Aboriginal Australians, New Zealanders, Pacific Islanders, Indians, Chinese, Argentinians, Colombians, Brazilians, Senegalese, Somalis, Spanish, French, Germans, Italians, and Irish each with their own language and communication strategies, working practices, leadership models, and decision-making processes that are shaped by their cultural norms and values. All of them were trying to navigate through the rules and norms of site culture. With such a mix of trades, professions, nationalities, ethnic backgrounds, and individual personalities on-site collaboration can become quite difficult as each group and their individual members behave according to the cultural norms, patterns, and values in which they have been socialised. How we communicate, lead, make decisions, determine value, manage time, create trust, build relationships, and reciprocate are all culturally determined and shape our thoughts, speech, and behaviour.

One of the difficulties of working in such a diverse environment is to understand how culture shapes behaviour and working practices. A helpful resource for this is the work of American business studies professor Erin Meyer who has developed online surveys based on her bestselling book *The Culture Map* (2014). These surveys allow people to identify themselves and some of their cultural practices according to eight dimensions: *Communicating*, *Evaluating*, *Leading*, *Deciding*, *Trusting*, *Disagreeing*, *Scheduling*, and *Persuading*. Meyer argues that each dimension is culturally determined and shapes how organisations, and their members, behave and interact with others, as individuals and members of the group or other groups on a daily basis.

If we take Meyer's models and apply them to each group involved in the build, what can be seen is that each contractor and sub-contractor has their own culture, in the shape of practices, values, norms, and rules that are adaptations of the national culture. The very nature of the work conducted by the individual subcontractors gives further shape to their culture and that of the industry as a whole. Trade specialists such as carpenters, plumbers, and electricians all have their own practices, tools, materials, communication styles, language, (which includes jargon) that they employ daily as part of their profession, and which forms a part of their identity, creating 'in-group' and 'out-group' boundaries (Byron 2002). I will address the importance of communication in more detail later in this chapter. Finally, the individuals who form the workforce of each organisation influence the culture of the companies they work for, which is shaped by their personality types, intellect, and life experiences, especially in Western societies, such as Australia where individualism is promoted. Conducting ethnographic research allows the researchers to uncover the explicit and implicit rules, rituals, and meanings that shape behaviours within the construction industry.

A number of concepts raised during the QMCA workshops in Brisbane, in 2019, centred on *diversity* and *communication*. The concerns raised around these concepts and the way that the participants spoke about them piqued my interest and led to the commencement of my ethnographic fieldwork – which I discuss in more detail in the following pages. One point of concern raised during the workshops centred on a lack of diversity within the construction industry. However, from my own experiences on construction sites, I found the industry in southeast Queensland to be incredibly diverse, which raises questions such as how is diversity being defined and how is it being measured?

Research conducted by Canadian anthropologist Anne Irwin is good example of how 'emic' and 'etic' views shape people's understanding of diversity or the lack of it. According to Irwin, in order to appreciate the diversity within organisations it is important to approach the study of diversity through two frameworks. The first is etic, the other is emic (2009: 495–496). When both the etic and the emic perspectives are applied, the diversity within organisations and teams becomes apparent. This approach led Irwin to notice factors such as age, marital status, rural and urban backgrounds, family backgrounds, i.e., military/civilian backgrounds, or personality traits such as sociability, work ethic, and sense of humour.

Despite the apparent low levels of diversity in the military unit she studied, Irwin argues that the unit was diverse. This diversity enhanced group cohesion. She continues that, from the emic perspective, what matters to the soldiers is the contribution that individual members of the unit make to the group as a whole, not whether the soldiers belong to a certain category determined by outsiders (2009: 504).

Understanding Diversity, Equity, and Inclusion (DEI) is essential for businesses that strive to be more equitable and inclusive and to attract and retain more staff. There are many reasons why DEI is important. One reason is that diverse teams are better at solving problems. Studies conducted by anthropologists in the US researching the effects of diversity amongst fellow anthropologists specialising in lab-based biological or physical anthropology (see Antón et al. 2018) have noted that diverse teams produce better science, are more creative, and are better equipped to solve a variety of problems. Diverse teams are also better at anticipating, recognising, and navigating a broader range of social and political issues, which can help to maintain harmony and cooperation (Hong & Page 2004; Apfelbaum et al. 2014; Antón et al. 2018).

Outside of academia, companies such as Deloitte (Bourke & Dillon 2018) and McKinsey & Company (Hovnanian et.al. 2020) have produced reports arguing that socially and culturally diverse teams are more innovative, more likely to anticipate change and are better at responding effectively to potential changes. Another significant reason to focus on DEI is that employees who feel excluded also tend to feel disenfranchised, which can have negative effects on their mental health and well-being. It can also affect how they communicate and collaborate with others, their motivation and productivity, and the group's cohesion, collaboration, and productivity. As was mentioned earlier in this chapter, while shared values are fundamental to these concepts, not all shared values are the same.

While it can be argued that the Queensland construction industry is diverse, that diversity is more cultural and professional than gender based. There can be no doubt that the industry is male dominated, and lacks gender diversity, especially on construction sites. I have not conducted ethnographic research that specifically examines gender and diversity within the construction industry, however during my time working onsites I have noticed the lack of female and transgender workers. One possible reason for the gender imbalance within the industry could be due to the nature of the work and the biological differences between the sexes, especially in relation to size, strength, and stamina (see Soh 2021). The industry is labour intensive and can involve a great deal of heavy, repetitive lifting, long hours, and the pressures created by time constraints, looming deadlines, and tight profit margins place high demands on productivity which can make the industry an unattractive career choice for many. It is an industry where size and strength often do matter and do make a difference to the work that is being carried out. Another more important reason centres around intrinsic interests that might be shaped by biological gender as well as cultural education (see Correll 2001). In Western societies, most women simply do not want to work in construction. Instead, women tend to be drawn to professions related to education, health, and well-being, often referred to as HEED (Health care, Elementary Education, and Domestic work) (Schulstok & Wikstrand 2020; Sassine & Hajj 2021), whereas professions that centre around physical activities, such as the manual trades, the military, and the emergency services or that require an interest in Science, Technology, Engineering and Mathematics subjects (STEM) tend to be male dominated (Su et al. 2009; Tellhed et al. 2017). Even in countries considered to be pioneers in gender equality, such as the Scandinavian countries, the division of labour is still based on gender, indicating that when given a choice women and men choose differently (Melkas & Anker 2003; Behrend et al. 2007; Schulstok & Wikstrand 2020). While an in-depth discussion of gender diversity falls outside of the scope of this chapter (see Chapter 4 for further detail) it is a subject matter that, I believe, requires and warrants more ethnographic research if organisations wish to understand exactly why there is a lack of gender diversity within the Australian construction industry.

Communication was another main point of concern raised during the QMCA workshops. The use of language such as jargon, slang, or gestures can signal identity and group membership in the construction community, but it can also create barriers and misunderstandings (Byron 2002). To collaborate effectively, people need to understand communication strategies across cultures. Erin Meyer, author of *The Culture Map* (2014), distinguishes between two main types of communication, 'low context' and 'high context'. In low-context cultures, such as many Western countries, especially Anglo-Saxon, Germanic, Nordic, and Slavic countries, speakers give and understand messages at face value. According to Meyer, good communication in low-context communication cultures is precise, simple, and clear, speakers usually repeat messages to clarify they have been understood. In high-context communication cultures, on the other hand, such as most East Asian countries, speakers employ sophisticated, nuanced, and layered communication styles. Messages are spoken and implied and the listener is expected to read between the lines. Another consideration is how people give feedback in different cultures. Low-context communication cultures tend to use 'systems of 'direct negative feedback', which means feedback is frank and to the point. They value honesty and do not soften their messages, which means that language is usually strong and to the point. They may give criticism in public, in front of colleagues. In contrast, high-context communication cultures favour an 'indirect negative feedback' strategy, which means feedback and criticism are conveyed subtly, in a soft and diluted manner. Diplomacy is paramount and messages are wrapped in positive form, i.e., negative feedback is delivered using positive language. In such cultures criticism is only given in private, never in front of others.

Australian culture favours a low-context communication strategy with a direct negative feedback strategy; however, Australian construction sites are a mix of low- and high-context communication strategies and direct and indirect feedback strategies, which can cause problems and misunderstandings between individual workers when the factors above are not taken into consideration and directly affects communication and collaboration as a result.

Ethnographic research allows researchers to discover subtle nuances that influence how people interact and communication in the construction industry which might go unnoticed by more traditional quantifiable research methodologies (see Chapter 20 for a further discussion). By observing the explicit and implicit rules and rituals that shape behaviour – such as the low and high of communication styles – ethnographers can understand how language, jargon, slang, or gestures can act as markers of identity and group inclusion within the construction community and how that affects collaboration and further communication. By immersing themselves in the setting, ethnographers can gain an in-depth understanding of these cultural dynamics and how they influence work processes and worker relationships. Employing ethnographic research has allowed me to gain a much better understanding of what life and culture is like on Queensland construction sites as the following case study will demonstrate.

## Case study: Queensland, Australia

While conducting ethnographies on a number of construction sites across southeast Queensland since 2019, I have come to understand some of the cultural practices and problems experienced by construction workers. On one occasion, I worked on a large multi-level construction site in Brisbane. It was mid-summer, and the weather was scorching and humid, reaching close to 40°C. I was working with a team of seven labourers tasked with cleaning the upper floor of a new building. This required the labourers to remove all construction waste left behind by the various trades that had passed through. The build was behind schedule and management were visibly stressed and under pressure. As soon as the clean-up commenced, the air filled with fine dust that stuck to everyone and everything. Safety glasses fogged up (due to the heat and humidity) and were quickly coated with the dust, which formed a paste that smeared when you tried to clean them. People had to frequently stop working to clean their glasses, so they could see what they were doing and where they were going. It slowed everyone down, and that started to anger one of the supervisors, who began micromanaging and shouting orders at everyone. While his actions made people work a little faster for a short time, it visibly angered some of the labourers and affected their mood and work. A few started complaining amongst themselves about the poor design of the glasses and masks, and how useless and dangerous they were. When they complained to the supervisor about the Personal Protective Equipment (PPE), he told them to get rid of the PPE if it bothered them so much. All he cared about, he said, was that they did the job they had been hired to do, without killing anyone. If they didn't want to do it, he could find someone else who 'wanted the money' and 'could follow orders.' Two of the labourers, a German and British worker, refused to carry on working and walked off site. The remainder continued with the job at hand, at a much slower pace. As the day progressed, the team tried different types of PPE and increased the ventilation in an attempt to remove some of the dust and cool the working area down. A task that was predicted to take no more than a couple of hours, ended up taking over ten hours to complete and caused a number of problems and conflicts between workers and management that could have been avoided.

Some of the workers who remained that day, and continued with the task, spoke angrily of feeling 'underappreciated', 'disposable', 'worthless', 'trapped', and 'helpless', amongst other things. They spoke about the low wages, the precarious nature of casual labour hire, how it made them feel, and how it affected their motivation and their relationships with those around them. Some spoke of not being able to take time off because they couldn't afford to, which made them feel mentally and physically drained, as they tried to balance finances and family relationships. An older Australian labourer told me how he dreaded Christmas and Easter because he was forced to take days off over the public holidays, which he didn't get paid for. His main concern was how he was going to pay his rent and bills when he was forced to take time off. He also spoke about how he worried about getting sick or injured. Most admitted to drinking too much in order to unwind as a result of these stresses, which further added to their financial problems and social relationships, especially with partners. One complained about how '*they*', meaning the foremen, supervisors, managers, and companies don't care about '*your*' health and safety, all they care about is '*you*' getting the job done as quickly possible and not causing '*them*' any problems. When I asked what kind of problems, he told me a story about a job he had been on where another labourer was asked to perform a task without the adequate PPE. The labourer had been on that particular site for a several weeks and was viewed as a 'good worker', a 'solid bloke', and a 'guy who turns up to put a shift in'. On this particular day he informed the supervisor that he didn't feel comfortable performing the task. The supervisor said 'ok' and found another labourer willing to finish of the task without PPE. The following day the 'solid bloke' was missing from site, when the labourer telling the story enquired about his whereabouts, he was told that he was not coming back because 'he couldn't follow directions'. It highlights several difficulties some construction workers face and sheds light on how cultural practices can affect their well-being, motivation, and their productivity. It should come as no surprise that employees with low motivation are less productive and produce work to a lower standard than their more motivated colleagues.

The problem I observed caused several issues and conflicts between workers and management that could have been avoided. It also effected the quality and efficiency of the work, a task that was predicted to take no more than a couple of hours, ended up taking over ten hours to complete. The morale and satisfaction of the workers was affected as well.

This small case study demonstrates the value of conducting an ethnographic to observe and experience first-hand some of the realities and problems faced by construction workers and labourers on construction sites. These challenges often go unnoticed or unaddressed when using surveys and interviews as the main method of data collection. By being close to the action I was able to capture how concepts such as 'values', especially those centred around professional worth, come into play and shape workers' motivations and productivity. I was also able to gain insights into how relationships were being formed and maintained amongst the labourers and how they related to and communicated with one another and with their supervisor and management teams onsite. I found that collaboration and communication were linked to how the workers related to others' actions and expectations (see Meyer 2014).

According to Sahlins (1972) human relationships can be divided into three main types: *generalised* reciprocity, *balanced* reciprocity, and *negative* reciprocity. The first refers to giving without expectations, such as amongst family and close friends, the second is giving with the expectation of an equal value in return, like when someone goes grocery shopping, and the third is giving with the intention of gaining a value over the other individual or group, such as in many unskilled and low paid positions. However, who we relate to and how we relate to them are cultural constructs. A common narrative amongst the labourers was of feeling undervalued as workers and as 'people', which some argued was visible by their working conditions, low hourly salaries, employment contracts, and the timeframes and expectations placed upon them to complete their tasks. These factors visibly affected their motivation and productivity.

The ethnography also shed light on the effects that time constraints and schedules have on construction workers. Construction in Australia, like in many other parts of the world, is notorious for its cost overruns and delays, often causing financial strain for all everyone involved in the build (see Flyvbjerg et al. 2003; Love et al. 2013; Ding et al. 2023). Numerous factors contribute to this problem, including market trends, risk allocation, inadequate project planning, poor project management, changes in scope, and unforeseen events such as weather conditions, material shortages, price hikes, and staff shortages (Peckiene et al. 2013).

By being onsite, I was able to observe how these factors influenced the construction site setting. For example, I noticed how the financial pressures led to the use of cheap and poor-quality PPE, which could result in division, communication breakdowns, and tension between workers and management. I also learned about the lives of the workers, their challenges balancing work and personal lives, and how they were viewed by others within the construction site setting. These few examples demonstrate the wealth of opportunity that ethnography can bring for understanding the construction site context, to provide the knowledge required for change.

## Conclusions

Ethnographic research has a strong and proven track record in allowing its practitioners to make sense of human groups and their cultures. Its techniques for gathering data, such as participant observation allow the ethnographer to gain deeper understandings of the human experience and the meanings people attach to those experiences. By embedding themselves with a particular group and participating in their daily activities the researcher gains a better understanding of group dynamics and the rules, norms and values that govern them. It requires the practitioner to create social relationships based on reciprocity, trust, and a number of shared values, which means they have to learn how these concepts are constructed and how they relate to each other from group to group. Armed with defined concepts, social theories, and models, ethnographers have the tools that allow them to see through the kaleidoscope of behaviours and order them into patterns which then enables them to make sense of the cultures that drive those behaviours.

From the start of my ethnographic research on construction sites across southeast Queensland it became apparent that there were multiple cultures co-existing, collaborating and often clashing with each other onsite. Working on-site daily revealed not only the diversity of workers from different national and ethnic cultural backgrounds, but also the different organisational subcultures that make up the construction industry in Queensland. The number of different sub-groups present onsite and individuals from different cultural groups with different values and communication strategies provided some insights into the complications that can arise from such a diverse ensemble and the importance of having shared values, practices, and communication strategies. The ethnographies highlighted the importance of understanding how reciprocal relationships are created and how they are inextricably linked to other concepts such as trust, communication, motivation, collaboration, and productivity.

It also became apparent very quickly that working on a construction site in any capacity is not for the faint hearted. The work can be, and often is, incredibly demanding, both physically and mentally, the working days are long and the constant pressures to be productive can often create an uncomfortable environment to work in, especially with the sub-tropical climate in southeast Queensland. The chances of injury are always present due to the nature of the work and the working environments, and while there is legislation and procedures in place to reduce accidents and injuries, workers are constantly getting minor injuries. These factors plus precarious nature of the work, rise in casual contracts, and the low wages offered to some workers make the construction industry unattractive and a hard industry to be employed in. Under the current cultural and business models contractors and subcontractors are on the back foot from the go. The constant pressure to be productive and to achieve the unachievable promotes toxic behaviour which can lead to toxic practices when left unchecked. These practices include cutting corners, breaking certain rules, trades arriving onsite in the wrong order due to the time changes and delays, and ultimately jobs and processes getting rushed. This often causes mistakes to be made and tempers to fray, affecting communication and collaboration as well as motivation, the standard of work, productivity, absenteeism, recruitment, and retention of skilled and unskilled staff, delaying projects, reducing profits and possibly damaging organisational reputations.

The good news is that many of the problems within the construction industry are cultural, which means they are human-made, and if they are human-made, they can be studied, analysed, understood, and changed. If the construction industry is serious about addressing some of the cultural challenges, they believe are negatively affecting the industry then I believe that employing qualitative research methodologies such as ethnographic research and combining them with more traditional quantitative methodologies is the most effective way to find long-lasting solutions to many of our common and not so common human-centric problems.

## References


ABC (Associated Builders and Contractors) (2023). "Construction workforce shortage tops half a million in 2023, says ABC." (Online 3 February 2023). https://www.abc.org/News-Media/News-Releases/construction-workforce-shortage-tops-half-a-million-in-2023-says-abc

ACA (Australian Constructors Association) (2022). "Disrupt or die: Transforming Australia's construction industry." https://www.constructors.com.au/wp-content/uploads/2022/11/Disrupt-or-die_November-2022.pdf

Antón, S. C., Malhi, R. S. & Fuentes, A. (2018). Race and diversity in U.S. biological anthropology: A decade of AAPA initiatives. *American Journal of Physical Anthropology*, 165, 158–180. https://doi.org/10.1002/ajpa.23382

Apfelbaum, E. P., Phillips, K. W. & Richeson, J. A. (2014). Rethinking the baseline in diversity research: Should we be explaining the effects of homogeneity? *Perspectives on Psychological Science*, 9(3), 235–244.

Aurora Marketing (2019). "Behind closed doors: Insights from evaluation panels." Report commissioned by Queensland Major Contractors Association (QMCA).

Behrend, T. S., Thompson, L. F., Meade, A. W., Grayson, M. S. & Newton, D. A. (2007, April). Gender differences in career choice influences. In 22nd Annual Meeting of the Society for Industrial and Organizational Psychology, New York, pp. 1–10.

Berger (Roland) (2022). "Economic uncertainty – Construction at risk despite its' solid fundamentals. Impacts on DACH region." https://content.rolandberger.com/hubfs/Construction_Trend_Radar.pdf?utm_campaign=22-0056_Sustainability-in-Fashion&utm_medium=email&_hsmi=229982114&_hsenc=p2ANqtz-_5_WkINBS1pe8Ue_0E1GG-6R2pLz-qlf3cSt0dIPHdTo9gi88iiJblm6A58lx1m2NyUuUO7UyMuDZj-alFLevvKyIbrQ&utm_content=229982114&utm_source=hs_automation

Beruguette, B. (2023). "Global construction industry insights." *Solutions Outlook*. Allianz. https://commercial.allianz.com/content/dam/onemarketing/commercial/commercial/reports/global-industry-solution-outlook-2023-construction.pdf

Bonnett, A. (2017). *The Idea of the West: Culture, Politics and History*. London: Bloomsbury Publishing.

Bourke, J. & Dillon, B. (2018). "The diversity and inclusion revolution: Eight powerful truths." Deloitte Development LLC (online). https://www2.deloitte.com/content/dam/insights/us/articles/4209_Diversity-and-inclusion-revolution/DI_Diversity-and-inclusion-revolution.pdf

Bruce, A. (2023). "Struggling UK economy buttressed by construction as other sectors flag." https://www.reuters.com/world/uk/struggling-uk-economy-buttressed-by-construction-other-sectors-flag-2023-04-18/

Byron, R. (2002). "Identity" in A.J. Barnard & J. Spencer (eds.) *Encyclopedia of Social and Cultural Anthropology*. London: Routledge, p. 441

Cohen, A. (2002). *Self-Consciousness: An Alternative Anthropology of Identity*. London: Routledge.

Correll, S. J. (2001). Gender and the career choice process: The role of biased self-assessments. *American Journal of Sociology*, 106(6), 1691–1730.

Crawford, D. (2003). Becoming a man-the views and experiences of some second-generation Australian males. *Electronic Journal of Sociology*, 7(3). https://sociology.org/ejs-archives/vol7.3/02_crawford.html

Demossier, M. (2012). What can anthropology bring to the debate on identity. *Free Football Research in an Enlarged Europe*. https://www.researchgate.net/profile/Marion-Demossier/publication/266166711_What_can_anthropology_bring_to_the_debate_on_identity/links/5772b71408aeef01a0b65d7a/What-can-anthropology-bring-to-the-debate-on-identity.pdf

Ding, M. J., Jie, F., Sisombat, S. & Bandlamudi, B. S. (2023). Impact of the skill shortage on the construction supply chain performance in Australia. *Civil Engineering Journal*, 9(2), 356–371. https://doi.org/10.28991/CEJ-2023-09-02-08; https://ro.ecu.edu.au/ecuworks2022-2026/2149/

Federal Ministry of the Interior and Community (2022). "The importance of the construction sector." https://www.bmi.bund.de/EN/topics/building-housing/building/construction-sector/importance-construction-sector/importance-construction-sector-node.html#:~:text=In%202019%2C%20Germany%27s%20construction%20industry,and%20in%20building%20completion%20work

Federal Statistical Office of Germany (2023). "National accounts at a glance 2022". https://www.destatis.de/EN/Themes/Economy/National-Accounts-Domestic-Product/_node.html#sprg246446

Flood, M. (2018). "Australian study reveals the dangers of 'toxic masculinity' to men and those around them." *The Conversation*. https://theconversation.com/australian-study-reveals-the-dangers-of-toxic-masculinity-to-men-and-those-around-them-104694

Flyvbjerg, B., Skamris Holm, M. K. & Buhl, S. L. (2003). How common and how large are cost overruns in transport infrastructure projects? *Transport Reviews*, 23(1), 71–88. https://doi.org/10.1080/01441640309904

Hofstede, G. (1980). Culture and organizations. *International Studies of Management & Organization*, 10(4), 15–41.

Hong, L. & Page, S. E. (2004). Groups of diverse problem solvers can outperform groups of high-ability problem solvers. *Proceedings of the National Academy of Sciences*, 101(46), 16385–16389.

Hovnanian, G., Kumar, A. & Luby, R. (2022). "Will a labor crunch derail plans to upgrade US infrastructure?" McKinsey & Company. https://www.mckinsey.com/industries/public-sector/our-insights/will-a-labor-crunch-derail-plans-to-upgrade-us-infrastructure

Irwin, A. (2009). Diversity in the Canadian forces: Lessons from Afghanistan. *Commonwealth & Comparative Politics*, 47(4), 494–505.

Jones, C. (2023). "Construction workforce shortage reaches half a million in 2023." In *International Construction.com* (16 February 2023. Online). https://www.international-construction.com/news/construction-workforce-shortage-reaches-half-a-million-in-2023/8026735.article

LeCompte, M. D. & Schensul, J. J. (2010). *Designing and Conducting Ethnographic Research: An Introduction* (Vol. 1). Lanham: Rowman Altamira.

Love, P. E., Wang, X., Sing, C. P. & Tiong, R. L. (2013). Determining the probability of project cost overruns. *Journal of Construction Engineering and Management*, 139(3), 321–330.

Madsbjerg, C. (2017). "Observation and listening." *The Creative Mind and Leadership Lecture Series*. Keller Center. Princeton University. https://www.youtube.com/watch?v=wLAbKu7rCjA&t=679s

McGrath-Champ, S., Rosewarne, S. & Rittau, Y. (2011). From one skill shortage to the next: The Australian construction industry and geographies of a global labour market. *Journal of Industrial Relations*, 53(4), 467–485. https://doi.org/10.1177/0022185611412897

Melkas, H. & Anker, R. (2003). *Towards Gender Equity in Japanese and Nordic Labour Markets: A Tale of Two Paths*. Geneva: International Labour Office.

Meyer, E. (2014). *The Culture Map: Breaking through the Invisible Boundaries of Global Business*. New York: Public Affairs.

Morris, M. W., Leung, K., Ames, D. & Lickel, B. (1999). Views from inside and outside: Integrating emic and etic insights about culture and justice judgment. *Academy of Management Review*, 24(4), 781–796. https://www.researchgate.net/profile/Michael-Morris-22/publication/271807725_Views_from_Inside_and_Outside_Integrating_Emic_and_Etic_Insights_about_Culture_and_Justice_Judgment/links/56377efa08ae9d3e0347c35f/Views-from-Inside-and-Outside-Integrating-Emic-and-Etic-Insights-about-Culture-and-Justice-Judgment.pdf

Muir, B. (2019). "Australian identity – Is it relevant anymore?" In *Courier Mail* (June 2019). https://www.couriermail.com.au/news/queensland/sunshine-coast/australian-identity-is-it-relevant-anymore/news-story/17d7912075f2511ab0aa8e3f1a86abe5

Office for National Statistics (2023). "Construction output in Great Britain: December 2022, new orders and Construction Output Price Indices, October to December 2022." https://www.ons.gov.uk/businessindustryandtrade/constructionindustry/bulletins/constructionoutputingreatbritain/constructionoutputingreatbritaindecember2022newordersandconstructionoutputpriceindicesoctobertodecember2022#:~:text=Total%20annual%20construction%20output%20increased,%25%20and%208.5%25%2C%20respectively

Peckiene, A., Komarovska, A. & Ustinovicius, L. (2013). Overview of risk allocation between construction parties. *Procedia Engineering*, 57, 889–894.

Sahlins, M. ([1972] 2017). *Stone Age Economics*. Oxfordshire: Taylor & Francis.

Salt, B. (2019). "What is Australian Culture now?" *The Weekend Australian Magazine*. https://www.theaustralian.com.au/weekend-australian-maga-zine/what-is-australian-culture-now/news-story/bd80f71dd063d6efce43b77da6d81d45

Sassine, J. & Hajj, J. (2021). Gender differences and career choice. https://www.researchgate.net/profile/Joey-Sassine/publication/358497347_Gender_Differences_and_Career_Choice_The_Role_of_Personality_Interest_Ability_and_Motivation_in_Choosing_to_Pursue_a_Career/links/6204f884075f695e892f64ee/Gender-Differences-and-Career-Choice-The-Role-of-Personality-Interest-Ability-and-Motivation-in-Choosing-to-Pursue-a-Career.pdf

Schulstok, T. & Wikstrand, F. (2020). "Gender equality and career guidance in a Nordic context" in *Career and Career Guidance in the Nordic Countries*. Leiden: Brill, pp. 51–64.

Soh, D. (2021). *The End of Gender: Debunking the Myths about Sex and Identity in Our Society*. New York: Simon and Schuster.

Su, R., Rounds, J. & Armstrong, P. I. (2009). Men and things, women and people: A meta-analysis of sex differences in interests. *Psychological Bulletin*, 135(6), 859–884. https://doi.org/10.1037/a0017364

Tellhed, U., Bäckström, M. & Björklund, F. (2017). Will I fit in and do well? The importance of social belongingness and self-efficacy for explaining gender differences in interests in STEM and HEED majors. *Sex Roles*, 77(1), 86–96.

Toby, X. (2013). "What is Australian culture?" *The Sydney Morning Herald* (June 2013). https://www.smh.com.au/entertainment/what-is-australian-culture-20130618-2ogqp.html

Triandis, H. C. (2001). Individualism-collectivism and personality. *Journal of Personality*, 69(6), 907–924.

Turk, Ž. (2023). Structured analysis of ICT adoption in the European construction industry. *International Journal of Construction Management*, 23(5), 756–762. https://www.tandfonline.com/doi/pdf/10.1080/15623599.2021.1925396

U.S. Bureau of Labor Statistics (2023). "Construction labor productivity." https://www.bls.gov/productivity/highlights/construction-labor-productivity.htm#:~:text=The%20construction%20sector%2C%20as%20defined,were%20attributable%20to%20this%20sector

Ward, R. B. (1956). "The ethos and influence of the Australian pastoral worker." PhD Thesis. https://openresearch-repository.anu.edu.au/bitstream/1885/9584/1/02Whole_Ward.pdf

Willis, O. (2019). "Traditional masculinity and mental health: Experts call for a gendered approach to treatment." *ABC News*. https://www.abc.net.au/news/health/2019-02-05/mens-mental-health-masculinity-gendered-psychology-guidelines/10768294

Winton, T. (2018). "About the boys: Tim Winton on how toxic masculinity is shackling men to misogyny." *The Guardian*. Australian Edition. https://www.theguardian.com/books/2018/apr/09/about-the-boys-tim-winton-on-how-toxic-masculinity-is-shackling-men-to-misogyny

# 3

# THE CULTURE OF CONSTRUCTION IN EASTERN SOCIETY

*Sarah Swider*

*Wayne State University, Detroit, MI, USA*

**Key takeaways**

- The Eastern society is culturally very diverse and so it its construction context.
- Four key cultural aspects provide insight into doing ethnography within an Eastern society context.
- A large part of the Eastern construction sector is based on an informal economy.
- Concepts like guanxi capture the essence of social relationships and power in a country like China.
- Status and authority influence social networks, especially when large shares of immigrants work in construction.
- Collective and individual divide between public and private differs from Western society.
- Reflection on the positionality of the (Western, female) researcher in the Eastern context is presented to encourage thinking about one's own position, safety and access to data.

## Introduction

Conducting ethnography is not easy, it is even more difficult in setting such as the construction industry. There is no one roadmap that one can follow because it is not a 'one-size-fits-all' endeavour. However, there are nuggets of knowledge that can be collected from different sources and put together to create a working guide for each unique project. This chapter presents elements of culture, specifically culture in the East, that shape the context the construction industry and how we do our research. The goal is to share experiential and scientific knowledge about important cultural considerations when conducting ethnographic studies in Asia. The hope is that this supports future researchers' design and efforts to carry out robust ethnographic studies in the East.

It is easy to assert that culture in the East is different than in the West, but difficult to define the culture in the East. In part, this is because it is too diverse. East Asia, South and Southeast Asia represent distinctive regions with unique cultures. East Asian cultures share roots in Confucianism.[1] In contrast, South and Southeast Asia are diverse in terms of their cultural practices and related differences observed in the construction industry. For example,

DOI: 10.1201/9781003379584-4

in Southeast Asia not only are there several countries in which women are a significant proportion of the workforce in construction but also there are several countries where this is not the case. In many of these countries, workers are both imported and exported in the construction industry, but in some Asian countries, the construction industry workforce is primarily drawn from the domestic workforce. In other Asian countries, it is almost exclusive immigrant labour.[2] Given the caveat of diversity in the East, the ideas and concepts in this chapter have been drawn from research conducted mostly in East Asia. They are being used as a loose proxy for Eastern Society.

On a macro-level, culture is an ambiguous concept with no established agreed upon definition or approach to study. One school of thought is that culture can be understood as the social characteristics that distinguish the members of one category of people from another (Hofstede, 2001). For others, the definition of culture centres on meaning, consisting of 'symbolic vehicles of meanings' including language and daily practices, beliefs, rituals, art forms and ceremonies (Geertz, 1973; Sewell, 2005). Nathan (1993) labels the two approaches positivistic and hermeneutic, respectively. In practice, both approaches to studying culture are valid, and each offers insights that can be integrated together to push forward our understanding of culture and its role in shaping society. In a somewhat similar formulation, Sewell (2005) conceptualises culture as both a theoretical category and as a 'concrete bounded body of beliefs and practices' and then suggests these two different conceptualisations are in dialectic relation to each other.

This chapter follows Sewell by drawing insights from both the positivistic and hermeneutic approaches to culture. It starts with the positivistic studies. Most notably, it draws from Choi et al. (2015) who examined culture and organisations in the construction industry and distilled a list of important factors that are central to understanding cultural differences. The list includes (1) group attachment and relations, (2) authority and status, (3) uncertainty and rules, (4) gender roles and assertiveness, (5) time and future orientation, (6) emotions and (7) caring for others. This study, and other like it, provided an inspired starting point. However, the list is too long to be covered in a meaningful way in a chapter. When faced with choosing which factors were most important, I resorted to using some on the list, creating a remix and adding something new.

Out of that process, four aspects of culture emerged. They will be the focus of this chapter. They are as follows: (1) the informal/formal segments, (2) relationships and rules, (3) authority and status and (4) public and private divide.[3] In some ways, these four aspects of culture derived from the larger list are arbitrary, but I found they are the most useful prisms through which we can explore the field. Each researcher setting up a new project that considers culture will be faced with the decision of which factors should be examined. The decisions are guided by the research question(s) and the specific characteristics of a project. I have chosen these four aspects of culture because they have been useful to me across different projects and sites. I have used them to help me make sense of the field, social interactions, organisations and macro-phenomena in the political economy.

## Four important cultural considerations

The next section lays out the four factors that can be used by ethnographers to map out the cultural context for construction research. They are relational and comparative by nature, but for the purposes of this chapter, they are applied to culture in the East. I will discuss each of the four aspects of cultured listed above, working from abstraction towards specificity, and from categories to symbolic meaning.

Also, while most literature on the construction industry and culture provide either organisation (Ankrah & Langford, 2005) or project (Zou et al., 2009) level of analysis, this chapter zooms out and considers national level analysis, within which the organisations and projects are embedded, and sometimes also zooms in and provides individual level analysis.

### *Informal and formal*

Scholars working in the West tend to focus on, and analyse, the formal sectors of the construction industry, with some notable exceptions.[4] The formal sector is our default because it is largest part of the industry in terms of size, labour force and production. However, in the East, the informal is a significant segment of the industry, and in some cases, it is larger than the formal sector of the industry. In most countries, both formal and informal economy exist, overlap and interact.

The informal sector of the economy, or of the construction sector more specifically, is the that part which operates outside of formal state laws and governance. The size of the informal economy varies across countries but about 60 per cent of all workers in the world (thus not only in construction) are working in the informal economy.[5] In emerging and developing countries the share of informal employment is just shy of 70 per cent of the employed population, in contrast, in 'developed nations' it is closer to 20 per cent.[6] Informal employment is also concentrated in specific industries like domestic help and construction, representing over 90 per cent of the construction industry workforce in places like Nepal, Indonesia and India (Chan & Pribadi, 2022; Chmutina & Rose, 2018).

Recognising and understanding that there is both a formal and informal sector in the construction industry is crucial if we want to produce informative and enlightening studies (Centeno & Portes, 2006; Chen, 2012). The size of the formal versus informal segment of the industry shapes how it is organised, the proportion of labour to machine, level of mechanisation and methods of construction. In terms of labour, informal employment usually falls outside of the realm of social benefits and the welfare system. Informal work is usually done without contracts or other forms of protection.

Most importantly, the culture that operates on jobsites, and in the construction industry more generally, is shaped by informality. Informality increases flexibility (Chmutina & Rose, 2018; Lee et al., 2020) and changes the relationship between actors like developers, contractors and workers (Hirslund, 2021; Wells & Jason, 2010). Informality of the construction industry changes worker's relationship to the work and their migratory patterns (Tutt et al., 2013). It also creates different sets of incentives on the jobsites (Chmutina & Rose, 2018) and even within the housing market, in determining what kind of housing is built (Burlotos et al., 2020; Lizarralde & Root, 2008).

The existing literature on the construction industry has proven that recognising and analysing how informality shapes the construction industry is fertile ground for researchers. For example, in Nepal, the implementation of building codes and regulations have been stymied not by a lack of knowledge of informal workers but by the way culture has shaped the process of transforming that knowledge into practice (Chmutina & Rose, 2018). In this case study, it was the belief about gender and practices around trust that prevented women workers from getting training and led many to hire trusted older contractors over younger, but better trained, contractors. Not only is the economy and industry operating in the informal sector, but often much of the labour is informal, and that informality shapes everything from labour hierarchies and gender inequalities

(Hirslund, 2022), to solidarity and labour struggles (Swider, 2016; Wells & Jason, 2010) and safety practices (Tutt et al., 2013).

The literature on the informal sector is large, but there are few studies that focus on informality and the construction industry. This lack of recognition of the informal sector or informal segments of the industry creates a large blind spot in our research. Acknowledgement of the informal should be a part of most studies in the construction industry, especially in the East and other parts of the world in which the informal is prevalent. Interrogation of the informal/formal divide will also enrich our understanding of how the larger political economy impacts the industry.

Finally, it is worth noting that economic activity in the informal economy, by standard definition, refers to practices and activities outside of state regulation, laws and rules. However, it does not mean that the activity or relations are unregulated. Rather, the mechanisms of regulation are different, and it is up to the ethnographer to help us identify these alternative mechanisms of regulation and how they shape the social relations and practices.

### *Relationships and rules*

Another important cultural axis in which the East differs from the West is how social systems are regulated. Western societies adhere more closely to a rule-based system, striving towards the ideal of transparency and impartiality. In contrast, Eastern culture, especially those influenced by Confucianism, are more relationship-based. Relationships have been described as the glue that bind people to others who are present in their life worlds. In the East, when trying to map out the context of any research field site, it is important to observe and identify the relationships that exist among social actors in the field. While the East may operate with a relationship-based system, rules still matter, they just matter differently.

One important difference in relationships is the structure. In the West, we tend to think of social relationships as taking the form of social networks. One person is linked to the next, creating webs of personal connections that are linked together (Granovetter, 1983). In contrast, Chinese social networks are described as taking the same form of 'ripples formed from a stone' thrown in the water, in which the individual stands in the centre and relationships are formed in concentric circles around the individual in the middle (Fei et al., 1992). These social relations tend to make a horizontal web of networks, in contrast to the vertical relations in social hierarchy (Hong & Engeström, 2004).

Social networks are extremely important because they produce social capital, which is conceptualised as resources acquired and accrued through relationships. Social capital can be transformed into information or social recognition, or it can be used to exert power on a group or individual, who in turn mobilise resources on behalf of the person with social capital (Bourdieu, 2018).

Scholars examine how social networks operate both as a resource and as a mechanism. As a resource, it provides structural social capital. Structural social capital relates to the properties of the social system and of the network of relations, specifically facilitating the flow of information and resources (Nahapiet & Ghoshal, 1998). In this case, the structure of the network matters. Ethnographers give attention to the density and connectivity of networks, to tie strength and centrality, and to network stability and size (Lefebvre et al., 2016). Also, given the overall structure of social networks is different, the role of these characteristics might vary. For example, in the West, it is weak ties that are most effective in helping people find jobs (Granovetter, 1983), but in East Asia, it is the strong ties that help people locate employment (Bian, 1997; Horak et al., 2019).

Social networks are also a mechanism through which people can solve problems, facilitate business, assign obligations and expectations and enforce sanctions. This is often referred to as relational or cognitive social capital (Chou, 2006). Relational social capital includes 'goods' that operate as mechanisms produced through social networks such as trust, obligation, respect, friendship and loyalty. They are used throughout the jobsites in different ways. For example, trust plays an important role in facilitating recruitment but is also used to exploit workers (Eade et al., 2007; Swider, 2016). Social networks can play different roles in shaping employment relations (Li et al., 2018).

In China, the network of social relations and the power which it contains is often referred to as guanxi. Some describe guanxi as a culturally complex concept, nested in a wide range of emotional states and social obligations, each of which defies direct translation into the English language. For others, its influence, especially in business, needs to be contextualised. It is a product of specific political and economic factors that encourage a preference for particularistic ties. The former argue it is a stable, enduring feature of China's national culture and a source of security and protection in an unpredictable society. In contrast, the latter argue that the shift towards open-market competition will diminish the importance of social ties such as guanxi, especially in privately owned enterprises (Guthrie, 2012).

While researchers may disagree on the future role of quanxi, they agree that guanxi has thus far played an important role in business and industry in China. Alon (2003) describes Chinese management style as relationship-based as compared to rule-based Western management. Park and Luo (2001) bring analysis to the next level by focusing on the role of guanxi in the firm on the organisational level. They show how businesses use guanxi (which can here be understood as cooperation and exchanging favours) with competitors and government representatives in order to overcome competitive disadvantage.

Bian's (2018) classification of guanxi as a five-level variable provides a lens that helps sort the complexities of social relationships and social exchanges that may be observed in fieldwork. He argues that the nature and the form of guanxi are dependent upon what kind of social tie is operating: (1) a tie of connectivity, (2) a sentimental tie, (3) an instrumental tie and (4) an instrumental particular tie or an obligational tie facilitating money and power exchanges. This approach, which must be historically and geographically contextualised, affords opportunities to enrich our understanding how these mechanisms operate in the construction industry across countries and cultures.

One issue with scholarship that recognises the primacy of social relationships and networks over rules and laws is that the latter is sometimes underestimated. The East is not without laws and rules, they are more or less followed, but often only with the grease provided by guanxi or by social networks more broadly speaking.

### *Authority and status*

At first glance, status and authority seem related to social networks. They both connect people and shape social relationships but do so in very different ways. The former is about power and relies on order and obedience, respect, adherence to rules, and top-down information flow. This is contrasted to the latter, which operates through discussion and negotiation, reciprocity, respect, mutual exchange of information and long-term trust. In the construction industry, even in other countries that are not influenced by Confucianism, there is sometimes a clash between what is called a 'clan culture' or social network relations and 'hierarchy culture' (Teravainen et al., 2018). In part, this may be because the workforce is often a large part immigrant or migrants.

Status and authority are vertical and usually hierarchical dyadic social ties which stand in contrast to horizontal web of relations that make up social networks. In many Eastern countries status and authority are shaped by Confucianism, which places a strong emphasis on 'wu-lun', otherwise known as the 'five relationships' that create a code of conduct based on hierarchical dyadic social ties (Child & Warner, 2014). The five relations are defined as being between (1) father and son, (2) husband and wife, (3) elder and younger brother, (4) friend and friend and (5) emperor and subjects. These five relations are all male-oriented and paternalistic, placing woman as secondary and/or in the subordinate role. In organisations, and in society at large, relations in social networks are complemented by relationships of authority and status.

There are both methodological and theoretical considerations that arise from the paternalistic and hierarchical nature of many Eastern cultures. Methodologically, examining status and authority makes visible the ethnographer and forces the issue of positionality. It also brings up issues about the level of analysis as status and authority are established not on the jobsite or in the industry but in society, established through society wide institutions and practices. Finally, theoretically, the lens of status and authority offer insights into stubbornly persistent questions in the field.

There are ongoing calls for ethnographers to engage in, and report on, researcher self-reflection, or reflexivity, but it is still not commonplace (Oswald & Dainty, 2020). This connects to the idea that we cannot enter the field ignorant of our power and authority or lack of it. Most likely, depending on who we are interacting with, our power and authority will change (Thiel, 2012). Reflectivity is our tool for dealing with these shifting settings and our place among varied social actors. Ethnographers must take note of their own gender, race, age and other status markers such as institutional affiliation or nationality. Doing so allows us to adequately assess our relationships to others and the consequences of those relations.

Positionality and reflexivity were central to navigating the field when doing my own field research on construction sites in China. It was through reflexivity that I came to understand that, for the most part, construction jobsites in China were quite safe for foreign female researchers. Workers were excited to talk with me since I was usually the only foreigner they had met and been able to communicate with (since I spoke Chinese). Their curiosity and questions drove our relationship in the beginning, but over time, I became familiar to them. I thought that once our relationship reached that plateau, where they were comfortable with me, they might cross lines of appropriate behaviour and things could become dangerous.

However, I learned they respected me as a student researcher, emphasised on researcher, and were protective of me as a student. Furthermore, they often felt they had a responsibility to represent China in a good light (since I was a foreigner). Finally, as one project manager explained, construction workers have relatively low social standing and understand that if an American woman researcher was harmed in anyway (physically or sexually) by a worker, they would face the wrath of both their own government and the powerful American government. For these reasons, I came to realise that contrary to my expectations, I could gain access and feel safe conducting research on jobsites.

In relation to developers and large contractors, my positionality played out quite differently. I had a much lower status in relation to them. Most saw me as a woman first, and my nationality did not raise my status, it only made me a novelty. As a student researcher, I carried no authority, and in totality, my positionality made me vulnerable. In some ways, I was a seen as a kind of 'prize' by developers and large contractors. As such, my positionality opened access as they welcomed me to business dinners and meetings, an exotic ornamental addition to the gathering.

However, I quickly learned that these developers and contractors had no fear of the government. They had close relations with high-ranking officials, which is what allowed them to operate as developers. Not only did they have high status and authority, but they also had unlimited cash and they were used to getting what they wanted. It was in their presence that I felt most vulnerable and sometimes even worried about my safety. In addition to my own safety and access to jobsites and research field sites, I realised that positionality also affected what methods would be (or would not be) successful when attempting to gain access to jobsites or companies and the quality of the data collected, especially through interviews.

Beyond the methodological issues authority and status shape cultural practices in the industry with age, gender and even the migrant/local divisions creating hierarchy and shaping how the construction industry is organised and operates.

The role of status and authority in shaping construction work is not unique to Eastern cultures, it is visible in other countries that use migrant labour in the industry. Haakestad and Friberg (2020) show how, in Norway, an increasing per cent of migrant workforce in the industry changes the status of workers, and in turn, the way authority operates on jobsites. They note that theoretically and analytically we tend to treat class and status separately but 'the status situation of workers will often reflect their material rewards and formal authority in production' (p. 644). Status and authority can also operate as reinforcing statuses. For example, gender on the jobsites advantage men through informal mechanisms such as homosocial loyalty and gendered opportunities.. (Chappell & Galea, 2017). Gender also operates to keep women in 'less skilled' or paid less in the same positions. Other studies point to the way that status and authority prevent workers from reporting safety concerns (Lestari et al., 2020; Loosemore & Lam, 2004; Paap, 2006).

Studies using qualitative methods, especially ethnography, should continue to include sections on positionality and reflexivity, which increases validity and reliability of the studies. Beyond methodology, scholars should use the status and authority lens to interrogate persistent questions and problems that remain in the field and to help us see what we do not see when we do not use the lens.

### *Public/private divide*

The public/private divide is important to understand how an industry or organisation, or group of workers and managers, is embedded into the broader society. The public/private divide should be placed in the cultural category of 'group attachment and relations' as outlined by Choi et al. (2015). Other ways to conceptualise this cultural category (group attachment and relations) include individual/collectivism (Hofstede, 1983), individualism/communitarianism (Trompenaars & Woolliams, 2002), embeddedness/autonomy (Schwartz, 2006) and collectivism/institutional (Brodbeck, Chhokar & House, 2007). I have chosen to use the public/private divide to capture the way that space shapes group attachment and relations because often this divide is quite different in Asia, and in the East more broadly, as compared to the West. How the public and private is defined and practiced shapes how space is constructed and, in turn, shapes how groups are embedded into the broader society. When we think about the public/private divide, on the sector and organisational level, we can see how it may influence the structure of organisation and how they are embedded in society. On the family and individual level, the public/private divide provides a lens to examine how space is produced in everyday life and work.

Evans (1996) opens the door to exploring state and society relations, in collecting cases that examine how they can create synergy rather than zero-sum outcomes in economic development. In the case of China, the state is clearly classified as public, and while there are private businesses, many businesses, especially larger companies, are state-owned and, hence, blur the traditional notion of public-private divide. In fact, on one jobsite, there are probably all sorts of businesses operating, including property developers which may be large state-owned enterprises, small private contractors with workers, and larger labour contractors that may be a village or township level public enterprise or operate as a private contractor.

In the East more generally, the prevalence of state-owned enterprises changes the landscape and blurs the public-private dichotomy that we are accustomed to in the West. On the macro state level, this is compounded as the divide is further blurred with states pushing strong industrial policies and setting the direction for the overall industry. In some places it opens interesting new approaches to construction, with public-private partnerships, but it can also create new spaces of exploitation and obscure the source of difficulties. On the organisational level, the types of businesses and their relationship to the state shape formal and informal practices, change the incentives and have other significant influences. In some situations, this kind of organisational structure creates synergies and does facilitate work and practices, but in other cases, these same organisational structures can be the source of corruption, inefficiencies, or other perverse behaviours that are counter to organisational goals.

On the individual level, the public/private divide is also complicated. Family and private life and work are often blended together in ways that are unfamiliar in the West. This is especially pronounced in the construction industry which often relies on migrant workers. In some cases, they migrate with families, but in other cases, the vast majority leave their families behind and migrate by themselves. When there are large populations of men migrating to the cities without their families, the lines between public and private can become blurred. As a result, workers are more likely to live and work on the jobsite, fusing the public and private and reproductive and productive realms of their lives. This can shape the management styles, worker's agency, exploitation, the rhythm of life and workplace struggles (Ngai & Smith, 2007).

Along these lines, Fei (2020) shows how 'compound labour regimes' instituted by Chinese construction companies in Africa shape the interactions among workers and shape worker struggles. At the same time, construction sites become 'melting pots' or 'boiling pots' of migrants from different places. The cultural differences among workers and between local project managers and different groups of workers often lead to communication problems, mismanagement and low productivity (Ling et al., 2013).

Families left behind mean that the productive work is separated across place and space from the reproductive work. Construction workers send remittances home to help fund the reproductive work and they themselves, find ways to traverse space to keep those relationships. Thus, these workers are linked to politics of mobility, sometimes across nations (Yeoh et al., 2017). In other sectors of the industry, workers take over public spaces and use them as physical 'labour markets', places where they can go and secure work in as day labourers. In some places, they may become much less public as labour bosses control the market and the space, determining who can come and how much they will pay for labour (Dhal, 2020; Swider, 2016; Theodore et al., 2006).

The sharp rupture of public and private space and practices in the industry pushes ethnographers to think about space in the tradition of Lefebvre (2014). We need to document how space is used and constructed, how it connects or separates construction workers on the jobsite and how it connects and separates them from their families, the public and the rest of society. Scholars have shown how remaking space is itself a process of shaping working bodies and social relationships among them (Lyon, 2013). The understanding of space put forward by the urban theorist, philosopher and sociologist, Lefebvre (2014), is central to thinking about how these bodies are embedded and/or dis-embedded, on the jobsite, in their communities and in society.

## Conclusion

In this chapter, I have outlined some of the important cultural considerations that ethnographers who are researching the construction industry could consider. I focused on informal/formal, relationships and rules, authority and status, and public and private. These elements of culture provide empirical examples of how cultural considerations are imperative to ethnographic research done in the East. Including such considerations and weaving cultural analysis into our ethnographic fieldwork will yield rich new data and understandings to the underlying processes that occur in this industry and help us answer persistent questions.

The focus on the informal/formal divide requires attention be given to understanding whether the industry and specific projects are operating in or across the informal and formal sectors of the economy. Also, whether the workforce is hired formally or informal. Attention to the structure of the industry and workforce is crucial because it impacts everything from the logics behind processes on the jobsites and worker safety to power relations between social actors.

Attention given to relationships and rules is also crucial as the relative importance and formality of rules varies across cultures. In some places, it is social relationships that play an important role in regulation in place of rules. Ethnographers, instead of assuming the rules are the rules, can pull back the curtain and look at social networks and how they operate to gain a better understanding of the industry and what is happening. Similarly, status and authority are configured differently across cultures, strongly impacting processes and relations.

Finally, the examination of the public/private divide is highlighted because social space is constructed differently across cultures. The relationship between public and private varies and can be blurred. The implications are far-reaching. For example, in the East, it is not uncommon for the workplace and home to be blurred with workers living on jobsites and at workplaces. Also, there is a much higher prevalence of state-owned enterprises and town and village enterprises, which operate quite differently than private businesses. While I chose to highlight these elements or arenas of culture, there are many more that have been identified by cultural scholars and may be equally or more important, depending on the project.

## Notes

1 This would include China, Japan, Mongolia, North Korea, South Korea and Taiwan.

2 South Asia consists of the countries Bangladesh, Bhutan, India, Maldives, Nepal, Pakistan and Sri Lanka, with Afghanistan also often included. Southeast Asia includes Brunei, Burma (Myanmar), Cambodia, Timor-Leste, Indonesia, Laos, Malaysia, the Philippines, Singapore, Thailand and Vietnam.

3 These four dimensions in some ways overlap and intersect with Hofstede's (1983) four cultural factors: (1) individualism/collectivism, (2) power distance, (3) uncertainty avoidance and (4) masculinity/femininity. His list was deduced by studying the national cultures of 50 countries.
4 See work done by Theodore et al. (2009) (and also Chappell & Galea, 2017; Portes & Sassen-Koob, 1987).
5 Informal Economy Database (worldbank.org).
6 ILO (2018).

## References


Alon, I. (2003). *Chinese Culture, Organizational Behavior, and International Business Management.* Greenwood Publishing Group, Santa Barbara.

Ankrah, N. A., & Langford, D. A. (2005). Architects and contractors: A comparative study of organizational cultures. *Construction Management and Economics*, 23(6), 595–607. https://doi.org/10.1080/01446190500126973

Bian, Y. (1997). Bringing strong ties back in: Indirect ties, network bridges, and job searches in China. *American Sociological Review*, 62(3), 366–385. https://doi.org/10.2307/2657311

Bian, Y. (2018). The prevalence and the increasing significance of guanxi. *The China Quarterly*, 235, 597–621.

Bourdieu, P. (2018). The forms of capital. In *The Sociology of Economic Life* (pp. 78–92). Routledge, Abingdon.

Burlotos, C., Kijewski-Correa, T. L., & Taflanidis, A. A. (2020). The housing market value chain: An integrated approach for mitigating risk in informal residential construction in Haiti. *Sustainability*, 12(19), 8006.

Centeno, M.A.; Portes, A. 2006. The informal economy in the shadow of the state. In: Fernandez-Kelly, P.; Sheffner, J., eds. Out of the shadows: political action and the informal economy in Latin America. University Park, PA: Pennsylvania State University Press: 23-48.

Chan, T.-K., & Pribadi, K. S. (2022). *Construction in Indonesia: Looking Back and Moving Forward.* Routledge, Abingdon.

Chappell, L., & Galea, N. (2017). Excavating informal institutional enforcement through 'rapid' ethnography: Lessons from the Australian construction industry. *Gender and Informal Institutions*, 67–89.

Chen, M. A. (2012). The informal economy: Definitions, theories and policies [WIEGO Working Paper No.1].WIEGO

Chhokar, J. S., Brodbeck, F. C., & House, R. J. (Eds.) (2007). *Culture and Leadership across the World: The GLOBE Book of In-depth Studies of 25 Societies*. Routledge, Abingdon.

Child, J., & Warner, M. (2014). Culture and management in China. In *Culture and Management in Asia* (pp. 24–47). Routledge, Abingdon.

Chmutina, K., & Rose, J. (2018). Building resilience: Knowledge, experience and perceptions among informal construction stakeholders. *International Journal of Disaster Risk Reduction*, 28, 158–164.

Choi, J. O., Gad, G. M., Shane, J. S., & Strong, K. C. (2015). Culture and organizational culture in the construction industry: A literature review. 5th International/11th Construction Specialty Conference.

Chou, Y. K. (2006). Three simple models of social capital and economic growth. *The Journal of Socio-Economics*, 35(5), 889–912.

Dhal, M. (2020). Labor stand: Face of precarious migrant construction workers in India. *Journal of Construction Engineering and Management*, 146(6), 04020048. https://doi.org/doi:10.1061/(ASCE)CO.1943-7862.0001761

Eade, J., Drinkwater, S., & Garapich, M. (2007). Class and ethnicity: Polish migrant workers in London: Full research report. ESRC End of Award Report, RES-000-22, 1294.

Evans, P. (1996). Introduction: Development strategies across the public-private divide. *World Development*, 24(6), IN1, 1033–1037.

Fei, D. (2020). Variegated work regimes of Chinese investment in Ethiopia. *World Development*, 135, 105049.

Fei, X., Hamilton, G. G., & Zheng, W. (1992). *From the soil: The Foundations of Chinese Society*. University of California Press, Oakland.

Geertz, C. (1973). *The Interpretation of Cultures* Basic Books, New York.

Granovetter, M. (1983). The strength of weak ties: A network theory revisited. *Sociological Theory*, 201–233.

Guthrie, D. (2012). *China and Globalization: The Social, Economic, and Political Transformation of Chinese Society.* Routledge, Abingdon.

Haakestad, H., & Friberg, J. H. (2020). Deskilling revisited: Labour migration, neo-Taylorism and the degradation of craft work in the Norwegian construction industry. *Economic and Industrial Democracy*, 41(3), 630–651. https://doi.org/10.1177/0143831x17735671

Hirslund, D. V. (2021). Brokering labour: The politics of markets in the Kathmandu construction industry. *Ethnography*, 22(4), 495–514. https://doi.org/10.1177/1466138119886601

Hirslund, D. V. (2022). A culture of informality? Fragmented solidarities among construction workers in Nepal. *Focaal*, 2022(93), 75–88.

Hofstede, G. (1983). National cultures in four dimensions: A research-based theory of cultural differences among nations. *International Studies of Management & Organization*, 13(1–2), 46–74.

Hofstede, G. (2001). *Culture's Consequences: Comparing Values, Behaviors, Institutions and Organizations across Nations*. Sage, Thousand Oaks.

Hong, J., & Engeström, Y. (2004). Changing principles of communication between Chinese managers and workers: Confucian authority chains and guanxi as social networking. *Management Communication Quarterly*, 17(4), 552–585.

Horak, S., Taube, M., Yang, I., & Restel, K. (2019). Two not of a kind: Social network theory and informal social networks in East Asia. *Asia Pacific Journal of Management*, 36, 349–372.

ILO (2018). *Women and Men in the Informal Economy: A Statistical Picture* (3rd ed.). International Labour Office, Geneva.

Lee, B. H., Swider, S., & Tilly, C.2020. "Informality in action: A relational look at informal work." *International Journal of Comparative Sociology*. 61(2-3): 91–100. Sage London

Lefebvre, H. (2014). The production of space (1991). In *The People, Place, and Space Reader* (pp. 289–293). Routledge, Abingdon.

Lefebvre, V., Sorenson, D., Henchion, M., & Gellynck, X. (2016). Social capital and knowledge sharing performance of learning networks. *International Journal of Information Management*, 36(4), 570–579.

Lestari, F., Sunindijo, R. Y., Loosemore, M., Kusminanti, Y., & Widanarko, B. (2020). A safety climate framework for improving health and safety in the Indonesian construction industry. *International Journal of Environmental Research and Public Health*, 17(20). https://doi.org/10.3390/ijerph17207462

Li, J., Wang, W., Sun, G., Jiang, Z., & Cheng, Z. (2018). Supervisor–subordinate guanxi and job satisfaction among migrant workers in China. *Social Indicators Research*, 139(1), 293–307. https://doi.org/10.1007/s11205-016-1471-6

Ling, F. Y. Y., Dulaimi, M. F., & Chua, M. (2013). Strategies for managing migrant construction workers from China, India, and the Philippines. *Journal of Professional Issues in Engineering Education and Practice*, 139(1), 19–26. https://doi.org/doi:10.1061/(ASCE)EI.1943-5541.0000124

Lizarralde, G., & Root, D. (2008). The informal construction sector and the inefficiency of low cost housing markets. *Construction Management and Economics*, 26(2), 103–113.

Loosemore, M., & Lam, A. S. Y. (2004). The locus of control: A determinant of opportunistic behaviour in construction health and safety. *Construction Management and Economics*, 22(4), 385–394. https://doi.org/10.1080/0144619042000239997

Lyon, D. (2013) The labour of refurbishment: the building and the body in space and time. In Pink, S. and Tutt, D. (eds) *Ethnographic Research in the Construction Industry*. London: Routledge.

Nahapiet, J., & Ghoshal, S. (1998). Social capital, intellectual capital, and the organizational advantage. *Academy of Management Review*, 23(2), 242–266.

Nathan, A. J. (1993). Is Chinese culture distinctive? – a review article. *The Journal of Asian Studies*, 52(4), 923–936.

Ngai, P., & Smith, C. (2007). Putting transnational labour process in its place: The dormitory labour regime in post-socialist China. *Work, Employment and Society*, 21(1), 27–45.
Oswald, D., & Dainty, A. (2020). Ethnographic research in the construction industry: A critical review. *Journal of Construction Engineering and Management*, 146(10). https://doi.org/10.1061/(ASCE)CO.1943-7862.0001917
Paap, K. (2006). *Working construction: Why white working-class men put themselves – and the labor movement – in harm's way*. Cornell University Press, Ithaca.
Park, S. H., & Luo, Y. (2001). Guanxi and organizational dynamics: Organizational networking in Chinese firms. *Strategic Management Journal*, 22(5), 455–477.
Portes, A., & Sassen-Koob, S. (1987). Making it underground: Comparative material on the informal sector in Western market economies. *American Journal of Sociology*, 93(1), 30–61.
Schwartz, S. H. (2006). A theory of cultural value orientations: Explication and applications. *Comparative Sociology*, 5, 136–182.
Sewell, William H., Jr, 2005, "The Concept(s) of Culture", in Practicing History: New Directions in Historical Writing After the Linguistic Turn, Gabrielle M. Spiegel (ed.), London: Routledge, 76–95.
Swider, S. (2016). *Building China: Informal Work and the New Precariat*. Cornell University Press, Ithaca.
Teravainen, V., Junnonen, J.-M., & Ali-Loytty, S. (2018). Organizational culture: Case of the Finnish construction industry. *Construction Economics and Building*, 18(1), 48–69.
Theodore, N., Valenzuela, A., & Meléndez, E. (2006). La esquina (the corner): Day laborers on the margins of New York's formal economy. *Working USA*, 9(4), 407–423.
Theodore, N., Valenzuela, A. Jr, & Meléndez, E. (2009). Worker centers: Defending labor standards for migrant workers in the informal economy. *International Journal of Manpower*, 30(5), 422–436.
Thiel, D. (2012). Ethnography and flux: Identity and epistemology in construction fieldwork. In *Ethnographic Research in the Construction Industry* (pp. 73–90). Routledge, Abingdon.
Trompenaars, F., & Woolliams, P. (2002). A new framework for managing change across cultures. *Journal of Change Management*, 3(4): 361–375.
Tutt, D., Pink, S., Dainty, A. R., & Gibb, A. (2013). Building networks to work: An ethnographic study of informal routes into the UK construction industry and pathways for migrant up-skilling. *Construction Management and Economics*, 31(10), 1025–1037.
Wells, J., & Jason, A. (2010). Employment relationships and organizing strategies in the informal construction sector. *African Studies Quarterly*, 11(2/3), 107.
Yeoh, B. S., Baey, G., Platt, M., & Wee, K. (2017). Bangladeshi Construction workers and the politics of (im) mobility in Singapore. *City*, 21(5), 641–649.
Zou, J., Zillante, G., & Coffey, V. (2009). Project culture in the Chinese construction industry: Perceptions of contractors. *Construction Economics and Building*, 9(2), 17–28.

# 4

# NEGOTIATING GENDER IN CONSTRUCTION ETHNOGRAPHY

*Natalie Galea*

*University of Melbourne, Melbourne, Australia*

**Key takeaways**

- Gender cannot be ignored especially when undertaking ethnography in male-dominated work sectors such as construction.
- An ethnographers' gender shapes access to the field, production of knowledge and participant-researcher relationships.
- A twinned researcher approach (woman/man) can provide rich insights into the gendered nature of the rules and practices in male-dominated work sectors.
- Researcher's positionality and reflexivity must consider the researchers' gender and the gendered context within which the research occurs.

## Introduction

Ethnography is a useful tool to understand covert resistance to gender equality in male-dominated sectors such as construction. It allows researchers to 'get under the skin' of an organisation, to observe the rules-in-use that preserves men's powerfulness and overrepresentation in the work setting. Through ethnography, researchers can observe how taken-for-granted and deeply embedded informal rules intersect with gender to maintain the gendered status quo in construction. This is because ethnography has the potential to make explicit the tacit, seemingly banal social dimensions of organisations (Ybema et al., 2009, p. 7) or the 'familiar strange' (Ferguson, 1993, p. 122). However, little consideration has been given in the literature about the ethnographer's gender and its influence on ethnographic research in the gendered space of the construction context. For example, whether an ethnographer's gender influences the participant-researcher relationship or the researcher's ability to slip-in under the radar and build a rapport with participants. Or whether the ethnographers' gender shapes the data collected. In this chapter, I draw on my experiences in the field as a woman and feminist ethnographer studying gender in the context of construction research. I compare my experience studying gender equality in the construction sector with my male 'research twin,' with whom I undertook the ethnography within the context of the Australian construction sector. I explore how gender dynamics affected my field work experience and research outcomes.

This chapter is broken into four sections. In Section I, I outline the importance of ethnography and how it intersects with feminist research ethics and gender. In Section II, I consider

DOI: 10.1201/9781003379584-5

how construction scholars have interacted with gender, I detail our rapid ethnographic case studies with a twinned researcher approach (man/woman and construction insider/outsider) and include my positionality statement. In Section III, I compare my experiences with that of my male research twin whom I debriefed with routinely while in the field. I conclude by emphasising that the construction context is not a gender-neutral space and encourage scholars who are applying ethnographic methods in the construction context to consider the gender dynamics of their research.

## Section I: The importance of ethnography in construction research

The strength of ethnography in the context of the construction sector lies in its ability to allow the researcher to step into the daily lives of participants, to unveil the complexities of the routine organisation of work. Ethnographic research reveals what a stranger would need to know if wanting to participate meaningfully in social groups within the construction sector (Creswell & Poth, 2017, p. 150). It is the embedded ethnographic method of enquiry that allows the researcher to 'see' first-hand what people do and say in a particular context. In the context of the construction sector for example, it helps to reveal what practices and norms are performed, and whether they are gendered, how work is organised and adhered to on a construction site, how these patterns of work affect the lives of construction workers and the gendered implications of such work structures (Hammersley, 2006).

Through a process of observing, occasionally participating, conversing, interviewing and closely examining documentary sources, ethnography allows the researcher to step into the contextual life of participants and interpret how things are viewed and understood from their perspective: the multiplicity of voices and interpretations (Hammersley & Atkinson, 1995; Ybema et al., 2009). Conducting an ethnography has the advantage of providing rich insight into participant's behaviour, the institutions they respect and enforce and the context they inhabit (Pink et al., 2010; Khan & Jerolmack, 2013; Jerolmack & Khan, 2014). It allows the researcher to witness naturally occurring events rather than simply relying on participants' accounts. To see first-hand, for example, how company rules and policies are understood, practised or ignored.

Traditionally ethnography has been seen as the result of the researcher's efforts to 'soberly describe' what he or she experiences through immersion in the field (Smit & Onwuegbuzie, 2018). Ethnographies conducted in workplaces have relied on an interpretative framework (Van Maanen, 2006; O'Reilly, 2012) where the researcher needs to have distance from the field to translate experiences and create meaning from their observations (Cunliffe & Karunanayake, 2013; Holck & Villeseche, 2023). In this way, ethnography demands that the researcher is consciously self-aware in the 'meaning making' process. Ethnographic research requires the researcher to recognise the effect they are having on the field, data collection and analysis (known as reflexivity). And what biases, privilege and power they as the researchers are bringing to the research (known as a researchers 'positionality') and how this might shape the knowledge claims they make (Ybema et al., 2009; Pink & Morgan, 2013).

### Feminist research ethics and ethnography

There are crossover points between ethnography and feminist research ethics including the importance of recognising reflexivity and positionality. A feminist informed approach to ethnography asks researchers to view their research through a gendered and power lens

(Harris, 1968; Krook & Squires, 2006; Ackerly & True, 2008, 2010, 2013). Feminist researchers are particularly conscious of the power relations which exist in the field between researcher and participants both theoretically and pragmatically (Weems, 2006). Feminist ethnography is political and active (McLaren, 2002). As Brooke Ackerly and Jacqui True (2010, p. 2) explain: 'A critical feminist perspective uses critical inquiry and reflection on social injustice by way of gender analysis, to transform, and not simply explain, the social order.'

In practice, this means committing to exploring absence, silence, difference and oppression and to finding new ways of knowing that will hopefully lead to new ways of doing. In this sense, a feminist ethnography operates from a normative foundation. Feminist scholars continue to challenge the idea of neutral and objective ethnographic research and ask researchers conducting ethnography to consider how social identities (such as gender, race, disability, age, class, ethnicity, sexuality, etc.) and personal attributes of the researcher influence the research process (Caplan, 1993; Speer, 2002; Daley, 2010; Haddow, 2022; Rodriguez & Ridgway, 2023). Feminist researchers are often, but not always, women.

Despite being conventionally used to explain the perceived differences between women and men, I understand gender to be an analytical tool that operates both as a 'category' tied to cultural symbolism of masculinity and femininity and a 'process' that organises social activities and power (Scott, 1986; Beckwith, 2005). As a category, gender represents the fluid, socially constructed and culturally specific characteristics, values and behaviours conceived of as masculine and feminine (Beckwith, 2005, p. 131).

Gender is embodied and enacted in everyday social interactions and processes. Bodies are intricately involved in the social reproduction of gender – the 'doing' of gender – as both objects and agents (West & Zimmerman, 1987; Connell, 2002). This is important in relation to ethnography as Holck and Villeseche (2023, p. 36) observe that 'body and bodily relations make the research of workplace ethnography much more complex, multi-layered and intersubjective, distorting and contracting the ideal of neutrality and objectivity embedded in (traditional) ethnographic research.'

While mainstream methodological textbooks have been largely silent on the importance of gender in conducting ethnography (Poulton, 2012), feminist scholars have long documented the complexities faced by women conducting research in male-dominated settings and cultures (Easterday et al., 1977; Bordo, 1991; Caplan, 1993; Olive & Thorpe, 2011). For example, women researchers have documented being subjected to sexualisation, sexualised humour, banter and flirting, sexist behaviour, not been taken seriously, or have men confide in them about personal vulnerabilities, as women researchers are seen as 'powerless and nonthreatening' (Easterday et al., 1977, p. 344) or holding an 'inferior and devalued position' (Gurney, 1985, p. 12).

For some women researchers, the problem of sexualisation from participants via comments about her appearance and questions about her private life leave them wanting to leave the field earlier than planned (Haddow, 2022, p. 317). Evasive action, laughing off approaches by participants and attempting to change the subject rather than directly challenging sexualisation and at the extreme end, engaging a male chaperone to accompany them, are some of the tactics employed by women researchers in response to being sexualised by participants in the field (Warren & Hackney, 2000; Haddow, 2022).

Gender and other social determinants have also shaped researchers ability to achieve 'insider status' through rapport and a position of proximity to participants in the field (Hodkinson, 2005; Haddow, 2022). According to Caplan (1993, p. 178) the role adopted by women researchers in the field is often shaped by positionality, i.e. 'Who we are for them? Who are they for us? (…).' Thus, women researchers conducting ethnography in male-dominated settings often find themselves being 'othered' or expected to perform according to gendered scripts like the women participants they may be studying (Mazzei & O'Brien, 2009).The process of conducting ethnography for women in male-dominated environments helps us to understand issues of women's 'representation and legitimatisation' (Haddow, 2022, p. 314). That is not to say that there is uniformity in how women researchers and women participants are responded to, rather it is a useful reminder of the power dynamics which disproportionately affect women in male-dominated spaces (Gurney, 1985; Puwar, 2004).

## Section II: Gender and ethnography in construction

Ethnographic studies on the construction sector to date have raised gender issues and drawn attention to the masculinity operating in and around construction practices (Applebaum, 1981; Paap, 1999; Mars, 2005, 2006; Thiel, 2007; Datta & Brickell, 2009), but few ethnographies have explicitly focused on gender (Iacuone, 2005; Faulkner, 2009; Galea, 2018; Galea et al., 2020). David Oswald and Andrew Dainty's (2020) review of ethnographic research in construction finds researchers lack reflexivity and self-scrutiny, overlooking critical reflection of their role and influence on studies. As Emma Poulton (2012) reminds us, 'a failure to consider the positioning, practices and performance of the gendered self in the gendered space' treats the construction sector as neutral space and ignores the well-documented and deeply embedded gender power relations that exist in construction workplaces (Galea, 2018; Turner et al., 2021).

While reflexivity is a prerequisite for a robust ethnographic practice, few scholars in construction research have specifically discussed how they have negotiated their gender and social identities while conducting ethnographic studies in construction workplaces (Paap, 2006; Lowstedt, 2015) or on the construction workforce (Gama Gato & Matyska, 2021). Kris Paap's (2006, p. 3) work is a notable exception. As an apprentice carpenter and ethnographer in the USA over a period of two years, Paap (2006) describes herself taking 'social beating[s]' while 'attempting to negotiate a constantly changing set of social conditions in which I, as a female on site, would be the object against which the men's masculinities would be measured and against which their masculinities would be proved.'

Reflexive rigour, while good practice, exposes inherent risks and challenges faced by researchers in the field (Oswald & Dainty, 2020). Reasons that construction scholars have shied away from self-reflexivity maybe that the very process challenges the notion that a researcher is objective and free from bias. Exactly this reflexivity reveals the researcher's bias, giving the researcher nowhere to hide. Sanitised ethnographic accounts also fail to reflect the impact on researchers. Rebecca Olive and Holly Thorpe (2011, p. 424) call on researchers to differentiate whether their reflective approaches are 'theoretical, methodological, or representational.' In my case study described next, I applied theoretical and methodological reflexivity. Methodologically this was undertaken via a debriefing session with my research partner, Adam and documenting patterns of absence, silence or othering in the field which links with theoretical reflexivity, through recognition that in some cases the informal gender rules applied to me and my research team. I detail this next, but first begin by describing my positionality.

## My positionality

I grew up in a construction family. My working-class parents owned and operated a successful plumbing business from our home. Despite familiarity with the construction sector, on leaving school I never considered a career in construction. Only after I had competed in the Olympics, where there is little financial security, did I decided to enrol in a construction management degree. This was my second bachelor's degree. I was 25 years old. I started working in construction immediately. I realise on reflection that my gender and difference was constantly pointed out to me. These reminders came often, seemingly in an offhand manner, and with little thought. I did not focus on them for long, I was eager to fit in and succeed in my construction career. I was employed by large construction companies and property development firms in Australia and the Middle East, including one of the companies I later studied in my PhD ethnography.

I felt my career progression was slow. I was often the only woman in the room. I frequently felt excluded from roles, opportunities and social events that my male peers were invited to. While working in the UAE and Qatar, I noted my racial privilege. Being white and Australian carried currency that gave me legitimacy amongst some. However, my gender continued to undermine my legitimacy as a construction professional. I came to this ethnographic study as a middle-aged middle-class PhD student with pre-existing social and cultural connections to the construction sector. I was, and am, a feminist working for social transformation and gender justice. Each one of these attributes fundamentally positions my scholarship.

I found that the process of ethnography made me feel accountable to my participants and peers and engaged in improving the conditions of women and men working in construction. Although I had worked in the construction sector for nearly two decades, throughout the research process I saw in operation for the first-time gendered dimensions to norms, practices and ways of working in construction. My identity shifted during my research. At the beginning of my PhD, I described myself as an industry 'insider.' But as the research progressed, I realised as a woman I had never been an 'insider.' I more accurately became to think of myself as an 'informed outsider.' I became comfortable observing my once familiar environment from a distance. That is not to say that I remained impervious to what I observed, heard and experienced during the research.

I experienced states of shock, anger, sadness, repulsion, exhaustion, frustration, exhilaration, delight and hopefulness in the field, during analysis and in the delivery of findings which inspired, challenged and frustrated me as a researcher and feminist. An example of the kind of confusing discomfort I felt during the research can be found in this excerpt from my fieldnotes. It details my interaction with an ex-colleague, a man of my age that I had known well enough to chat with casually and at company events. He had agreed to be interviewed when we met by chance on site the day before:

> Just after lunch and in a hushed tone, conscious not to disturb others in the open planned office I approached Tom for the interview. Tom did not look up from the printed spreadsheet he was reading. He ignored me. His finger running along the document. His hand made a fist which he brought towards his ear, like he was going to punch someone, and he yelled at the top of his voice, 'What the fuck!' After a pause and still not turning to look at me, he said, 'Yes. I will be available this afternoon, but I have other things to do right this moment.' I could feel my eyes grow wider as I attempted to keep control of the rest of my face. Embarrassed and uncomfortable from the confrontation, I began to retreat back to my desk, 'Okay. Later then,' I said. As I walked away avoiding eye contact with anyone else in the office, Tom's voice lowered, 'After 4.30pm I will be ready.' I don't think he looked at me the entire encounter.

Sometimes as an ethnographer you are spontaneously jolted into self-reflexivity. In this case, I was left feeling embarrassed and put in my place, but also elated. Instead of changing his behaviour, I had been treated to an example of hyper masculine command and control behaviour where I was not left wondering who had power. During my ethnography, I was reminded regularly of my humanity. I found myself consciously negotiating and suppressing strong emotions, particularly when participants' experiences chimed with my own, for example, when women participants described in interviews the favouritism given to male colleagues by managers and their frustration in the face of discrimination. Trying to fit long hours and trying to fit in my care responsibilities that included supporting my unwell father, left me tired and often emotional. These experiences brought into sharp focus how rules and practices in construction are gendered or have gendered effects.

Going into the field with a 'twin' enhanced reflexive rigour. These debriefing sessions were recorded and transcribed and became a key part of our analysis. Whilst it wasn't the original intention, it emerged quickly through the debriefing sessions that we were able to compare and see how the experiences of woman and man researchers in the construction setting differ. Before I describe how the debriefing with my research twin worked, I will briefly outline our rapid ethnographic research project.

## Section III: Twinned researchers

We applied a rapid ethnographic approach to study why the number of women working in professional construction roles in Australia had plateaued despite the existence of formal company gender equality policies? Applying a feminist institutionalist lens, our study paid attention to how the 'rules-in-use' on construction sites and in construction companies intersected with gender (see Lowndes, 2020). Rules shape behaviour and exist in two forms: formal rules, which are written down and more easily identifiable, and informal rules that operate through practices, norms, and narratives. Informal rules are harder to detect and study due to their widely accepted nature. Ethnography allowed our team to observe informal rules in practice and their gendered dimensions (Galea, 2018; Lowndes, 2020).

Rapid ethnography is a focused ethnographic probe, a time-limited study within a discrete organisation or community (Muecke, 1994, p. 199). In our case, we studied two large construction companies operating in Australia. We undertook a document analysis of company policies, attended company events in head office as participant observers, shadowed construction professionals on six construction sites around Australia and interviewed construction professionals and senior management (see Galea et al., 2020 for more details on our application of rapid ethnography).

One the notable challenges we encountered using ethnography to study gender relations on construction sites was obtaining ethics approval from our university. We did not want to disclose to participants explicitly that we were studying gender. This was to minimise participants' altering their behaviour due to our presence as we were studying the embedded taken-for-granted gendered rules in operation in construction. But our university ethics committee made the disclosure of our research intent a condition of our ethics approval (see for more details Chappell, 2020, p. 131). In practice, participants probably did adjust their behaviour when we first began to shadow and observe them, but for many participants, the pace of construction work meant they often forgot our presence to a point.

We chose a twinned researcher approach to our study, attending the events and sites in pairs, although each researcher worked independently throughout the day, interviewing, observing, and shadowing. A critical component of our methodology involved the two researchers debriefing at the end of each site visit/event. The twinned researchers were a construction insider and a construction outsider and for the most part, one of our team was a woman and the other a man. I was the construction insider and woman, and my twin for much of the field work was Adam. An early career researcher, Adam was a sociologist who studied masculinities. Adam had never worked in construction sector. Adam and I represented 'twins' because we shared the same theoretical background – masculinities and feminist institutionalism. At times, members from the larger research team also pulled on their steel cap boots and joined Adam or I on site to shadow construction workers.

We chose a twinned approach due to the documented challenges women researchers faced 'blending in' into a male-dominated environment and we expected that a 'gendered logic of appropriateness' (Chappell, 2006) may shape some participants behaviour in the field depending on whether they were shadowed by a woman or a man. Although we were also curious to see if it would be easier for me as the construction insider to build quicker rapport, as I understood construction and spoke the same language as participants.

We found that Adam and I observed informal rules, norms and practices that were gendered or had gendered consequences that advantaged men including performing to hyper-masculine scripts, practicing presenteeism, total availability and homosociality (where older men sponsor younger men – see Galea et al., 2023). Yet our gender gained us discrete access to different types of gendered interactions and notably different informal gendered rules because of our gender and the gendered nature of construction.

## Blending in and being 'one of the boys'

Adam was made to feel like 'one of the boys' by participants. Men freely swore vigorously in front of him without a filter. As Adam's fieldnotes show, swearing was often performative, a demonstration of the participants hypermasculinity.

> Howard asked Doug to attend a meeting for him while he was away. Doug questioned Howard's decision, but Howard replied, 'I want you to go to the meeting with the plumbing subcontractor. I want you to put some c*nt in the meeting.' He used the word 'c*nt' to mean, a bit of mongrel, hard-nosed edge. Everybody had a laugh at that. Doug looked like he really liked that idea of him being a c*nt and being asked to be a c*nt. I had observed Doug during the week; he walked around the site office with his chest puffed out wearing a big moustache. He looked like a big peacock.
>
> *(Adam, Fieldnotes)*

Adam was given a nickname by participants to his face and was made to feel like 'one of the boys.' As Adam explains in one of our debriefing session:

> Being a male researcher, my experiences were a little different to yours in that I really felt like I got brought into the team … Tom, he had nicknames for everybody. Like endearing nicknames like his other foreman, Josh Cook, was called 'Cookie' and everybody had a nickname. And within half an hour he was calling me 'Ads'. I guess that was an interesting thing in becoming part of team, becoming accepted you get given a nickname.
>
> *(Adam, Debriefing session)*

Men did swear in front of me, but they rarely used the word c*nt and if they did so, an apology swiftly followed. Class and context appeared to play a role in whether men apologised to me for swearing. Blue collar (wage) workers and tradespeople apologised more often for swearing in front of me than white collar (salaried) construction professionals who seemed to not care as much about my presence. As my fieldnotes show, apologies were so prevalent, I started to count the number of times a day I was apologised to:

> The level of swearing in the office … they punctuate their sentences with 'f*ck', where onsite, I was apologised to four times today by blue collar construction workers. Each time a man swore and saw that I was in earshot, he mumbled an apology. The white-collar guys did not seem to care if they swore in front of me.
>
> *(Natalie, Fieldnotes)*

Unlike Adam, I was never given a nickname that I knew about. Rather I experienced three incidents of pushback from male participants which Adam did not. I have already mentioned the first incident earlier when discussing my positionality. The second was the passive resistance from a group of men on site including a junior site engineer named Paul, the site manager and design manager. Paul had recently been awarded a 'construction cadet of the year' and had nominated himself to be shadowed. Despite sitting within a metre of each other and working closely within a small site shed, the site manager and design manager did not engage with me at all over the entire week. If I made eye contact, they would look away. They never introduced themselves nor spoke to me the entire week. Their behaviour aside, it was challenging to build rapport with Paul who spent almost the entire day sitting at his computer and working on the same document. When I asked Paul questions, his responses were short and instructional, describing to me the formal rules. Paul agreed to doing an interview with me, but never got back to me with a suitable time. At 5 pm, Paul turned to me and said, 'So, are you leaving?' It was clear from their behaviour that these three men did not want researchers on their site. Maybe they were suspicious. While I respect their position, I was still left frustrated by the tedium of the day and my inability to build a rapport. I was exposed to silence and distance. It gave me a sense of how hard it must be to break into that inner circle or clique on site.

The third notable pushback towards me occurred in public in front of Adam and an entire construction team. It was the end of the week, and Adam and I were observing a project team meeting as described:

> At the end of the meeting the Project Manager went around the room and asked everyone in the team to share something from their week. When it came to me, the project manager said, 'Natalie?' suggesting that I speak. I thanked the team and said if anyone wanted to be involved in an interview they could let me know. When I finished, the foreman sitting across the table from me said something which I didn't hear in full except for the words 'shut the f*ck up'. His comments were met with a gasp of shock and then embarrassed laughter from the team. I looked at him and then at the project manager and asked, 'Sorry, what was that?' The project manager said, 'Oh, don't worry. I don't understand him either.' I asked the cadet I had been shadowing all day who was sitting beside me to repeat what had been said, he said, 'No, no, I wouldn't dare repeat that.'

Adam also had not heard the foreman's statement. These three experiences left me feeling 'othered' and like a 'stranger spoiling the good atmosphere' (Holck, 2018, p. 6). It also gave me and Adam a direct insight into the tolerated antagonistic practices and behaviours in construction towards women, that act to keep construction a male space and make women feel like they are unwelcome 'space invaders' (Puwar, 2004).

Overall, it was easier for Adam to be accepted by participants who were for the most part men, and slide in under the radar in the male-dominated construction setting. I found that the researchers' gender was more important in relation to the participant-researcher relationship than the researchers' 'insider' status. Maybe if our team was composed of two men, one a construction insider and the other a construction outsider, the insider status might have mattered more. Maybe if our team were undertaking a traditional ethnography and in the field for longer things might have been different. But I found that sliding in under the radar and even building rapport was more challenging for myself and the women in our team conducting a rapid ethnography, no matter whether they were the industry insider or outsider.

## Sexualising women

On site, we were both exposed to sexualised commentary and sexist and pornographic graffiti (see Galea, 2018, p. 188 for graffiti). On occasions, Adam and I were in attendance to observe the same meetings. One project team meeting we attended featured the site team discussing their 'Christmas video' which would be screened at the company Christmas party. The conversation was light, jovial and the intent was humour, yet a video of a site team orgy in the same format as a pornographic video was inferred.

> The project manager said, 'There's suggestions that we ... do something like the Dave Hughes [Australian comedian] type video: What tradies say.' And someone suggested, 'Well why don't we get one of the foremen pissed and they can talk about the project on the video?' And then someone else asked what wins, and the project manager said, 'Well it seems that cross-dressing wins.' Basically, after a lot of light discussion, including a suggestion that they video-record a mass orgy and one of the guys said, 'Well, you know, I'll be right for, if it's only a four-minute video, I'll be right for four minutes,' said one of the foremen. The other foreman said, 'I can record it in one hand.' Acting out holding a camcorder. Like a porno. There was lots of discussions about this and laughter.
>
> *(Adam's fieldnotes)*

Participants took the sexualisation of women further with Adam. They were comfortable talking about their sexual conquests, showing pornographic movies in front of him on their iPhone and as the fieldnotes show below, perving and commenting on women on/around the construction site. It was assumed by participants that all men were heterosexual and interested in women in their gym attire.

> These were primarily with the white-collar guys from the office I guess ... Two times in particular when we drove back onto site ... there was women exercising on the [precinct]. So groups of women jogging around doing their exercise. And we'd drive past them. And there was always comments made. Some of the comments were things like, 'Oh, can you slow down a little bit?' 'You don't see that on-site every day.' Like real women as sexual

> objects … I guess, as a researcher, my job is to try and float into the background a little bit and not jump up and down at that point and say anything. Its just I guess you have to laugh along a little bit and not like jump up and down at that point and say anything.
>
> *(Adam, debrief)*

Like other women researchers, the sexual innuendos from participants in the form of humour were directed at me, which I laughed off.

> 'I forgot to bring the ethics form to have you sign it,' even though I'd already sent it to him via email [before the day]. Steve joked and said, 'Is that the form that says you're allowed to molest me?' Straight from the get-go, as soon as I entered that work area, within the first five minutes I had heard jokes with sexual innuendo, which were said in 'good' humour and I responded accordingly with a laugh, but it was 6.35 am in the morning.
>
> *(Natalie, fieldnotes)*

But unlike Haddow (2022) and others, participants did not make sexual advances towards me. Being on site for short periods of time and being a middle-aged woman may have had something to do with this. At times, when I was shadowing younger women on site, I stood back and walked at a distance behind them. This gave me the opportunity to observe the behaviours of others on the site. I was able to blend in only to a point, when stopped by polite social conventions associated with gender.

> I stood back and what I saw was akin to willy-willy[1] sweeping through the site. As Jenny walked through the site, there was a lot of reaction. There was head-turning. There was checking-out. Men congregated. There was noise and chatter as she approached. One guy stuck out his tongue as she walked by as if sexually provoked. The interesting thing is the focus was all on her, not on me as the observer. When we got to the sub-contractor, I think he was really quite shocked and pleased with himself to meet Jenny, and was very polite. 'After you,' said the sub-contractor to both of us, which was a bit uncomfortable for me because I was really trying to stay back and observe the interaction. I said to him, 'It's okay, you can go on and you don't have to worry about letting me [a woman] go before you.'
>
> *(Natalie, fieldnotes)*

One of the challenges Adam and I found we had to negotiate was how we responded to behaviours and practices that perpetuate women's disadvantage in the construction setting. At times participants realised that we were in the room, while at other times participants appeared not to register our existence. As my fieldnotes explain:

> In terms of how we behaved as researchers, we laughed along with them. There was no looking shock[ed] or anything from us. We giggled with the group … particularly around the time of the discussions about site orgies and holding the camcorder in their hand, with one hand, a la porno. At that moment, I could feel [name of project manager] looking at me and just dying as it was happening.
>
> *(Natalie, Fieldnotes)*

Laughing off sexual innuendos in my case, and not commenting on derogatory commentary about women, made Adam and I feel like we were in a double bind. Our aim as ethnographers was to blend in and extract the taken-for-granted and embedded gendered rules and practices that act to maintain men's overrepresentation in the construction sector. Yet as feminist researchers, we felt complicit in the performance of these rules and practices that perpetuate gender inequality in the construction setting. I found that there was an opportunity to ask participants about sexist comments and practices observed in fieldwork. This process of referring back to these practices, narratives and norms allowed me to delve deeper in the field of enquiry. However, I did not challenge perpetrators of sexist practices in the field, as I did not see it was my place to. I saw it was my role to build evidence of these misogynistic practices and deliberately share these findings back to the construction sector and provide recommendations for change. To answer Olive and Thorpe's (2011, p. 426) question whether we must bite our 'feminist tongue' to collect quality data in ethnography. My response is, sometimes, but only to a point. We must not forget that post research knowledge transfer is as important and powerful as data collection.

## Women's unfair advantage and men's mental health

Having twinned researchers revealed that men in our study were more comfortable talking about their antagonism towards gender equality measures to a man and talking about the impact that construction work had on their health and well-being to a woman. I recognise that rules about gender may have enabled and constrained men and women's interactions with Adam and myself.

Male participants explicitly shared with Adam their concerns about 'reverse sexism' and their perception that a focus on gender equality was giving women in construction an 'unfair advantage.' As our debriefing conversation below shows, men did not share these beliefs with me, most likely because of my gender.

***Me:*** And in terms of reverse sexism?

***Adam:*** Yep. So nearly every bloke I spoke to basically gave that narrative about women in construction.

***Me:*** I never get this narrative. That's probably because they find out I was a woman in construction, maybe.

***Adam:*** Or maybe just the fact that you're a woman. I don't think they would say that to a woman because there's too much political incorrectness inherent in what they're saying.

*(Natalie and Adam debriefing)*

From my first day on site, men shared with me the effect that construction work pressure had on them, their health and their personal relationships. Below are my fieldnotes from the first day when I was shadowing a site superintendent and at 'smoko'[2] he revealed to me very openly, that he was having panic attacks on his way into work each day:

> The long hours on site were one thing but today it was the openness of an older man, a site manager that caught me off guard. On our way to smoko in a nearby food court, the site

> manager disclosed to me that he was suffering panic attacks each morning on his drive into work. He was also drinking more heavily than normal and had recently taken up smoking again. I was surprised by his frankness. His manner was direct and affable.
>
> *(Natalie, fieldnotes)*

Daily I was confronted by the stories from men about for example their poor mental health, marriage breakdowns and substance abuse. Which is interesting because when I worked in the construction sector as a construction professional, male colleagues did not share their vulnerabilities with me, so I was quite taken aback when they did as in my role as researcher. Men tended to avoid detailed discussion with Adam about these personal issues. It is not unusual according to the literature; women researchers have remarked that in the field they have often been expected to undertake caring roles leaving them feeling emotionally exhausted (Haddow, 2022).

Unlike male participants, women shared openly with Adam and I about their personal challenges including ire from their male colleagues who feel they are receiving an unfair advantage and their frustration with persistent discrimination faced during their construction career. Given my positionality, I consciously asked Adam to shadow and interview more women participants as often their experiences upset me. While Rebekah Wilson (2010, p. 139) explained she 'smiled through gritted teeth' as she endured sexist comments. I observed that despite my years of work in construction, strangely my response was more surprise and sadness at how normative a tolerance of sexism and discrimination was. I found that the process of ethnography made me feel both accountable to my participants and peers and engaged in improving the conditions of women and men working in construction – a position familiar to other researchers (Olive & Thorpe, 2011, p. 426).

## Conclusions

Gender cannot be ignored when undertaking ethnography in the deeply male-dominated sector of construction. I found that applying a twinned (woman/man and insider/outsider) rapid ethnographic method allowed us to see clearly how gender intersects with the rules-in-use that operate within construction. While our research team observed consistent informal rules and practices that were gendered or had gendered consequences, our positionality and gender as ethnographers clearly influenced our access to the field, shaped our relationships with participants and provided us with discrete exposure to different gendered interactions and rules.

I found that a researchers' gender was more important in relation to the participant-researcher relationship than the researchers' 'insider' status. As a woman studying ethnography, my experiences reflected the well-documented experiences of women working in construction. I was 'othered,' subject to sexual innuendo and antagonism from male participants. Male participants also seemed more open to me about their vulnerabilities and the impact on construction work on their health, family and intimate relationships.

By contrast, my male research twin was accepted, blended in, exposed to different performances of masculinity relational to him and other men, and different masculine practices and narratives that were discriminatory towards women and not corporately appropriate such as sexual commentary of women and expressing concern about the unfair advantage women now have in construction. Like other feminist scholars, we were left with a dilemma as to how to navigate these uncomfortable, sexist and richly gendered interactions without feeling complicit in their performance. While there is an opportunity for these uncomfortable interactions to form part of the research inquiry with participants, it seems particularly challenging for a woman researcher in a male-dominated context to push back on these behaviours and practices while in the field, as it might result in antagonism or expulsion from the field. Rather, I focused these interactions when I was writing up the research and sharing the research findings with the construction sector. While it is acknowledged that not every ethnographer can enter the field with a research twin, it was a useful methodological approach to studying gender in the construction context. Had either Adam or I conducted this research alone, it would have eschewed the research findings leaving the rich, tacit gendered differences hidden below the surface. These findings add to the body of literatures on women ethnographers studying in male-dominated work settings.

One clear limitation of this chapter is that it reflects binary notions of gender in relation to the researchers. Future research needs to consider the researcher-participant relationship from a gender non-binary perspective and consider the experiences of people with different gender identities conducting ethnographies. This chapter offers a new way of considering gender and ethnography in the construction setting, reinforcing the importance of reflexivity in construction ethnography. It calls on researchers to give more critical thought to their roles in the field and to add more layers of complexity to this very position. The significance of this chapter is that it asks ethnographers in construction research to consider the different ways gender intersects with ethnography – most notably through researchers themselves. To recognise that the construction context being study is not a gender-neutral setting and cannot be treated so by scholars. Ethnographers conducting research in the construction sector need to be reflexive of how their gender and positionality and the gendered context of construction intersect to shape the creation of knowledge.

## Acknowledgement

I would like to recognise the generosity of Dr Adam Rogan, my research pair, without whom this research and chapter would not have been possible.

## Notes

1 Australian term for a whirlwind or dust storm.
2 Australian term for morning tea.

## References

Ackerly, B., & True, J. (2008). Reflexivity in practice: Power and ethics in feminist research on international relations. *International Studies Review*, 10(4), 693–707.

Ackerly, B., & True, J. (2010). *Doing Feminist Research in Political and Social Science*. London: Palgrave Macmillan.

Ackerly, B., & True, J. (2013). Methods and methodologies. In G. Waylen, K. Celis, J. Kantola, & S. L. Weldon (Eds.), *The Oxford Handbook of Gender and Politics* (pp. 135–159). Oxford: Oxford University Press.

Applebaum, H. A. (1981). *Royal Blue: The Culture of Construction Workers*. New York: Holt Rinehart & Winston.

Beckwith, K. (2005). A common language of gender? *Politics & Gender*, 1(1), 128–137.

Bordo, S. (1991). *Unbearable Weight: Feminism, Western Culture and the Body*. Oakland: University of California Press.

Caplan, P. (2013). Learning gender: Fieldwork in a Tanzanian coastal village (pp. 168–181), In Bell, B, Caplan, P and Karim, WJ. (Eds) Gendered Fields: Women, Men & Ethnography. Routledge.

Chappell, L. (2006). Comparing political Institutions: Revealing the gendered 'Logic of appropriateness.' *Politics & Gender*, 2(02), 223–235. https://doi.org/10.1017/S1743923X06221044

Chappell, L. (2020). Doing elite interviews in feminist research: Confessions of a born-again observationist. In P. Wadds, N. Apoifis, S. Schmeidl, & K. Spurway (Eds.), *Navigating fieldwork in the social sciences: Stories of danger, risk and reward*. Springer Nature. (pp. 129–145). https://link.springer.com/chapter/10.1007/978-3-030-46855-2_7

Connell, R. W. (2002). *Gender*. Cambridge: Polity Press in association with Blackwell Publishers.

Creswell, J. W., & Poth, C. N. (2017). *Qualitative Inquiry and Research Design: Choosing among Five Approaches*. Thousand Oaks: Sage.

Cunliffe, A. L., & Karunanayake, G. (2013). Working within hyphen-spaces in ethnographic research: Implications for research identities and practice. *Organizational Research Methods*, 16(3), 364–392. https://doi.org/10.1177/1094428113489353

Daley, A. (2010). reflections on reflexivity and critical reflection as critical research practices. *Affilia*, 25(1), 68–82. https://doi.org/10.1177/0886109909354981

Datta, A., & Brickell, K. (2009). "We have a little bit more finesse, as a nation": Constructing the Polish worker in London's building sites. *Antipode*, 41(3), 439–464. https://doi.org/10.1111/j.1467-8330.2009.00682.x

Easterday, L., Papademas, D., Schorr, L., & Valentine, C. (1977). The making of a female researcher: Role problems in field work. *Urban Life*, 6(3), 333–348. https://doi.org/10.1177/089124167700600305

Faulkner, W. (2009). Doing gender in engineering workplace cultures. II. Gender in/authenticity and the in/visibility paradox. *Engineering Studies*, 1(3), 169–189. https://doi.org/10.1080/19378620903225059

Ferguson, K. E. (1993). *The Man Question: Visions of Subjectivity in Feminist Theory*. Oakland: University of California Press.

Galea, N. (2018). *Built for Men: Institutional Privilege in the Australian Construction Industry [PhD]*. University of New South Wales.

Galea, N., Powell, A., Loosemore, M., & Chappell, L. (2020). The gendered dimensions of informal institutions in the Australian construction industry. *Gender, Work & Organization*, 27, 1214–1213. https://doi.org/10.1111/gwao.12458

Galea, N., Powell, A., & Salignac, F. (2023). The role of homosociality in maintaining men's powerfulness in construction companies. *Construction Management and Economics*, 41(2), 172–182. https://doi.org/10.1080/01446193.2022.2151025

Gama Gato, L., & Matyska, A. (2021). Gendered encounters in mobility: Women researching migrant construction workers. *Ethnography*, 146613812110382. https://doi.org/10.1177/14661381211038290

Gurney, J. N. (1985). Not one of the guys: The female researcher in a male-dominated setting. *Qualitative Sociology*, 8(1), 42–62. https://doi.org/10.1007/BF00987013

Haddow, K. (2022). 'Lasses are much easier to get on with': The gendered labour of a female ethnographer in an all-male group. *Qualitative Research*, 22(2), 313–327. https://doi.org/10.1177/1468794120984089

Hammersley, M. (2006). Ethnography: Problems and prospects. *Ethnography and Education*, 1(1), 3–14. https://doi.org/10.1080/17457820500512697

Hammersley, M., & Atkinson, P. (1995). Insider accounts: Listening and asking questions. In S. Ybema, D. Yanow, H. Wels, & F. Kamsteeg (Eds.), *Ethnography: Principles in Practice* (pp. 124–156). Thousand Oaks: Sage.

Harris, A. (1968). *Next Wave Cultures: Feminism, Subcultures, Activism*. Abingdon: Routledge.

Hodkinson, P. (2005). 'Insider Research' in the study of youth cultures. *Journal of Youth Studies*, 8(2), 131–149. https://doi.org/10.1080/13676260500149238

Holck, L. (2018). Affective ethnography: Reflections on the application of "useful" research on workplace diversity. *Qualitative Research in Organizations and Management: An International Journal*, 13(3), 218–234. https://doi.org/10.1108/QROM-11-2016-1456

Holck, L., & Villeseche, F. (2023). Affective ethnography: Interpreting body language in the workplace fieldwork. In *Methodological Approaches for Workplace Research and Management* (pp. 36–53). Abingdon: Routledge.

Iacuone, D. (2005). "Real men are tough guys": Hegemonic masculinity and safety in the construction industry. *The Journal of Men's Studies*, 13(2), 247–266.

Jerolmack, C., & Khan, S. (2014). Talk is cheap: Ethnography and the attitudinal fallacy. *Sociological Methods & Research*, 43(2), 178–209.

Khan, S., & Jerolmack, C. (2013). Saying meritocracy and doing privilege. *The Sociological Quarterly*, 54(1), 9–19.

Krook, M. L., & Squires, J. (2006). Gender quotas in British politics: Multiple approaches and methods in feminist research. *British Politics*, 1, 44–66.

Lowndes, V. (2020). How are political Institutions gendered? *Political Studies*, 68(3), 543–564. https://doi.org/10.1177/0032321719867667

Lowstedt, M. (2015). 'Taking off my glasses in order to see': Exploring practice on a building site using self-reflexive ethnography. *Construction Management and Economics*, 33(5–6, SI), 404–414. https://doi.org/10.1080/01446193.2015.1066021

Mars, G. (2005). Locating causes of accidents in the social organisation of building workers and some wider implications: An approach from cultural theory. *International Journal of Nuclear Knowledge Management*, 1(3), 255–269.

Mazzei, J., & O'Brien, E. E. (2009). you got it, so when do you flaunt it? Building rapport, intersectionality, and the strategic deployment of gender in the field. *Journal of Contemporary Ethnography*, 38(3), 358–383. https://doi.org/10.1177/0891241608330456

McLaren, M. A. (2002). *Feminism, Foucault, and Embodied Subjectivity*. Albany: State University of New York Press.

Muecke, M. A. (1994). On the evaluation of ethnographies. in Morse, J (Ed.) *Critical Issues in Qualitative Research Methods*, Thousand Oaks, CA. Sage. (pp 187–209).

Olive, R., & Thorpe, H. (2011). Negotiating the 'F-word' in the field: Doing feminist ethnography in action sport cultures. *Sociology of Sport Journal*, 28(4), 421–440.

O'Reilly, K. (2012). *Ethnographic Methods*. Abingdon: Routledge.

Oswald, D., & Dainty, A. (2020). Ethnographic research in the construction industry: A critical review. *Journal of Construction Engineering and Management*, 146(10). https://doi.org/10.1061/(ASCE)CO.1943-7862.0001917

Paap, K. (1999). *Masculinity under construction: Gender, class, and race in the construction industry* [Doctoral Thesis]. The University of Wisconsin, Madison, USA.

Paap, K. (2006). *Masculinity Under Construction: Gender, Class, and Race in the Construction Industry*. Ithaca: Cornell University Press. http://forward.library.wisconsin.edu/catalog/ocm50264785

Pink, S., & Morgan, J. (2013). Short-term ethnography: Intense routes to knowing. *Symbolic Interaction*, 36(3), 351–361.

Pink, S., Tutt, D., Dainty, A., & Gibb, A. (2010). Ethnographic methodologies for construction research: Knowing, practice and interventions. *Building Research & Information*, 38(6), 647–659.

Poulton, E. (2012). "If you had balls, you'd be one of us!" Doing gendered research: Methodological reflections on being a female academic researcher in the hyper-masculine subculture of "Football hooliganism". *Sociological Research Online*, 17(4), 4.

Puwar, N. (2004). *Space Invaders: Race, Gender and Bodies Out of Place*. Oxford: Berg Publishers.

Rodriguez, J. K., & Ridgway, M. (2023). Intersectional reflexivity: Fieldwork experiences of ethnic minority women researchers. *Gender, Work & Organization*, 30(4), 1273–1295. https://doi.org/10.1111/gwao.12977

Scott, J. W. (1986). Gender: A useful category of historical analysis. *The American Historical Review*, 91(5), 1053–1075.

Smit, B., & Onwuegbuzie, A. J. (2018). Observations in qualitative inquiry: When what you see is not what you see. *International Journal of Qualitative Methods*, 17(1), 160940691881676. https://doi.org/10.1177/1609406918816766

Speer, S. A. (2002). What can conversation analysis contribute to feminist methodology? Putting reflexivity into practice. *Discourse & Society*, 13(6), 783–803. https://doi.org/10.1177/0957926502013006757

Thiel, D. (2007). Class in construction: London building workers, dirty work and physical cultures. *The British Journal of Sociology*, 58(2), 227–251. https://doi.org/10.1111/j.1468-4446.2007.00149.x
Turner, M., Zhang, R. P., Holdsworth, S., & Myla, M. (2021). Taking a Broader Approach to Women's Retention in Construction: Incorporating the University Domain. *Proceedings of the 37th Annual ARCOM Conference*, 188–197.
Van Maanen, J. (2006). Ethnography then and now. *Qualitative Research in Organizations and Management: An International Journal*, 1(1), 13–21. https://doi.org/10.1108/17465640610666615
Warren, C., & Hackney, J. (2000). *Gender Issues in Ethnography*. Thousand Oaks: Sage.
Weems, L. (2006). Unsettling politics, locating ethics: Representations of reciprocity in postpositivist inquiry. *Qualitative Inquiry*, 12(5), 994–1011. https://doi.org/10.1177/1077800406288628
West, C., & Zimmerman, D. H. (1987). Doing gender. *Gender and Society*, 1(2), 125–151.
Wilson, R. (2010). Raising the curtain on survey work. In R. Ryan-Flood, & R. Gill (Eds.), *Secrecy and Silence in the Research Process: Feminist Reflections* (pp. 188–199). Abingdon: Routledge.
Ybema, S., Wels, H., & Yanow, D. (2009). *Organizational Ethnography: Studying the Complexity of Everyday Life*. Thousand Oaks: Sage. http://digital.casalini.it/9781446248188.

## SECTION II

# Ethnographic research considerations for the construction context

# 5

# ETHNOGRAPHIC METHODS IN CONSTRUCTION

*David Oswald*

*RMIT University, Melbourne, Australia*

**Key takeaways**

- This chapter provides a historic overview of the use of ethnography in the construction.
- It highlights the methodological debate of the mid-1990s, and the calls of late 2000s, as key milestones in the emergence of ethnographic research in construction.
- It notes that with a rise and diversification of ethnographic research, greater discussion and debate on assessing ethnography in construction are needed.

## Introduction

The previous chapters within this book have introduced how ethnographic research in the construction industry has been performed globally across Eastern (see Swider, Chapter 3) and Western cultures (see Porter, Chapter 2). While ethnographic research in construction is growing globally, it is still a minor pursuit, when compared to more traditional methods (Oswald & Dainty, 2020). The study of management within the construction industry began as early as the 1930s, but it was not until the 1980s when the field began to rapidly develop with a growth in academic literature. This growth only accelerated into the 2000s, and brought with it a diversity of research approaches, including ethnography.

Some of the initial research in construction management appeared in the journal *Transactions of the American Society of Civil Engineers* (Vol. 98, Issue 2) with, for instance, Allen (1933), Holcomb and Mueller (1933), Hopkins (1933) and Bush and Greensfelder (1933) contributing early work. There are a few further examples of studies on the management of construction (but not many) that followed in the 1940s, 1950s, 1960s, and 1970s (see, for example, Thieme, 1943; Rositsky, 1958; Nicholls, 1963). It was into the 1980s that the mainstream journals known in the field today, such as *Construction Management & Economics* (Issue 1 in 1983), were established. These journals published more regularly throughout the 1980s and 1990s. For example, in the *Journal of Construction Engineering and Management*, when comparing articles in the early-mid-1980s with the late-1990s, the number of authors publishing more than doubled and the number of publications almost doubled (see Pietroforte & Stefani, 2004). The growth of the field led to some diversification in research methods and debate on the domination of quantitative approaches in the 1990s. As a result, there is no one

DOI: 10.1201/9781003379584-7

single scientific paradigm adopted within the field of construction management (see Ballard & Howell, 2003).

The construction management field is at the intersection of natural and social sciences (Love et al., 2002). There are many differences between natural and social sciences. For example, in the natural sciences, there is study into facts that are independent of what people think or say about them; whereas in the social sciences, peoples' perceptions (e.g. construction professionals) are studied. A field, such as construction management, that is on the intersection of natural and social sciences, is therefore subject to many potential ways to perform research, with no single methodological approach to all studies within the field. While this may be the case, the field has traditionally been (as still is) dominated with positivism and quantitative approaches (see Oswald & Dainty, 2020; Sherratt & Leicht, 2020).

The early ethnographic examples in construction are rare and sporadic. For instance, there was initial work by Sykes (1969a,b) on the construction of a hydroelectric plant in Scotland; Clegg (1975) on power rule and domination; Applebaum (1982) on construction management methods and then Bresnen (1990) on organising construction and matrix management. These examples introduced classic ethnographic research characteristics to the construction setting, which typically involves many months or years of being embedded within a cultural setting, participating, observing, and writing field notes. Despite these early adopters – Clegg (1975), Sykes (1969a,b), Applebaum (1982), and Bresnen (1990) – there was a reluctance to accept ethnographic research within the field of construction management. As Rooke and Seymour (2002, p. 7) explained: 'There is, indeed, some resistance to ethnographic studies among the construction management community and there has been something of a struggle to gain acceptance of them. Part of the problem is that they just don't look to engineers like research reports. They do not commence with set objectives, or present sets of numerical results.'

This chapter returns to a key methodological debate of the mid-1990s as a starting point for the growth of ethnographic methods in the construction industry. It then presents the various forms of ethnographic research that have emerged within the construction sector and highlights a key future challenge for the field.

## A pivotal methodological debate

In the mid-1990s there was methodological debate in the journal *Construction Management & Economics* that was initiated by Seymour and Rooke (1995, 1997). They argued that: 'if researchers are to have a role in changing the culture of the industry, then the culture of research must change also' (1995, p. 511). Their work, in a response to a ten-year review of articles within the journal (see Betts & Lansley, 1993), explained that there was a need to reconsider the conceptions of 'theory,' 'rigour,' and 'objectivity,' with one particular meaning of 'theory' and one particular way of arriving at 'theory' having been accepted without question (see Seymour et al., 1997). They pointed out that within construction management research the 'objects of study' are people and that people attribute meaning both to what they do and the presence of the researcher. Within research, meaning is intersubjectively created and managed between the researcher and the researched within an iterative and interpretive process. Considering this, they offered another path for researchers, stating: 'Our suggested alternative is to concentrate upon the interpretative methods that researchers and managers use to make sense of the world. This approach yields an investigation that is primarily concerned with meaning rather than causality, and produces an account that recognizes the respective viewpoints of practitioners in the process. It is our belief that such an account better reflects the realities of construction management as a practice' (1997, p. 118).

Seymour and Rooke's (1995, 1997) suggestions were met with a firm reply. Runeson (1997) described their argument as 'anti-scientific' and that 'contrary to Seymour et al.'s claims, positivist research methods are our best insurance against bad research' (p. 299). He went onto explain that their suggestions had two fundamental drawbacks: that no theory is being tested or developed, and the results are subjective and biased. Runeson (1997) stated: '… as scientific research it suffers from two serious draw-backs: there is no theory either being tested or developed and the results are so subjective and so biased in favour of construction managers and against designers, subcontractors and trade unions as to be totally meaningless. This has nothing to do with the question of whether information was derived from survey questionnaires and subjected to all kinds of descriptive statistics disguised as analyses, or from in-depth interpretative interviews presented some other way. Either way, it is bad research, and there is a lot of it around' (p. 300).

Runeson (1997) also responded to Seymour et al.'s (1997) notion that the objects of study are people (and therefore interpretative methods better reflect realities of construction management practice). He argued that: 'the most important consequence of this sweeping reinterpretation of scientific philosophy is that medicine, dentistry and all social sciences deal with people and therefore, by application of this logic, cannot accommodate causal relationships.' Before concluding that: 'A positivist approach does not guarantee progress. It does not guarantee good research, because there will always be poor researchers. It does not guarantee objectivity, because there can probably never be any completely objective research. However, positivism currently offers the best way to reduce subjectivity and to discipline undisciplined researchers and the results, in terms of advancement of science, are nothing short of spectacular' (p. 302).

There were various others that reacted to the debate, such as Raftery et al. (1997) and Root et al. (1997) as well as Edum-Fotwe et al. (1997), who reflected that 'the discipline is gradually witnessing a polarisation of research orientation into rational and interpretive schools of thought' (p. 447). They suggested that both perspectives could complement each other and produce a more balanced outcome. It was not long before Seymour et al. (1998) replied to Runeson's (1997) critiques by clarifying their positions and responding to the points raised. One of their points was in response to Runeson's (1997) example of medicine when discussing people as the object of research. Seymour and Rooke (1998) highlighted that while general practitioners (GPs) may prescribe a medicine (e.g. a sleeping pill), they also have to draw upon knowledge that was not learnt from scientific laws or from employing the hypothetico-deductive methods. This could include when they have to sympathetically listen to patients or enquire about possible forms of support through family and friends. Seymour and Rooke (1998) stated: 'Some of the orders of knowledge that doctors (and construction managers) draw on in their working lives can be referred to as folk psychology, common sense, practical reasoning and know-how. These are vital topics for study but also vital resources which we employ in our own studies. They are not accessible to the application of a single method' (p. 110).

They concluded by reinforcing that another pathway for researchers in the construction industry was a focus on the study of meaning: '…we observe that the debate concerning the academic credentials of construction management as a discipline has, usually implicitly, hinged on the natural sciences as exemplars. Runeson has performed a great service in making the case explicit. However, science, as it is understood by positivism, is not the only form of academic discipline. There exists an academic tradition of equal rigour and for many purposes of greater appropriateness to the study of social conduct. Central to this tradition is the study of meaning' (p. 112).

This was perhaps the most notable debate in the history of the journal (Leiringer & Dainty, 2017) and one that created more opportunity for research methods from a social science perspective in the field of construction management. A decade on from the debate, an analysis of every 2006 paper published in the journal *Construction Management & Economics* was undertaken, with Dainty (2007) reporting that only 8.4% of articles used purely qualitative methods. Of these qualitative papers, virtually all relied exclusively on semi-structured interviews (Dainty, 2007). This lack of diversification in types of qualitative methods and over-reliance of interview techniques can be criticised since qualitative researchers have a range of traditional, contemporary, and novel methodologies and methods available to them for project studies (see Sergeeva et al., 2022). More recently, a study of articles within the journal of *Construction Engineering and Management* found similar results, with 74% using primarily quantitative data and 18% using a mix of quantitative and qualitative data, leaving 8% of other approaches, including the primary use of qualitative data (see Sherratt & Leicht, 2020).

Thus, while there has been wider acceptance of social sciences approaches within the construction industry since the debate of the mid-1990s, positivism and quantitative approaches remain dominant within the field. Further, of the research using qualitative methods, most relied on semi-structured interviewing and not other approaches, such as ethnography.

A critical review between 2011 and 2019 found that there were only 57 articles that used ethnographic methods within leading construction journals (see Oswald & Dainty, 2020). Previous studies have found that qualitative work remains less common in construction management literature, and ethnographic work is only one qualitative approach[1] to choose from. However, it is an approach that has grown in popularity recently and one that has diversified with various forms of ethnographic research emerging.

## Ethnographic approaches used in construction research

The debate of the mid-1990s paved the way for wider acceptance of social science research within the field of construction management. One methodological approach used by researchers in the social sciences is that of ethnographic methods. However, by the turn of the millennium, there were still only a few examples of full-length ethnographic studies (e.g. Clegg, 1975; Applebaum, 1982; Bresnen, 1990) in the field of construction management. This was highlighted by Rooke and Seymour (2002) and later reflected by the work of Dainty (2007), as noted above.

Rooke and Seymore (2002) argued that 'The construction process has been virtually ignored. There has been little attempt to specify the realities of site work. Thus, part of the solution, we believe, is more and better ethnographies.' The authors themselves would attest to their own advice by publishing ethnographic work in the years that followed (see Rooke et al., 2003, 2004). Their ethnographic data collection involved unstructured interviews, direct observations (e.g. of meetings), and field notes.

Other ethnographic work in the early 2000s included interviewing techniques (such as Dainty et al.'s, 2000a, ethnographic interviewing and use of computer software to aid textual analysis) and participant observations (such as Gherardi and Nicolini's, 2002, three months of shadowing a trainee). In some cases, ethnographic interviewing was the main approach used, as in the work by Dainty et al. (2000b). In other cases, interviews were used alongside site observations over several site visits for periods of months as in the study by Shields and West (2003). Interviews have also been used in longitudinal ethnographic work, with, for example, Demain and Fruchter (2006) undertaking fieldwork for two weeks in a structural design office, before making several return visits to the site for a two-year period to observe further and interview participants. Their work was one of a few ethnographic construction industry studies in the mid-2000s (see also, for example, Shields & West, 2003; Iszatt-White, 2007) that emerged alongside the work led by Rooke (see Rooke et al., 2003, 2004; Rooke & Clarke, 2005; Rooke & Kagioglou, 2007). Though it was not until around years 2008–2010 when there were several calls for greater use of ethnographic methods within the construction industry. See, for example:

- '... It is recommended that an ethnographic study of the whole bidding process should be carried out' (Laryea & Hughes, 2008, p. 920)
- '... the present study should preferably be complemented by more ethnographic research approaches' (Styhre, 2009, p. 998)
- '... there is a strong methodological implication of the above discussion that points to the value of adopting approaches to research that—like more interpretative and ethnographic methods—are able to tap into the "lived experience"' (Bresnen, 2009, p. 932)
- '... Ethnographic theory-building methods provide a tremendous opportunity for construction research to address these issues in their full complexity' (Phelps & Horman, 2010, p. 64)
- '... Ethnographic methodologies developed in social anthropology and sociology hold considerable promise for addressing practical, problem-based research concerned with the construction site' (Pink et al., 2010, p. 647)

A proliferation of ethnographic research in construction followed (Tutt & Pink, 2019; Oswald & Dainty, 2020). This increase in popularity can also be attributed to the book 'Ethnographic research in the construction industry' by Pink et al. (2013). This book demonstrated, through ethnographic case studies, how theoretically informed ethnographic research could generate fresh insights, challenge common perceptions, and help us understand industry problems. It highlighted the potential value in undertaking an ethnographic approach in construction and provided 'an indication of the trajectories along which future work could travel' (p. 3).

Around 2010, the shift away from classic forms of ethnography was already being noted. See, for example: '... Already there was ethnographic work shifting from research that could be described as "classic" ethnography' (Pink et al., 2010, p. 658). Since the calls for further ethnographic work and Pink et al.'s book, the diversification of ethnographic research in construction has been even more obvious – see Table 5.1. This diversification from traditional 'classic' forms of ethnography to a range of approaches including audio, visual, and other practice has stretched the definition of ethnography (Tutt & Pink, 2019). Regardless of the form, it has been argued by Oswald and Dainty (2020) that to remain truly ethnographic, the researcher should be in a position where it is possible to provide thick description (see Geertz, 1973). It could also be contended that despite the form of ethnographic approach, there should be deep awareness of the context and that observation should still remain a key method, as a central characteristic of ethnography.

**TABLE 5.1** An overview of ethnographic examples since 2011, demonstrating a diversification of approaches

| *Ethnographic approach included* | *Example case(s)* |
|---|---|
| Observations, interviews, and documentary data | Laryea (2011, 2013); Cao et al., (2015); Ramalingam and Mahalingam, (2018); van den Berg et al. (2020) |
| A multi-sited approach | Tutt et al. (2013a, b); Jia et al. (2017); Hampton et al. (2019); Oswald et al. (2022) |
| An intense short-term approach | Pink and Morgan (2013); Pink et al. (2014); Pink et al. (2016); Galea et al. (2020); Brett et al. (2022) |
| A long-term single sited approach | Oswald et al. (2015); Oswald et al. (2018a, b, c); Oswald et al. (2019a, b, c); Oswald et al. (2020) |
| Using multiple researcher roles | Löwstedt (2015) |
| An autoethnography | Kanjanabootra & Corbitt (2016); Grosse (2019) |
| A retrospective ethnography | Ness (2012) |
| A longitudinal approach | Koch and Schultz (2019) |
| An ethnographic-action research approach | Zanen et al. (2013); olde Scholtenhuis et al. (2016) |
| The use of video alongside observations | Lingard et al. (2015); Mäki & Kerosuo (2015); Arana & Wittek (2016); Pink et al. (2016) |
| The use of photography alongside observations | Judson and Maller (2014); Mäki and Kerosuo (2015); Hamid and Tutt (2019) |
| Non-participant observations | Aboagye-Nimo et al. (2015); Ikioda (2016) |
| Participant observation with focus groups | Oswald and Lingard (2019); Lingard and Oswald (2020) |
| Ethnographic interviewing | Berteaux and Javernick-Will (2015) |
| A digital ethnographic approach | Dossick et al. (2019); Pink et al. (2016) |
| Used as a pilot or exploratory study | Cao et al. (2015); Spence et al. (2018) |
| Shadowing management | Mäki and Kerosuo (2015); Klitgaard et al. (2021) |

Ethnographic interviewing and other forms of interviewing techniques used alongside observation within an ethnographic approach have remained a popular approach since the early 2000s by construction researchers. For example, Teo and Loosemore (2014) performed 24 semi-structured interviews alongside ethnographic observations of protest activities; van den Berg et al. (2020) completed semi-structured interviews to complement 250 hours of observation and documentary data; and Grytnes et al. (2020) undertook 34 semi-structured interviews alongside eight months of ethnographic fieldwork on two large construction projects.

Ethnographic interviewing typically takes place in a specific location or setting that is under study. This means there is an opportunity to observe the actions, practices and objects that the participants refer to during the interview. The interviews ask questions that are open and aim to generate long and detailed accounts from participants. These descriptive types of questioning in a 'tell me about …' style allow participants to speak of the issues pertinent to them and draw out explicit and implicit cultural knowledge (Phelps & Horman, 2010). They go beyond traditional interviews in the sense that they are not only used to collect data but also to start ordering and categorising that data (Phelps & Horman, 2010).

Interviews can be carried out in both long-term and short-term ethnographic research. For instance, Baarts (2009) worked as an apprentice full-time for seven months in a gang of 30 construction workers. This embedment, participation, and observation of a particular cultural setting for a prolonged period, while writing field notes, represent characteristics associated with 'classic' forms of ethnography. In the last few weeks of the research fieldwork, Baarts collected 23 semi-structured interviews to complement her field notes. A shorter-term fieldwork study by Koch (2013) observed the workday of a group of carpenters for ten days, with supplementary semi-structured interviews.

As the boundaries and definition of ethnography have become stretched within construction management research, challenges emerge particularly around judging ethnographic research. It is clear that ethnographic research should not be judged through a positivist lens (Oswald & Dainty, 2020), despite this being the dominant research paradigm within the field. As Rooke et al. (1997) put it during the mid-1990s debate: 'Different research methods are required for different research purposes and are to be evaluated according to different criteria. These criteria must be made explicit.' Therefore, instead of positivist frameworks for assessing research, such as validity, reliability and generalisability, other forms of assessment should be considered.

For example, in journals on qualitative research, a criterion of substantive contribution, aesthetic, merit, reflexivity, impact, and expression of a reality has been proposed (see Richardson, 2000). In books on ethnography, it has been proposed that there is a focus on plausibility and credibility, the amount of evidence, the kind of evidence and the claims made (Hammersley, 1991, 1992). In broader organisational studies, validity has been described in terms of authenticity, plausibility, and criticality (see Golden-Biddle & Locke, 1993), and it has been argued that the test of truth is not in the presentation of an ever-greater number of data extracts or frequency counts of the codes and themes, but on the 'construction of convincing text in which the authenticity of the author's field experience is made accessible to the reader' (see Jarzabkowski et al., 2014, p. 284).

Within the field of construction management, aside from sporadic publications on topics of methodological rigour, such as evaluating research (see Rooke & Kagioglou, 2007) and on validity (see Phelps & Horman, 2009), there has been little debate or discussion on how ethnographic research should be assessed. Oswald and Dainty (2020, p. 1) stated that: 'greater levels of appreciation and consistency in judging ethnographic research, and greater reflexivity on the influence of the researcher's position and influence on the practices studied, are prerequisites for robust ethnographic practice.'

The lack of appreciation, consistency, and clarity becomes increasing important as the definition and boundaries of ethnographic research become stretched in construction with more digital, video, short-term, and intense forms rising in popularity (as opposed to long-term 'classic' forms). While sometimes criticised by classic ethnographers as a second-rate approach to ethnography (see Galea et al., 2020), rapid and short intense forms of ethnographic research in construction have been advocated as an alternative to classic approaches (Pink et al., 2010). It can be a practical solution to time-consuming fieldwork challenges (Galea et al., 2020) and is characterised by forms of intensity that lead to deep and valid ways of knowing (Pink & Morgan, 2013).

This book discusses a range of various approaches in Section III, including short intense forms of ethnographic research (see Chapter 10) and longer-term approaches (see Chapter 11). There is also discussion on: multi-sited ethnography (see Chapter 12), inter-organisational ethnography (see Chapter 13), ethnographic-action research (see Chapter 14), ethnographic video interviewing (see Chapter 15) online ethnographic approaches (see Chapter 16), and autoethnography (see Chapter 17),

Prior to such discussion on the various ethnographic approaches, this section (II) continues to more broadly consider methodology for construction ethnographies. Having provided an overview of the various ethnographic methods that have been employed within the construction industry in this chapter, the following chapters will focus on methodological considerations and reflections for doing and demonstrating robust ethnographic research (in Section II). These include: discussing ethics for doing industry-embedded research (see Chapter 6), reflecting on research philosophy relevant to construction ethnographies (see Chapter 7), explaining how contributions can be demonstrated from ethnographic research (see Chapter 8) and considering ethnographic research designs, for future construction settings that include robots and smart machines, in Construction 5.0 (see Chapter 9).

## Conclusions

Ethnographic research is a complementary approach to traditional positivist studies within the construction industry. Following the methodological debate of the mid-1990s (see Seymour & Rooke, 1995; Seymour et al., 1997), the calls for greater ethnographic use in the late 2000s, and publication of Pink et al.'s (2013) book, ethnography within the construction industry has begun to have a rise in popularity and a diversification of approaches. While ethnography is still very much a minor pursuit within the field, this greater adoption by researchers indicates a wider acceptance and understanding of ethnographic work within the field of construction management. This chapter has introduced the various forms of ethnographic research that have been performed in the construction industry. While these different ethnographic approaches offer flexibility (which is important for doing ethnographic research as it is undertaken in a natural setting), there are challenges with judging diverse ethnographic approaches. It is recommended that there is further debate and discussion for assessing ethnographic research within the field of construction management.

## Note

1 It should be noted that there are rare occasions where quantitative data are, and can be, used in ethnographic work (see Pole & Morrison, 2003).

## References

Aboagye-Nimo, E., Raiden, A., King, A., & Tietze, S. (2015). Using tacit knowledge in training and accident prevention. *Proceedings of the Institution of Civil Engineers-Management, Procurement and Law*, 168(5), 232–240.

Allen, W. (1933). Highway construction management. *Transactions of the American Society of Civil Engineers*, 98(2), 192–206.

Applebaum, H. A. (1982). Construction management: Traditional versus bureaucratic methods. *Anthropological Quarterly*, 55(4), 224–234.

Arana, M. M., & Wittek, R. P. (2016). Community resilience: Sustained cooperation and space usage in collective housing. *Building Research & Information*, 44(7), 764–774.

Baarts, C. (2009). Collective individualism: The informal and emergent dynamics of practising safety in a high-risk work environment. *Construction Management and Economics*, 27(10), 949–957.

Ballard, G., & Howell, G. A. (2003). Competing construction management paradigms. In *Construction Research Congress: Wind of Change: Integration and Innovation* (pp. 1–8, Vol. 1).

Berteaux, F., & Javernick-Will, A. (2015). Adaptation and integration for multinational project-based organizations. *Journal of Management in Engineering*, 31(6), 04015008.

Betts, M., & Lansley, P. (1993) Construction management and economics: Review of the first ten years. *Construction, Management and Economics*, 11, 221–245

Bresnen, M. (1990). *Organising Construction: Project Organisation and Matrix Management*. London: Routledge.

Bresnen, M. (2009). Living the dream? Understanding partnering as emergent practice. *Construction Management and Economics*, 27(10), 923–933.

Brett, R., Thomson, D., & Dainty, A. (2022). Exploring craft in construction with short-term ethnography: Reflections on a researcher's prior insight. *Construction Management and Economics*, 40(5), 359–373.

Bush, E., & Greensfelder, A. (1933). Bush on construction management. *Transactions of the American Society of Civil Engineers*, 98(2), 215–217.

Cao, D., Wang, G., Li, H., Skitmore, M., Huang, T., & Zhang, W. (2015). Practices and effectiveness of building information modelling in construction projects in China. *Automation in Construction*, 49, 113–122.

Clegg, S. (1975) *Power, Rule and Domination: A Critical and Empirical Understanding of Power in Sociological Theory and Organizational Life*. Oxfordshire: Taylor & Francis Group.

Dainty, A., (2007) A review and critique of construction management research methods. In *Proceedings of Construction Management and Economics 25th Anniversary Conference*, University of Reading, July (pp. 1533–1543).

Dainty, A. R., Bagilhole, B. M., & Neale, R. H. (2000a). Computer aided analysis of qualitative data in construction management research. *Building Research & Information*, 28(4), 226–233.

Dainty, A. R., Bagilhole, B. M., & Neale, R. H. (2000b). A grounded theory of women's career underachievement in large UK construction companies. *Construction Management & Economics*, 18(2), 239–250.

Demain, P., & Fruchter, R. (2006). An ethnographic study of design knowledge reuse in the architecture, engineering, and construction industry. *Research in Engineering Design*, 16(4), 184–195.

Dossick, C., Osburn, L., & Neff, G. (2019). Innovation through practice: The messy work of making technology useful for architecture, engineering and construction teams. *Engineering, Construction and Architectural Management*. https://doi.org/10.1108/ECAM-12-2017-0272

Edum-Fotwe, F. T., Thorpe, A., McCaffer, R., & Price, A. D. F. (1997) Developing construction management research: A balance of perspectives. In P. Stephenson (Ed.), *13th Annual ARCOM Conference*, 15–17 September 1997, King's College, Cambridge. Association of Researchers in Construction Management, Vol. 2, 447–454.

Galea, N., Powell, A., Loosemore, M., & Chappell, L. (2020). The gendered dimensions of informal institutions in the Australian construction industry. *Gender, Work & Organization*, 27(6), 1214–1231.

Geertz, C. (1973). Thick description: Toward an interpretive theory of culture. In *The Interpretation of Cultures: Selected Essays* (pp. 3–30). New York: Basic.

Gherardi, S., & Nicolini, D. (2002). Learning the trade: A culture of safety in practice. *Organization*, 9(2), 191–223.

Golden-Biddle, K., & Locke, K. (1993). Appealing work: An investigation of how ethnographic texts convince. *Organization Science*, 4(4), 595–616.

Grosse, H. (2019). An insider's point of view: Autoethnography in the construction industry. *Construction Management and Economics*, 37(9), 481–498.

Grytnes, R., Tutt, D. E., & Andersen, L. P. S. (2020). Developing safety cooperation in construction: Between facilitating independence and tightening the grip. *Construction Management and Economics*, 38(11), 977–992.

Hamid, W., & Tutt, D. (2019). "Thrown away like a banana leaf": Precarity of labour and precarity of place for Tamil migrant construction workers in Singapore. *Construction Management and Economics*, 37(9), 513–536.

Hammersley, M. (1991). *Reading Ethnographic Research*. London: Longman.

Hammersley, M. (1992). *What's Wrong with Ethnography?* London: Routledge.

Hampton, P., Chinyio, E. A., & Riva, S. (2019). Framing stress and associated behaviours at work: An ethnography study in the United Kingdom. *Engineering, Construction and Architectural Management*, 26(11), 2566–2580. https://doi.org/10.1108/ECAM-10-2018-0432

Holcomb, A., & Mueller, C. (1933). Holcomb on highway construction management. *Transactions of the American Society of Civil Engineers*, 98(2), 207–210.

Hopkins, R. (1933). Construction management. *Transactions of the American Society of Civil Engineers*, 98(2), 211–214.

Ikioda, F. (2016). The impact of road construction on market and street trading in Lagos. *Journal of Transport Geography*, 55, 175–181.

Iszatt-White, M. (2007). Catching them at it: An ethnography of rule violation. *Ethnography*, 8(4), 445–465.

Jarzabkowski, P., Bednarek, R., & Lê, J. K. (2014). Producing persuasive findings: Demystifying ethnographic textwork in strategy and organization research. *Strategic Organization*, 12(4), 274–287.

Jia, A. Y., Rowlinson, S., Loosemore, M., Xu, M., Li, B., & Gibb, A. (2017). Institutions and institutional logics in construction safety management: The case of climatic heat stress. *Construction Management and Economics*, 35(6), 338–367.

Judson, E. P., & Maller, C. (2014). Housing renovations and energy efficiency: Insights from homeowners' practices. *Building Research & Information*, 42(4), 501–511.

Kanjanabootra, S., & Corbitt, B. (2016). Reproducing knowledge in construction expertise: A reflexive theory, critical approach. *Construction Management and Economics*, 34(7–8), 561–577.

Klitgaard, A., Gottlieb, S. C., & Svidt, K. (2021). The researcher as audience and storyteller: Challenges and opportunities of impression management in ethnographic studies. *Construction Management and Economics*, 39(5), 383–397.

Koch, C. (2013). From crew to country? Local and national construction safety cultures in Denmark. *Construction Management and Economics*, 31(6), 691–703.

Koch, C., & Schultz, C. S. (2019). The production of defects in construction–an agency dissonance. *Construction Management and Economics*, 37(9), 499–512.

Laryea, S. (2011). Quality of tender documents: Case studies from the UK. *Construction Management and Economics*, 29(3), 275–286.

Laryea, S. (2013). Nature of tender review meetings. *Journal of Construction Engineering and Management*, 139(8), 927–940.

Laryea, S., & Hughes, W. (2008). How contractors price risk in bids: Theory and practice. *Construction Management and Economics*, 26(9), 911–924.

Leiringer, R., & Dainty, A. (2017). On the right to disagree, healthy debates and moving forward. *Construction Management and Economics*, 35(7), 383–384.

Lingard, H., & Oswald, D. (2020). Safety at the front line: Social negotiation of work and safety at the principal contractor–subcontractor interface. *Journal of Construction Engineering and Management*, 146(4), 04020024.

Lingard, H., Pink, S., Harley, J., & Edirisinghe, R. (2015). Looking and learning: Using participatory video to improve health and safety in the construction industry. *Construction Management and Economics*, 33(9), 740–751.

Love, P., Holt, G., & Heng, L. (2002). Triangulation in construction management research. *Engineering, Construction and Architectural Management*, 9(4), 294–303.

Löwstedt, M. (2015). "Taking off my glasses in order to see": Exploring practice on a building site using self-reflexive ethnography. *Construction Management and Economics*, 33(5–6), 404–414.

Mäki, T., & Kerosuo, H. (2015). Site managers' daily work and the uses of building information modelling in construction site management. *Construction Management and Economics*, 33(3), 163–175.

Ness, K. (2012). "Constructing masculinity in the building trades": Most jobs in the construction industry can be done by women. *Gender, Work & Organization*, 19(6), 654–676.

Nicholls, R. L. (1963). Operations research in construction planning. *Journal of the Construction Division*, 89(2), 59–74.

olde Scholtenhuis, L. L., Hartmann, T., & Dorée, A. G. (2016). Testing the value of 4D visualizations for enhancing mindfulness in utility reconstruction works. *Journal of Construction Engineering and Management*, 142(7), 04016015.

Oswald, D., Ahiaga-Dagbui, D., Sherratt, F., & Smith, S. (2020) An industry structured for unsafety? An exploration of the cost-safety conundrum in construction project deliver. *Safety Science*, 122, 104535

Oswald, D., & Dainty, A. (2020). Ethnographic research in the construction industry: A critical review. *Journal of Construction Engineering & Management*, 146(10), 03120003.

Oswald, D., & Lingard, H. (2019). Development of a frontline H&S leadership maturity model in the construction industry. *Safety Science*, 118, 674–686

Oswald, D., Lingard, H., & Zhang, R. P. (2022). How transactional and transformational safety leadership behaviours are demonstrated within the construction industry. *Construction Management and Economics*, 40(5), 374–390.

Oswald, D., Sherratt, F., & Smith, S. (2015). An ethnographic study of an accident investigation. *Procedia Manufacturing*, 3, 1788–1795.

Oswald, D., Sherratt, F., & Smith, S. (2018a). Problems with safety observation reporting: A construction industry case study. *Safety Science*, 107, 35–45.

Oswald, D., Sherratt, F., Smith, S., & Dainty, A. (2018b). An exploration into the implications of the "compensation culture" on construction safety. *Safety Science*, 109, 294–302.

Oswald, D., Sherratt, F., Smith, S., & Hallowell, M. (2018c). Exploring safety management challenges for multi-national construction workforces: A UK case study. *Construction Management and Economics*, 36(5), 291–301.

Oswald, D., Sherratt, F., & Smith, S. (2019a). Managing production pressures through dangerous informality: A case study. *Engineering, Construction and Architectural Management*, 26(11), 2581–2596.

Oswald, D., Sherratt, F., & Smith, S. (2019b). How safety rewards can help and hinder: A case study. *Professional Safety*, 64(7), 27–31.

Oswald, D., Wade, F., Sherratt, F., & Smith, S. (2019c). Communicating health and safety on a multinational construction project: Challenges and strategies. *Journal of Construction Engineering & Management*, 145(4), 04019017.

Phelps, A. F., & Horman, M. J. (2010). Ethnographic theory-building research in construction. *Journal of Construction Engineering and Management*, 136(1), 58–65.

Pietroforte, R., & Stefani, T. (2004). ASCE journal of construction engineering and management: Review of the years 1983–2000. *Journal of Construction Engineering & Management*, 130(3), 440–448.

Pink, S., Lingard, H., & Harley, J. (2016). Digital pedagogy for safety: The construction site as a collaborative learning environment. *Video Journal of Education and Pedagogy*, 1(1), 1–15.

Pink, S., & Morgan, J. (2013). Short-term ethnography: Intense routes to knowing. *Symbolic Interaction*, 36(3), 351–361.

Pink, S., Morgan, J., & Dainty, A. (2014). Safety in movement: Mobile workers, mobile media. *Mobile Media & Communication*, 2(3), 335–351.

Pink, S., Tutt, D., Dainty, A., & Gibb, A. (2010). Ethnographic methodologies for construction research: Knowing, practice and interventions. *Building Research and Information*, 38(6), 647–659.

Pink, S., Tutt, D., & Dainty, A. (2013). *Ethnographic Research in the Construction Industry*. London: Routledge.

Pole, C., & Morrison, M. (2003). *Ethnography for Education*. Berkshire: Open Universities Press, McGraw-Hill Education.

Raftery, J., McGeorge, D., & Walters, M. (1997). Breaking up methodological monopolies: A multi paradigm approach to construction management research. *Construction Management and Economics*, 15(3), 291–297.

Ramalingam, S., & Mahalingam, A. (2018). Knowledge coordination in transnational engineering projects: A practice-based study. *Construction Management and Economics*, 36(12), 700–715.

Richardson, L. (2000). Evaluating ethnography. *Qualitative Inquiry*, 6(2), 253–255.

Rooke, J., & Clark, L. (2005). Learning, knowledge and authority on site: A case study of safety practice. *Building Research & Information*, 33(6), 561–570.

Rooke, J. A., & Kagioglou, M. (2007). Criteria for evaluating research: The unique adequacy requirement of methods. *Construction Management and Economics*, 25(9), 979–987.

Rooke, J., & Seymour, D. (2002) Ethnography in the construction industry: Competing bodies of knowledge in civil engineering. In R. Fellows & D. Seymour (Eds.), *Perspectives on Culture in Construction*. Rotterdam: CIB.

Rooke, J., Seymour, D., & Crook, D. (1997) Preserving methodological consistency: A reply to Raftery, McGeorge and Walters. *Construction Management and Economics*, 15, 491–494.

Rooke, J., Seymour, D., & Fellows, R. (2003). The claims culture: A taxonomy of attitudes in the industry. *Construction Management & Economics*, 21(2), 167–174.

Rooke, J., Seymour, D., & Fellows, R. (2004). Planning for claims: An ethnography of industry culture. *Construction Management and Economics*, 22(6), 655–662.

Root, D., Fellows, R., & Hancock, M. (1997). Quantitative versus qualitative or positivism and interactionism – A reflection of ideology in the current methodological debate. *Journal of Construction Procurement*, 3(2), 34–44.

Rositsky, M. (1958). Concerning the new forms of industrial and construction management in the Czechoslovak Republic. *Problems in Economics*, 1(8), 103–104.

Runeson, G. (1997). The role of theory in construction management research: Comment. *Construction Management and Economics*, 15(1), 299–302.

Sergeeva, N., Ninan, J., & Oswald, D. (2022). Call for papers for the special paper collection: Novel research methodologies, methods and data in project studies. *Project Leadership and Society*, 3, 1–3.

Seymour, D., Crook, J., & Rooke, J. (1998) The role of theory in construction management: Reply to Runeson. *Construction Management and Economics*, 16, 109–112.

Seymour, D., & Rooke, J. (1995). The culture of the industry and the culture of research. *Construction Management and Economics*, 13(6), 511–523.

Seymour, D., Rooke, J., & Crook, J. (1997). The role of theory in construction management: A call for debate. *Construction Management and Economics*, 15(1), 117–119.

Sherratt, F., & Leicht, R. (2020). Unpacking ontological perspectives in CEM research: Everything is biased. *Journal of Construction Engineering & Management*, 146(2), 04019101–2.

Shields, R., & West, K. (2003). Innovation in clean-room construction: A case study of co-operation between firms. *Construction Management and Economics*, 21(4), 337–344.

Spence, A., Goulden, M., Leygue, C., Banks, N., Bedwell, B., Jewell, M., Yang, R., & Ferguson, E. (2018). Digital energy visualizations in the workplace: The e-Genie tool. *Building Research & Information*, 46(3), 272–283.

Styhre, A. (2009). Tacit knowledge in rock construction work: A study and a critique of the use of the term. *Construction Management and Economics*, 27(10), 995–1003.

Sykes, A. J. M. (1969a) Navvies: Their work attitudes. *Sociology*, 3, 21–35.

Sykes, A. J. M. (1969b) Navvies: Their social relations. *Sociology*, 3, 157–172.

Teo, M. M. M., & Loosemore, M. (2014). The role of core protest group members in sustaining protest against controversial construction and engineering projects. *Habitat International*, 44, 41–49.

Thieme, F. E. (1943). When should construction equipment be retired from service? *Journal of Forestry*, 41(6), 398–401.

Tutt, D., & Pink, S. (2019). Refiguring global construction challenges through ethnography. *Construction Management and Economics*, 37(9), 475–480.

Tutt, D., Pink, S., Dainty, A. R., & Gibb, A. (2013a). Building networks to work: An ethnographic study of informal routes into the UK construction industry and pathways for migrant up-skilling. *Construction Management and Economics*, 31(10), 1025–1037.

Tutt, D., Pink, S., Dainty, A. R., & Gibb, A. (2013b). 'In the air' and below the horizon: Migrant workers in UK construction and the practice-based nature of learning and communicating OHS. *Construction Management and Economics*, 31(6), 515–527.

van den Berg, M., Voordijk, H., & Adriaanse, A. (2020). Recovering building elements for reuse (or not) – ethnographic insights into selective demolition practices. *Journal of Cleaner Production*, 256, 120332.

Zanen, P. P. A., Hartmann, T., Al-Jibouri, S. H., & Heijmans, H. W. N. (2013). Using 4D CAD to visualize the impacts of highway construction on the public. *Automation in Construction*, 32, 136–144.

# 6

# DOING ETHICAL RESEARCH IN AN INDUSTRY CONTEXT

*Jan Hayes and Sarah Maslen*

*RMIT University, Melbourne, Australia*

*University of Canberra, Canberra, Australia*

**Key takeaways**

- University research ethics can be bureaucratic, but there are important considerations behind what may appear to be a box ticking exercise.
- Research ethics is not only about fieldwork. It starts with choosing an appropriate research question, including something with societal value.
- Core ethical principles in dealing with research participants are minimising harm, preserving privacy, respecting personal autonomy, honouring commitments and dealing with people fairly.
- As a minimum, this means obtaining informed consent, keeping data secure and anonymising participants in reported results.
- Complexities can arise in industry settings where corporate values differ, but preplanning can anticipate most difficulties.

## Introduction

Let's start with a couple of descriptions of some research fieldwork.

> You have approached Developer A to conduct interviews on their FirstBuild site. They have agreed by email that you can interview workers at the site. You arrive on site only to find that Constructor B did not know you were coming. After several phone calls, the site manager for Constructor B reluctantly agrees to let you on site and make people available for interview. You find yourself speaking to people in the site office where others coming in and out of the office can hear what your interviewees are saying. Do you continue? In a quiet moment, one interviewee tells you they are being bullied at work and wants your help to deal with the situation. What do you do? The following week, Developer A rings you and asks for copies of the interview transcripts as they have heard you have been told things that are not true and they want to educate the workers that gave you the incorrect information. Do you provide the information as requested?
>
> You have approached Developer B to conduct fieldwork on their NewBuild site. Developer B has agreed to the observational aspect of the research and has put you in contact

DOI: 10.1201/9781003379584-8

with a member of the site safety team who has organised your access to the worksite. While you are there, you witness two supervisors teasing an apprentice using sexist language. None of the staff who witness this raise it as a concern. Do you speak up, or remain silent given your observational role and concerns about maintaining access to your field site?

While these are fictional scenarios, researchers regularly face these types of situations during qualitative fieldwork on construction sites when conducting research with industry partners. The role of research ethics processes is to help researchers anticipate these types of situations and give them the tools to deal with these issues if/when they arise.

Historically, research ethics has focused on values concerning how the people being studied or from whom data is obtained (which may be the same or different groups) should be treated by researchers. Such values include 'minimizing harm, preserving privacy, respecting personal autonomy, honouring commitments such as promises, and equity in dealing with different people' (Hammersley, 2018). But contemporary research ethics extends beyond questions of informed consent and protecting sensitive data (Beauchemin et al., 2022). As Beauchemin et al. (2022) point out, an ethical issue can be defined as any situation that may compromise, in whole or in part, at least one value considered legitimate and desirable. This extends consideration of research ethics beyond a narrow focus on consent and confidentiality to include much bigger questions such as whose interests the research serves. This is particularly relevant in research that is conducted in partnership with industry players who may have quite different considerations in mind.

The focus of this chapter is to address ethical issues specifically linked to research practice (i.e., the doing of a specific research project) since this directly impacts participants. There are other ethical issues that also impact researcher integrity in a broad sense that we do not address here. For example, Drolet et al. (2022) identified distributive injustices within academia regarding research funding and the culture of individualism within academia as two of the top ten ethical concerns of researchers. Such structural issues have a real impact on researchers and research integrity but are beyond the scope of the chapter.

The chapter starts by setting out the origins of current university processes for managing research ethics, some key cases where research ethics have been challenged, and an appeal for researchers to adopt an ethics in practice approach, rather than being focused on ethics bureaucracy. Practical considerations in research ethics linked to construction industry fieldwork are then canvassed in six key areas before some concluding remarks.

## Background

The concept of modern research ethics stems from the middle of the twentieth century and the establishment of the Nuremberg Code in 1947 in response to experiments conducted by Nazi doctors during the Second World War (Shuster, 1997). The Code established principles for ethical medical research which are still accepted today including requiring voluntary informed consent from participants, minimising harm to participants and research on human participants only when the results are for the good of society and unprocurable by other methods (Shuster, 1997).

Uptake of ethical research principles in a university setting took some time to become established. Laboratory-based psychology experiments, in particular, have been called out for the harm done to participants on more than one occasion. In 1963, the well-known Milgram experiment showed how far many research participants were prepared to go in following instructions, when they apparently continued to give increasingly strong electric shocks to other participants who they thought were giving incorrect answers to questions. In fact, the entire arrangement was designed to study those who were receiving the instructions. No shocks were being administered and the people apparently screaming in response were research team members acting. Despite having university ethics approval, the experiment left some participants 'traumatized by how far they had been willing to go rather than question authority' (Klitzman, 2015: 13). In this and other similar cases, researchers had judged their experiments to be of minimal risk since they did not expect the studies to cause any physical harm. This was not challenged by university ethics processes at the time and yet considerable psychological distress ensued (Klitzman, 2015).

Regulatory change in the US was finally driven by the notorious Tuskegee syphilis study in which poor African American men were deliberately infected with syphilis between 1932 and 1972. When this came to public attention in 1974, the resultant outrage led to the passing of the National Research Act which put in place processes that remain largely unchanged today. Similar requirements were formally introduced in Canada, South Africa, the UK and Australia among others (Hammett et al., 2022). Systems for managing research ethics are now typically based around completion of an application by researchers that is then considered by a formal ethics committee.

Even though researchers require approval from instructional research ethics committees and strict protocols surround the operation of such committees, harm to participants because of medical research still arises. Examples include unauthorised procedures being carried out on patients, deliberately exposing people to toxins without their consent and using samples taken from participants (e.g. blood) for research beyond that which participants have agreed to (Klitzman, 2015). Despite its origins in medical and psychology research, and some doubts about the effectiveness of such systems, modern ethics requirements have spread into the social sciences and now apply to all research using human participants.

Universities in many countries have formal research ethics governance processes in place. While such processes are typically grounded in the three principles of voluntary informed consent, minimising harm to participants (and researchers) and justifiable research, the overarching purpose is to manage risk to the university, primarily the potential for financial and/or reputational damage. The high level of bureaucratisation required to comply with such systems can lead to 'mindless rule following or frustrated resistance' (Allen, 2008: 105). At worst, ethics committees 'adopt uncritical positivist approaches, transpose bio-medical and quantitative methods-bases philosophies onto other disciples and methodological approaches' (Hammett et al., 2022: 583). While there is a significant literature critical of institutional ethics processes in the social sciences, at least one empirical study of research ethics suggests that criticism may be somewhat overblown and reflect researchers resistance to control as much as anything else (Jeanes, 2017). Having said that, Tesar et al. (2021) note that we are in need of an ethics that turns attention to the institution as well as researchers, but such considerations are beyond the scope of this chapter.

Despite the adoption of such processes to manage research ethics including within the social sciences, what makes for ethical research has occasionally been challenged very publicly by ethnographic research. One such 'test' came from Laud Humphreys' work that culminated in the book *Tearoom Trade* (1976). This sociological study was undertaken in the US in the 1960s and involved participant observation and a follow-up survey with men who frequented public toilets (called 'tearooms') for sex with other men that they do not know. This kind of anonymous sex happens all over the world, but until the late 1960s it was a rarely studied form of social interaction. Humphreys posed as a lookout to befriend the men and observe their behaviour.

Part of the problem of this research is that it was covert. Humphreys did not tell the men he was a researcher, so they assumed that he was there for the same reason as them. He did not lie about why he was there, but as he was not asked directly he simply did not provide that information. Sometimes it is concluded that research needs to be covert to collect the necessary data. But this was not the only part of the research. Humphries also collected car number plate details, deceived the police that he was conducting 'market research' to obtain the home addresses of drivers and used this information to visit the homes of the observed men under the premise of a neutral social survey to establish the profiles of men who were visiting the tearooms.

In many ways, this aspect of the study was what gave such striking findings: we know from this that most men who visit tearooms were leading otherwise conservative, and straight, lifestyles. But this deception is incredibly problematic from a social research perspective. Participants certainly did not give voluntary informed consent and even though Humphreys was careful to protect the men's confidentiality, the information he gained could have been incredibly harmful. The activity he was documenting was illegal, so the police could have demanded he release the information (Warwick, 1973).

Another recent ethnographic study that attracted debate over research ethics came from Alice Goffman and her study of the experiences of the 6th Street Boys and their encounters with the law (Goffman, 2014). Like *Tearoom Trade*, while originally received favourably by critics the book went on to receive significant criticism related to the accuracy of accounts, Goffman's criminal activity and her positionality. In the course of Goffman's participant observation, one of the 6th Street Boys was shot dead. In the aftermath, Goffman reports to have been the driver of a vehicle with armed passengers on a hunt for the killer, which is a felony (Lubet, 2015). Conspiracy to commit murder aside, Goffman's embeddedness in this community led to role slippage. She wrote in her book that in this moment she was not driving as part of the data collection, but because she 'wanted Chuck's killer to die'. Critics also raised the ethics of a middle-class white woman speaking for a marginalised black community which many felt had the feel of an exoticised 'jungle story' (Zussman, 2016).

These examples may seem extreme, but researchers active in more conventional social setting such as construction workplaces can still encounter major ethical questions and dilemmas as the vignettes at the start of the chapter illustrate. While there remains an extensive literature about research ethics in medicine and human behaviour, there is little discussion about this topic in the construction research literature. Stewart et al. (2017) point out that while human participation in construction management research is critical, their review of the literature suggests that construction management researchers are little concerned with human research ethics. They argue that increased attention to research ethics would be not only morally appropriate but also objectively useful in improving research outcomes. Their plea does not seem to have been picked up by the construction management research community with major citation indices listing only a single citation.

The challenge for construction industry researchers is to ensure that they adopt an attitude of ethics-in-practice and do not fall back on a procedural ethics approach with the sole goal of achieving one-off administrative approval. As Chiumento et al. (2020) warn, the documents produced to meet such requirements are abstracted from the research context and do not prepare researchers well for the potential ethical issues that they may come across in practice. A more situated and reflexive research ethics is required that recognises ethics as something that must continually be negotiated between researchers and participants (Hammett et al., 2022). It is beyond the scope of this chapter to make recommendations for change to university ethics processes, but we hope to encourage researchers in construction to think beyond institutional hoop jumping.

## Key considerations for ethical research in construction

### *Establishing the research question*

In construction as elsewhere, all research is about generating new knowledge. This could be new knowledge for its own sake or knowledge that is to be put into practice to develop new policies or otherwise impacting ways of working. Considerations of truth, justifiability and relevance in research need to consider what Hammersley (2018) calls the epistemic values of the research, i.e., the values inherent in the knowledge that the research is intended to produce. In practice, research topics emerge from our areas of interests, things we notice in previous research and wish to explore further and, increasingly, what we believe we can get funded. However research topics emerge, researchers must be cognisant of the social value of what they propose to do and whose interests will be served by the results as these are important ethical considerations. Working with industry partners has many advantages not least of which include access to funding and research participants. These benefits come with potential downsides that must be managed. Industry interests are not always the same as those of researchers and researchers would do well to be aware of this from the beginning of any relationship. This does not mean that industry people are wrong or unethical, but rather that their activities are informed by different values and hence potentially different ethical imperatives.

In the field of safety, for example, it is now widely accepted that accidents are organisational phenomena (Reason, 1997) meaning that the actions of individuals working in the complex sociotechnical system that comprises the construction sector are driven by the structures, practices and interactions that take place within the system. An ethnographic study of the actions of people on site may be useful in identifying such issues as where work is being done in ways that are different to that set out in procedures (i.e., the classic 'work as imagined' vis-a-vis 'work as done' divide as described by Hollnagel, 2014). Ethical issues arise when the cause of the divide, also referred to as the compliance gap, is linked to normative traits of frontline workers such as ignorance, laziness or stupidity. This approach shifts responsibility for noncompliance onto frontline workers; there is no space for questions about the validity of the procedures themselves or other factors that are driving non-compliant behaviour.

Research that focuses on finding an explanation for behaviour, rather than blaming individual frontline workers, is much more likely to form the basis of meaningful strategies for improving safety at work. Yet this may not be the lens that industry partners have in mind in the process of co-designing a set of research questions. Seeking explanations for behaviour directs researcher attention to issues such as remuneration arrangements, social norms in the workplace and organisational structures. These things are significantly the purview of senior management and yet, despite this, we find that by far the majority of empirical work in safety science focuses on the actions of frontline workers (Hayes & Maslen, 2022).

This tension between industry perspectives on research questions and research ethics leads us to reflect on selecting field sites. Commonly, ethnographers seem to gravitate towards sites that are 'on the front line', i.e., where construction operations take place. When you are designing your next research project, ask yourself: are we researching frontline workers because they have the data we need or because it is too hard to get to management? Think carefully about research design and which groups you should be targeting for study. Don't do what's easy – try to do what will make a difference, remembering that the social value of the research you chose to do is a key aspect of undertaking ethical research.

### *Dealing with gatekeepers*

Once the research question is established and the research method including the preferred group of research participants has been identified, the next challenge is to identify specific individuals who are members of that group and invite them to participate. Ethical conduct of research is about ensuring the rights of all participants are protected. University ethics protocols are predicated on the idea that there is a power imbalance between researchers and human participants that favours the researchers. In practice when working with industry, power relationships are more complex. Even gaining access to workplaces can be difficult and the power relationships within the workplace setting require careful negotiation.

Ethics considerations generally require that individuals are fully informed about the research in general and specifically what will happen to the data they contribute. Such knowledge allows them to give their informed consent regarding participation (or decline to take part). The standard requirements for ethical research are that the participants' identities are kept confidential and the content of what they contribute to the research is used in a way that is not identifiable to them.

It is important to understand that there are two reasons for confidentiality. The first relates to the consequences of others knowing about participants' involvement and the research value of minimising harm. In the professional setting of construction research, potential adverse consequences can relate to financial harm if, for example, a participant's employer discovers that they have said something controversial or critical and so chooses to take action against them. Researchers must remember that confidentiality is also seen as a deontological value, i.e., it has value in and of itself. Honesty on the part of researchers is another such deontological value and so promised confidentiality should be maintained regardless of the consequences, i.e., the fact that you cannot see that there are any significant consequences is not sufficient justification for breaking confidentiality.

Ethics processes usually assume that researchers will be approaching possible research participants directly, but for workplace-based observations or interviewee recruitment this is often not the case. Rather, researchers need to convince a gatekeeper of some kind that access should be granted. This involves several potential ethical issues. Firstly, in an environment where time is money, many organisations will not give access unless they can see a direct, short-term benefit to them of allowing researchers to approach their workers. Negotiating access in this situation means being able to articulate what's in it for them. Appeals to altruism, i.e., the societal benefit of the research, are unlikely to be successful on their own. Strategies that we have used to convince companies to give us access have included a quid pro quo of running a small number of training sessions for them (scheduled so data collection is not impacted) and/or offering to give the company a benchmark of their performance on the research topic (or something closely related) against the broader industry (for projects where data from multiple companies is collected) as well as giving them access to consolidated research results. Some companies may wish to use their involvement and their academic relationship for promotional purposes which can be acceptable within reason provided it is not used to coerce individuals to become research participants.

Observational work in an industry context does not happen as it does in other aspects of social life. Ethnographic work can have different orientations, whether it focuses on the experiences of a smaller group of people as a microcosm for a broader social experience, or a much larger group of participants to observe the tactics people develop to navigate situations (Katz, 2018). In both cases, it is common for researchers to find their place in the identities and situations that they study either organically or without organisational oversight in the sense that they choose who to study and are not watched as they do so. This is rarely the case in an industry context. In the office environment, this is likely an issue of corporate comfort with having a non-staff member hanging around unchaperoned. On site, ethnographers are given access to specific spaces and are also subject to oversight as any other guest on a site would be for safety reasons as much as organisational control.

For observational work in industry, researchers need to consider how this gatekeeping opens some doors and closes others. In having personnel from the organisation screen what events and spaces to observe, the research process as a knowledge-making exercise is constrained by this filtering from an interested third party, rather than the research lens being researcher-led. Researchers should reflect on these questions of access: what am I being shown here? Does this give access to data that allows me to answer my research question? Do I have the sense that there are fields of activity that are out of bounds? Why am I being introduced to these people/sites, rather than others? Who/what is being excluded? Such oversight also affects the anonymity of the research participants to the organisational leadership. People with the power to hire and fire know who they have put in the researcher's orbit. Such individuals can often open conversations about events and people in a way that seems casual, and yet can violate confidentiality. Researchers need to watch for the openings of these conversations and stick to their commitments to individual research participants.

For interview-based research, access can and should be structured so that the company does not find out from the researchers who has agreed to participate. (If participants chose to tell their colleagues, that is their prerogative.) Combining this requirement with privacy provisions regarding giving out contact information may mean that a company representative circulates an invitation for participation in research and invites those who are interested to contact researchers directly. This means that researchers only receive the contact details of those who chose to contact them and the company does not know who has volunteered to participate. We have successfully used such a scheme in multiple contexts.

Some companies request access to research data. In our experience, it is best never to promise access to raw data. It is very difficult to completely anonymise fieldnotes or transcripts if readers know possible interviewees personally and can recognise experiences described and language used. In any case, the value of raw data to a company is limited.

Something else to be negotiated is whether or not the research requires the permission of the interviewee's employer. Is company permission required for people to speak to you? On many occasions, this is not required. If individuals are being asked about their professional views and experiences on a topic, or about typical practices across an industry sector, then many people can talk to researchers as professional individuals. In this case, they are not speaking on behalf of their employing organisation, but if this is what the research design requires then broader organisational approval may be required. On the other hand, in addition to the ethics of individual confidentiality, if your research design requires you to have copies of company documentation, such as procedures, you may find that the company itself requires you to sign an agreement stating that you will use the information for research purposes only, not publish it and/or not give it to third parties. Such agreements must be scrutinised carefully before signing as they are rarely written with research in mind and may be written in a way that inhibits publication of research findings.

In the age of social media, data can also come from online spaces. In the ethnographic method, researchers often combine the methods of participant observation and interviewing with analysis of archival data, meaning publicly available content online that was generated without the researcher's involvement (Kozinets, 2010). Archival data presents ethical concerns with respect to informed consent. The public character of content online makes it difficult, if not impossible, to anonymise the data. Mining of this data for research purposes is a widespread practice, and those researchers who do this argue that the publicly available nature of the information negates the need for informed consent (Pace & Livingston, 2005; Kozinets, 2010). But this is a tricky area. Users typically do not anticipate their content being used for research purposes and may not appreciate its use for these purposes especially in cases of poor de-identification raising red flags among some ethics committees and journal editors. Use of online data for research also sometimes breaches the terms of use of online platforms that the researcher may be using to access the data. This is a live area of debate that we cannot fully do justice to here, but we encourage researchers who are considering adopting this research method to deeply engage with the ethical considerations that are raised by use of archival data and the most current recommendations in navigating the challenges it presents.

### *In the field*

Once people have agreed to participate, the question arises as to how the research data collection should take place. This includes making choices about role adoption, dealing with 'guilty knowledge', and the coping with unexpected unethical comments expressed by participants, which may not be related to the researched phenomenon.

First, in the case of ethnographic work, roles are likely adopted at the worksite itself. In participant observation, for example, it is common for researchers to adopt a role as part of the culture or context. This can involve the doing of the work or practice, which happens mostly where an ethnographer is already involved in a given practice, taking a given workplace as a research site in a second step, as in the case of research on martial arts (Foster, 2015) or wildland firefighting (Desmond, 2008). It can also involve adopting a novice role (Maslen, 2020) or an entry-level position in the field site such as that of a lab technician to study the doing of scientific work (Latour & Woolgar, 1979). In organisational studies, the door into observation can involve role adoption more akin to consultancy, where ethnographers are shown around and allowed to sit in on meetings with the expectation of feedback in exchange for their research participation (Hopkins & Maslen, 2015). It can also involve non-participant observation, as in sitting in a room trying to observe and engage as little as possible in the work (Hayes, 2013).

In the case of role adoption, participant observers can face ethical challenges due to role conflict, one explanation for Goffman's criminal behaviour described earlier. Signals that this is taking place include becoming involved in practices that, as a researcher, you would challenge, or struggling with how or even whether to articulate commentary over an area of organisation or practice due to not wanting to share 'bad news' in the context of the relationships formed and the researcher's place in the organisation. Unless you are mindfully adopting an action research approach (with the aim of intervening in practice and observing the impact), it is important to remember that even in participant observation the researcher is a researcher first and should avoid being drawn in to inter- or intra-organisational politics, or changing the reporting of the research to fit with organisational expectations. In a construction context this might include reporting on unsafe practices, or less than ideal attitudes to safety by managers, or even commenting on harassment observed in the workplace. Ethical researchers must report on what they find. Ultimately, it is up to the organisation as to how they respond.

Second, regardless of role adoption, observation can also raise ethical challenges with respect to what the literature calls 'guilty knowledge'. This can take a number of forms that have differing implications for researchers. The term was originally coined to refer to knowledge of illegal activities (Fetterman, 1983). Ethnographers such as Goffman and Humphreys whose work we discussed earlier set out to study social groups that are known to be conducting illegal activities, but it is also not unknown for researchers in more mainstream contexts to be made aware of actions going on within a workplace that might be against the law such as sexual harassment or fraud. Researchers must decide for themselves on their actions in accordance with their own moral codes (and in some cases legal responsibilities). In the case of participants who reveal a history of harassment, one strategy is to refer individuals to specialist support services.

Guilty knowledge also refers to knowledge incidentally acquired about things that are not illegal but breach the social norms of the fieldwork setting (e.g. Puttick's, 2017 study of teachers where the researcher was aware of one teacher's poor performance in the classroom). Researchers generally have an obligation not to share such knowledge even in cases where they may be asked by a member of the organisation to pass an opinion about someone or something. When asked directly, a researcher can always point out that, in their role as researcher, making a comment would not be appropriate.

A third situation which can also put researchers under pressure is when research participants express views that are antithetical to that of the researcher, particularly if this is unexpected (e.g. an informal non-work-related conversation where someone makes a highly racist comment). You must always remember why you are in the workplace and not be tempted to argue the point or try to educate research participants.

For interviews, it is often convenient to do these at the worksite too if there is a quiet place available where interviewees can express views without being overheard or interrupted and recordings can be made without excessive background noise. It may also be appropriate to meet office workers away from the workplace where both confidentiality of participation and the content of the conversation can be more easily managed. In our experience, most professional interviewees have no problem with others knowing that they are meeting an academic researcher (and this is entirely up to them), but they are keen for conversations to be confidential. We have also found that sometimes it is easier to undertake research interviews in the evening after office staff have departed for the day as operational and site-based people may have more time to talk.

As described above, the autonomy of research participants must be respected, so the research must be described to them in a way that gives them the opportunity to give informed consent. This may mean that research questions and topics that rely on academic or discipline-specific jargon will require translation into plain language. It is also possible that research questions require some generalisation in material for participants so as not to bias the data collected.

In interviewing, it is important to ensure that all participants understand what is required of them if they choose to participate and what will happen to the data that the researchers collect before formal data collection begins. In many countries this is set out in a plain language participant information sheet and an associated consent form which the participant must sign. Sometimes getting signed forms can be problematic. For fieldworkers, using paper and pens at work may be a poor cultural fit and it is acceptable to get participants to acknowledge at the beginning of a video or audio recording that they are aware recording is happening and consent to it.

Another situation in which getting signed forms can be problematic is in conversations with senior professionals. We have had managers refuse to sign such consent forms without legal review and in this case have again fallen back on oral consent (Hopkins & Maslen, 2015). In the case of interviewing, ethics committees can require that oral consent is captured on the recording following a script. When conducting online interviews, we always get participants to acknowledge and consent to recording as even when the return of signed form is promised, busy professionals sometimes fail to deliver. In all cases, the most important consideration is that informed consent is obtained and recorded in some form.

In the case of observational research, the concept of informed consent is much more grey and may even be waived. There are typically too many people in a setting for it to be feasible to get consent forms from all involved. This is also beyond the requirements of ethics committees. In lieu of this, a strategy that we have used in the past is for the researcher's status to be public knowledge, and for written consent only to be sought in more contained observations of practice, e.g. when shadowing.

In a professional setting such as most research in construction, interviewees are best placed to manage their own risk of harm to some extent but no matter what the stakes, researchers must not identify participants or link specific data to them without their explicit permission. A recent example of this from our own work involves writing about the role of being a standards committee chair (Hayes & Maslen, 2023). While our published work does not include the name of the individual, everyone in the relevant sector will know who we are writing about so publication was only possible with the direct and explicit approval of the individual involved. This was done in two steps. Firstly, we discussed the idea of the paper with the person involved to see if they had any fundamental objections. Once we had in principle endorsement, the analysis was further developed, and the written work was sent to them for review and approval before being submitted to a journal.

### *Stay safe*

We have reflected at length on how to ensure that research participants are not harmed. Research ethics is also about protecting the researcher from harm. This is a very real issue given that construction sites are dangerous places so be careful as you move around and obey all instructions. If you feel unsure or unsafe, stop.

If appropriate, researchers should volunteer to undertake a site induction. A site induction is so much more than an administrative hurdle keeping you away from the real data. In addition to supporting understanding of how to be safe on site, attending an induction is another source of research data that can, in itself, provide useful insights about cultural aspects of the organisational setting (it gives rise to questions such as: what assumptions underlie the session? How does the organisation speak to its employees/contractors/visitors?).

Safe considerations for fieldwork on building sites include wearing the appropriate personal protective equipment (PPE). Researchers should view this not as adherence to rules, but another opportunity for research insight as they place themselves in the material world of their research participants. It is best to ask beforehand what is required so that researchers can arrive wearing the right shoes and clothing and have any other necessary equipment such as hard hats and safety glasses to hand. Often organisations will arrange this equipment themselves, providing access to this equipment from their own storerooms. Being dressed for the part may also be an element of fitting in and establishing credibility with research participants for both interviews and observation-based fieldwork, though there may also be scenarios where workers generally do not wear their PPE requiring researchers to balance their personal risks with the value of fitting in. In general, we would advise compliance with PPE requirements irrespective of worker behaviour.

Other risks to researchers should also be explicitly identified and managed. This includes consideration of where interviews take place and whether one or two researchers should attend. An example of this is sending researchers to conduct interviews in people's homes. We prefer such interviews to be done in pairs. If this is not practical, the minimum requirement is for the researcher to text details to a colleague before they enter someone's home and check in again once the interview is completed. This is an especially important consideration for research students who may be more vulnerable and less experienced at dealing with tricky situations.

The potential for psychological harm linked to fieldwork should also be taken seriously, even with apparently benign topics. This is perhaps best managed by a common reflexive practice within the research team where trustful relationships can provide support in the first instance.

### *Reporting research results*

For industry-funded projects, delivering a report of the research results will likely be a contractual requirement, but it is worth noting that ethical research includes consideration of appropriate reporting of results to participants in some form. Essentially, researchers should avoid possible perceptions of inappropriate power relationships, for example, interviewing workers and then reporting findings only back to management.

Industry and academia have different cultural requirements regarding written communication which researchers working with industry also need to navigate in an ethical way. Industry is likely to ask for plain language in written deliverables, but this typically means adopting industry-based jargon, rather than academic writing tropes. Writing so that readers understand research findings is important, but researchers should be aware of the potential for self-censoring when it comes to reporting findings that industry players may find difficult. In our view, this is one key difference between the role of a consultant and the role of an academic researcher. Researchers must retain independence and report the findings of research as they see them. This is quite different to the role of consultants who are hired to meet their clients' needs.

When writing about qualitative research data, it is a common practice for participants to be identified by pseudonyms. This is the primary method used to ensure participants are not identifiable, even to their colleagues who may read research outputs (Heaton, 2022). In ethnographic studies, where research outputs may talk about individuals at length, pseudonyms usually take the form of replacement names essentially to allow results to be deidentified without being depersonalised. In some interview-based studies, direct quotes from participants are simply identified by a unique number. On other occasions, where sub-groups within the participant pool are important to interpretation of the data, more complex schemes can be used, e.g. allocation of a code that identifies the participant as an employee of a particular type of organisation, a particular education level or professional role or similar always providing that the group is large enough so that membership of the group does not identify the individual. If results are reported as an organisational level, similar considerations apply to allocating pseudonyms to differentiate between participating organisations. For complex cases, Heaton's recent review (2022) provides useful guidance.

### *Higher degree by research students*

The ethical considerations outlined to date apply to all researchers from students to early career researchers and senior professorial staff, but there are some specific considerations that apply to students. Research students are inexperienced, and, as we have already described, there are many ethical issues that are not addressed in the narrow confines of formal university processes. Students need to be supported to identify and manage ethical issues as they arise in both framing and conducting their research. Keeping a research journal can be a useful way to encourage reflection on research practice.

Since many research students also work part time in industry, questions of role conflict are particularly pertinent. The issue of guilty knowledge discussed above becomes even more complex as something that is peripheral knowledge in one role may be highly pertinent in another, and so taking no action might be an appropriate response in one role and completely unacceptable in another. Supervisors should be vigilant in addressing these issues with students throughout their fieldwork so that ethical dilemmas are resolved mindfully.

Students are also sometimes unprepared for the emotional impact of data that they may collect. This can be compounded by the fact that student fieldwork is usually done alone. Questions of student safety can arise (see, for example, Quickfall, 2022) that supervisors should do their best to anticipate and so avoid.

## Conclusions

It is important to follow ethical principles in conduct of research in construction and provided these are understood in the early stages of planning a research project, this is usually straightforward. Most qualitative research in construction does not pose significant risk of harm to either participants or researchers and the autonomy of participants is usually simple to respect. The kinds of ethical dilemmas that arise are mostly able to be anticipated and so strategies can be developed in advance.

Qualitative research is often iterative with the details of empirical work being adjusted in response to early data collection and analysis. With this in mind, ethics must be understood as a key aspect of ongoing research practice. The likelihood of harm to participants or researchers is usually very low provided ethical principles are front of mind throughout data collection and analysis. Problems may arise if the initial ethics application and approval is seen as a hurdle to be passed only to be forgotten once the real work begins.

The chapter began by raising a series of ethical questions that researchers may face in the field. We hope that readers will now have the skills to recognise that these are indeed ethical questions and have a greater understanding of how to find an ethical way forward in any given empirical domain.

## References


Allen, G. (2008). Getting beyond form filling: The role of institutional governance in human research ethics. *Journal of Academic Ethics*, 6, 105–116. https://doi.org/10.1007/s10805-008-9057-9

Beauchemin, É., Côté, L. P., Drolet, M. J., & Williams-Jones, B. (2022). Conceptualising ethical issues in the conduct of research: Results from a critical and systematic literature review. *Journal of Academic Ethics*, 20, 335–358. https://doi.org/10.1007/s10805-021-09411-7

Chiumento, A., Rahman, A., & Frith, L. (2020). Writing to template: Researchers' negotiation of procedural research ethics. *Social Science & Medicine*, 255. https://doi.org/10.1016/j.socscimed.2020.112980

Desmond, M. (2008). *On the Fireline: Living and Dying with Wildland Firefighters*. Chicago: University of Chicago Press.

Drolet, M.-J., Rose-Derouin, E., Leblanc, J.-C., Ruest, M., & Williams-Jones, B. (2022). Ethical issues in research: Perceptions of researchers, research ethics board members and research ethics experts. *Journal of Academic Ethics*. https://doi.org/10.1007/s10805-022-09455-3

Fetterman, D. M. (1983). Guilty knowledge, dirty hands, and other ethical dilemmas: The hazards of contract research. *Human Organization*, 42(3 (Fall 1983)), 214–224.

Foster, D. (2015). Fighters who don't fight: The case of aikido and somatic metaphorism. *Qualitative Sociology*, 38(2), 165–183. https://doi.org/10.1007/s11133-015-9305-4

Goffman, A. (2014). *On the Run: Fugitive Life in an American City*. Chicago: University of Chicago Press.

Hammersley, M. (2018). Values in social research. In R. Iphofen & M. Tolich (Eds.), *The SAGE Handbook of Qualitative Research Ethics*. Thousand Oaks: Sage.

Hammett, D., Jackson, L., & Bramley, R. (2022). Beyond 'do no harm'? On the need for a dynamic approach to research ethics. *Area*, 54, 582–590. https://doi.org/10.1111/area.12795

Hayes, J. (2013). *Operational Decision-Making in High-Hazard Organizations: Drawing a Line in the Sand*. Chesterfield: Ashgate.

Hayes, J., & Maslen, S. (2022). Putting senior management in focus: A critical reflection on the direction of safety science research. In K. P. Gould & C. Macrae (Eds.), *Inside Hazardous Technological Systems: Methodological Foundations, Challenges and Future Directions*. Boca Raton: CRC Press.

Hayes, J., & Maslen, S. (2023). Dynamics of the room: expert wrangling in the process of standard formation. *Environment Systems and Decisions*. https://doi.org/10.1007/s10669-023-09945-8

Heaton, J. (2022). "*Pseudonyms are used throughout": A footnote, unpacked. *Qualitative Inquiry*, 28(1), 123–132. https://doi.org/10.1177/10778004211048379

Hollnagel, E. (2014). *Safety I and Safety II: The Past and Future of Safety Management*. Chesterfield: Ashgate.

Hopkins, A., & Maslen, S. (2015). *Risky Rewards: How Company Bonuses Affect Safety*. Chesterfield: Ashgate.

Humphreys, L. (1976). *Tearoom Trade: Impersonal Sex in Public Places*. Abingdon: Routledge.

Jeanes, E. (2017). Are we ethical? Approaches to ethics in management and organisation research. *Organization*, 24(2), 174–197. https://doi.org/10.1177/1350508416656930

Katz, J. (2018). On becoming an ethnographer. *Journal of Contemporary Ethnography*, 48(1), 15–60. https://doi.org/10.1177/0891241618777801

Klitzman, R. (2015). *The Ethics Police? The Struggle to Make Human Research Safe*. Oxford: Oxford University Press.

Kozinets, R. V. (2010). *Netnography: Doing Ethnographic Research Online*. Thousand Oaks: Sage.

Latour, B., & Woolgar, S. (1979). *Laboratory Life: The Construction of Scientific Facts*. Princeton: Princeton University Press.

Lubet, S. (2015). Ethics on the Run. *The New Rambler Review, May 26, 2015*, Northwestern Public Law Research Paper No. 15-34. https://ssrn.com/abstract=2611742

Maslen, S. (2020). Between narrative and practice: Storytelling as a way of knowing how to be in nature. *Ethnography*, 23(4), 433–449. https://doi.org/10.1177/1466138120923707

Pace, L. A., & Livingston, M. M. (2005). Protecting human subjects in internet research. *International Journal of Business Ethics and Organization Studies*, 10, 35–41.

Puttick, S. (2017). Performativity, guilty knowledge, and ethnographic intervention. *Ethnography and Education*, 12(1), 49–63. https://doi.org/10.1080/17457823.2015.1110039

Quickfall, A. (2022). Reflecting on ethical processes and dilemmas in doctoral research. *Education Science*, 12, 751. https://doi.org/10.3390/educsci12110751

Reason, J. (1997). *Managing the Risks of Organizational Accidents*. Chesterfield: Ashgate.

Shuster, E. (1997). Fifty years later: The significance of the Nuremberg Code. *The New England Journal of Medicine*, 337, 1436–1440. https://doi.org/10.1056/NEJM199711133372006

Stewart, I., Fenn, P., & Aminian, E. (2017). Human research ethics – Is construction management research concerned? *Construction Management and Economics*, 35(11–12), 665–675. https://doi.org/10.1080/01446193.2017.1315151

Tesar, M., Peters, M., & Jackson, L. (2021). The ethical academy? The university as an ethical system. *Educational Philosophy and Theory*, 53(5), 419–425. https://doi.org/10.1080/00131857.2021.1884977

Warwick, D. P. (1973). Tearoom trade: Means & ends in social research. *The Hastings Center Studies*, 1(1), 27–38. https://doi.org/10.2307/3527471

Zussman, R. (2016). Alice's adventures in wonderland: On the run and its critics. *Society*, 53(4 (Aug 2016)), 436–443. https://doi.org/10.1007/s12115-016-0039-z

# 7

# PHILOSOPHICAL REFLECTIONS ON ETHNOGRAPHY IN CONSTRUCTION RESEARCH

*Hans Voordijk*

*Civil Engineering and Management, University of Twente, Enschede, The Netherlands*

**Key takeaways**

- Three relevant philosophical foundations of ethnographic studies in construction research are phenomenology, hermeneutics and critical research.
- Phenomenology is relevant for examining lived experience of those working in construction practice and the way certain concepts appear to the consciousness of construction experts and practitioners.
- Hermeneutics is relevant for the unique contexts of construction projects that require context-sensitive action of practitioners involved.
- Critical ethnography is relevant for providing insights in preconditions and obstacles for technology implementation.

## Introduction

Ethnography originally emerged from anthropology to explore exotic and unfamiliar cultures (O'Reilly, 2009). While classic ethnography seeks to describe whole cultures, ethnography explores nowadays also particular phenomena in their natural settings (Katz & Csordas, 2003). Ethnography provides descriptions to experience these cultures or phenomena first hand. In studying these cultures or phenomena, ethnography draws on philosophical sources, such as phenomenology, hermeneutics and critical research.

Studying the lifeworlds or lived experiences is an important part of ethnographic research. Phenomenology, the study of these experiences, is therefore a first important philosophical basis for ethnographic research (Laverty, 2003). According to the German philosopher Heidegger, individual history, culture and traditions influence a person's understanding of any experience. He focuses on the relation between individuals and their lifeworlds (Neubauer et al., 2019). Based on this Heideggerian perspective, examples of phenomenological studies in construction research will be provided.

Ethnography also recognises multiple voices, views and subjective truths. Hermeneutics, the science of the interpretation of language, whether written or spoken, is therefore discussed as a

DOI: 10.1201/9781003379584-9

second important philosophical basis for ethnographic research. Heidegger's notions of 'being-in-the-world' and 'thrownness' and the 'hermeneutics circle' of Gadamer (1976), one of the major representatives of contemporary hermeneutics, will be discussed. The relevance of these concepts for ethnographic research in construction research will be addressed and examples of studies using the hermeneutic approach provided.

Subsequently, based on the Heideggerian concept of 'being-in-the-world', ethnographic approaches when investigating different ways of engagement in general and in construction organisations and projects are discussed. Different ways of how humans engage with the world range from complete immersion to detachment. Heidegger's tool analysis and the concept of a temporary breakdown of Sandberg and Tsoukas (2011) make it possible to understand these different ways of human engagement. Examples of ethnographic studies in construction research investigating different ways of engagement are provided.

Besides phenomenology and hermeneutics, critical research is a third important philosophical basis for ethnographic research. Critical researchers challenge taken-for-granted assumptions, beliefs, ideologies and discourses (Richardson, 2005). They recognise that a researcher and a research participant are part of a wider set of social, economic and political relationships (Myers, 1997). It is discussed that, based on Habermas, the impact of systems is represented in critical ethnography as a struggle between system and the lifeworld. This critical ethnographic approach is discussed in studies on construction projects and organisations implementing and using digital technologies.

## Phenomenology

### *Descriptive and interpretative phenomenology*

Everyday experiences, the so-called 'life worlds' or lived experiences, are key in both ethnography and phenomenology (Heidegger, 1977). In phenomenology, a phenomenon is described by exploring it from the perspective of those who have experienced it (Neubauer et al., 2019). A phenomenon is anything that appears to someone in their conscious experience (Moran, 2002). This can be a real or imagined object, a feeling, idea or emotion which presents itself to consciousness (Van Manen, 1997). Phenomenology is the study of these experiences. By examining these experiences as they are subjectively lived, the aim is to better understand these experiences (Laverty, 2003). Two major approaches in phenomenology are Husserl's descriptive and Heidegger's interpretive phenomenology.

Husserl (1970), the founder of the descriptive phenomenology, believed that these experiences that people have cannot be reduced completely to a measurable object. Husserl tried to fill the gap between objectivism and these subjective experiences by describing the 'objects' of consciousness (Chari, 2014). Consciousness is always consciousness about 'something'. This 'something' is the 'object' of consciousness. This 'object' can be a physical object or abstract concept. Human thinking and perceiving are directed towards these objects of consciousness. Descriptive phenomenology examines experiences of human beings by focusing on these objects of consciousness or phenomena that appear to the consciousness (Laverty, 2003). It tries to identify the common features, or universal *essences*, which people have experienced about a certain phenomenon (Neubauer et al., 2019).

To identify the universal essences of a phenomenon requires the researcher to set aside – or bracket off – one's own beliefs, assumptions, preconceptions and expectations in order to focus on the participants' experience of a phenomenon (Lopez & Willis, 2004). This *bracketing*, although difficult to achieve, requires transcendence from the natural attitude of everyday way of perceiving the world (Aagaard, 2017). This bracketing is a critical assumption of the descriptive phenomenology of Husserl.

For Heidegger, founder of the interpretive phenomenology, an individual's conscious experience of a phenomenon is not separate from the world. An individual's consciousness is formed by historically lived experiences including an individual history and the culture in which a person was raised (ibid., 2004). Heidegger's term lifeworld refers therefore to the idea that individuals' realities are always influenced by the world in which they live (Lopez & Willis, 2004). He focuses on the relation between individuals and their lifeworlds (Neubauer et al., 2019). Humans cannot experience a phenomenon without referring to their background understandings. Culture and traditions influence a person's understanding of an experience. For that reason, Heidegger's interpretative phenomenology does not just describe experiences, but is also concerned with their interpretation to understand the meaning of experiences.

Heidegger (2008) denies the possibility of "presuppositionless apprehending of something presented to us" (pp. 191–192). We can never be free of assumptions nor have a fully detached reflection (Gill, 2014). Thereby, Heidegger disputes Husserl's idea of bracketing presuppositions to articulate an essence of a phenomenon. Understanding of a certain context is not possible by a researcher suspending her or his prejudices (Myers, 1997). This means that ethnographic researchers must become aware of their prejudices and make them explicit in the research process.

### *Studying lived experiences in construction research*

So, how does phenomenology inform construction researchers? This section illustrates based on previous ethnographic work, what phenomenological elements surface in construction ethnographies.

Studies on lived experiences, key in both ethnography and phenomenology, have been fruitfully conducted within construction research. Based on descriptive phenomenology Samuel (2020), for example, examines the lived experience of project delivery team members who are assigned simultaneously to multiple project delivery teams. This study aims to provide a better understanding of team members' self-leadership abilities; how they interact, collaborate and/or prioritise their project activities. Based on individual descriptions of lived experiences of team members universal meanings or, in other words, the essences of these experience are derived. In this way, Samuel provides insight into the phenomenon of the teaming experience as lived by members, who simultaneously serve on, or lead, multiple project delivery teams. Knowledge is gained of how the phenomenon of these multiple project delivery teams affects the collaborative processes that are necessary for successful outcomes. To this end, he asks team members about personal experiences, perceptions and comprehensions of the shared leadership environment while serving simultaneously in multiple teams.

Further, Venselaar and Gruis (2016) also used the approach of descriptive phenomenology. They studied the phenomenon of the emergence of supply chain partnering in a Dutch housing association that implements supply chain partnering. Their aim is to uncover intra-organisational dynamics in implementing supply chain partnering through an in-depth phenomenological study inside an organisation. The analysis of individual descriptions of the 'lived experiences' of those working in the same context and involved in supply partnering includes highlighting significant statements and describing the 'essence' of these experiences. Insights are generated in experiences participants have in common that could be beneficial to the whole supply chain.

Based on interpretative phenomenology, Wimalaratne and Kulatunga (2022) aim to create richer understandings on the phenomenon of buildability within construction projects. Buildability is the optimal integration of construction knowledge and experience at various project stages to achieve the overall project goals. Through interviews, perceptions from industry experts with regard to their lived experiences concerning buildability are obtained. Knowledge about the buildability concept as it appears to the consciousness of the experts are interpreted following the interpretive phenomenological perspective of describing what an expert perceives, senses and knows in one's immediate awareness and experience. It is attempted to understand the hidden deeper meanings underlying these subjective experiences.

Another example of a study based on interpretative phenomenology in construction research is presented in Akmam Syed Zakaria et al. (2018), who present key factors influencing the decision to adopt industrialised building systems. The aim of this study is to understand patterns of interaction between construction professionals making decisions on the adoption of these systems. In semi-structured face-to-face interviews with experienced construction stakeholders from across the industry feelings, attitudes and perceptions of the interviewees are explored. Insights are offered into the complexity of an adoption decision and how many interconnected 'aspects' influence this. The interpretative phenomenological approach explained the multiple interpretations of reality held by different professional during this decision-making process.

## Hermeneutics

### *The 'hermeneutics circle' of Gadamer*

As an interpretive methodology, ethnographic research recognises the multiplicity of meaning systems of participants of a research project (Sharp, 2019). Interpretivism views reality as a multiple and interconnected 'system' of meanings, which are socially constructed (Berger & Luckman, 2003). Interpretivist research seeks to capture these meanings through interaction with the participants (Hudson & Ozanne, 1988). This capturing of meanings through interpretive ethnography is based on humans continually interpreting their reality and researchers interpreting these interpretations. Hermeneutics or the science of interpretation is therefore an important philosophical basis for ethnographic research (Myers, 1997).

Originally, hermeneutics was associated with the interpretation of biblical texts (Turk, 2001b). In the 19th century, systematic procedures were developed that would lead to correct interpretations of texts by studying the context, in which they were written. In the 20th century, based on Heidegger, Gadamer (1976) developed the idea of the 'hermeneutics circle'. This circle refers to the dialectic between the understanding of the text as a whole and the interpretation of its parts. Gadamer explains it as a circular relationship where "the anticipation of meaning in which the whole is envisaged becomes explicit understanding in that the parts, that are determined by the whole, themselves also determine this whole" (ibid., p. 117). The awareness of a dialogue between the text and the interpreter is regarded as the basis for contemporary hermeneutics or interpretation of any text (Howarth, 1997; Myers, 1997).

For Heidegger and Gadamer all understanding and interpretation is rooted in hermeneutics, forming an inescapable part of 'being-in-the-world' or 'Dasein' (Heidegger, 1977; Aagaard, 2017). 'Dasein' is a German word which can be translated as 'being there', the 'human way of being' or 'everyday human existence' (Mitcham, 2003). Da-sein emphasises that a human finds itself 'thrown' into a situation – a typical Heideggerian concept (Turk, 2001a; Dias, 2006).

Humans being socialised into certain practices, prior to any theoretical understanding, is one of the main themes of Heidegger's philosophy (Dias, 2006). 'Thrown into a situation' or 'being-in-the-world' means that human beings always have the notion of these 'pre-theoretical' shared agreements or practices (Dias, 2006). These practices provide the ability to act pre-reflectively (i.e., without reflecting or thinking) and are the basis for everyday action when 'thrown' into a situation. Gadamer follows the same reasoning by stating that historically and culturally unbound 'objectivity' cannot exist. He rejects 'truth' interpreted through methodical, objective investigation. History and 'culture' of the interpreter always provide the starting point for the 'hermeneutic circle'. Understanding is part of the 'being' of the interpreter (Howarth, 1997).

Consequently, prejudice and prior knowledge play an important part in our understanding, while positivist social science considers these as a source of bias and therefore a hindrance to true knowledge (Myers, 1997). Objectivity, according to positivism, is best attained if social scientists adopt a value-free position and do not let biases interfere with their analysis (ibid., 1997). By contrast, Gadamer recognises that prejudice and prior knowledge are the necessary starting point of understanding and cannot be thought of as being removable. "Prejudices are not necessarily unjustified and erroneous, so that they inevitably distort the truth. In fact the historicity of our existence entails that prejudices … are simply conditions whereby we experience something - whereby what we encounter says something to us" (Gadamer in Kearney & Rainwater, 1996, p. 115).

The explicit recognition of prejudice is central to Gadamerian interpretation and the nature of understanding (Howarth, 1997). Human understanding is always interpretive. The task of interpretive ethnography then becomes one of distinguishing between "true prejudices, by which we understand, from the false ones by which we misunderstand" (Gadamer, 1976, p. 124).

### *Hermeneutics and thrownness in construction research*

So, how does hermeutics inform construction researchers? This section illustrates based on previous ethnographic work, what hermeneutic elements surface in construction ethnographies.

The idea of the hermeneutic circle can also be applied to the way in which we understand a construction project or organisation. It provides a point of departure for the dynamics of understanding, in which the researcher must attempt to 'fuse horizons' of different individuals (Howarth, 1997). The meaning of the data is the result of this joint process by which researcher and participants have fused 'horizons' and of the way certain situations are interpreted or understood. This dynamic that shapes understanding is encapsulated in the hermeneutic circle, which moves "constantly from the whole to the part and back to the whole" (Gadamer, 1976, p. 117). In other words, the more interviews we conduct and the more information we gather, the more we understand the construction project or organisation as a whole and its constituent parts. This hermeneutic process continues until contradictions and oppositions in the construction project or organisation or between organisations longer appear strange but make sense.

Heidegger's being-in-the-world and notion of 'thrownness' are relevant for ethnographic research in construction projects. 'Thrownness' can be compared to the context-sensitivity of actions and interpretations taking place in a construction project. Dias (2006) shows that each construction project has unique features that require context-sensitive action of practitioners. Turk (2001b) discusses the 'blindness' that would be experienced if an engineer relies on a physical, mathematical or conceptual model of either the product or process he or she is involved in, rather than recognising his 'thrownness' in a given context, with its particular idiosyncrasies. Engineers must ensure that outputs of their models are judged in the light of specific contexts. In other words, engineers are committed primarily to 'being-in-the-world' rather than to 'being-in-a-model-of-the-world' (Dias, 2006). As such, an engineer can be seen as being thrown into a context.

Turk (2001b) also analysed digital modelling standards from a hermeneutical perspective. According to Turk, these standards are not representations of objective reality but the result of (inter-) subjective interpretations. A strict definition of the semantics, the linguistic meaning of applied terms, is usually considered as a positive feature of digital modelling standards. This feature, however, does not prevent the possibility of different interpretations (Turk, 2001b). Voordijk et al. (2022) showed that, given these different interpretations of reality by the various disciplines involved in construction, defining a common ontology is difficult: "there is not one reality out there to check the models against" (Turk, 2001b, p. 90). A digital model is part of a particular user's lifeworld.

## Different ways of human engagement

### *Different subject-object distinctions*

Heidegger's 'being-in-the-world' suggests that there is a continuum of ways of human engagement. It can be interpreted as a form of being of entwinement: human beings are never separated but always entwined with others and things in a more or less holistic whole (Sandberg & Tsoukas, 2011, p. 343). This entwinement ranges from complete immersion to detachment. Sandberg and Tsoukas (2011) state that in particular ethnographic studies of organisational life have shown cases of practitioners immersed in organisational practices in a holistic manner. Acting immersed means that we are, in first instance, not separated with the objects and other human beings that surround us (Hartmann, 2012). This is contrary to the reflective analysis of a detached observer which occurs in situations where one is inspecting, designing or building a tool or equipment (Turk, 2001b).

Immersed acting or acting without reflection uses intuition and common sense and is based on being 'thrown into a situation' (Turk, 2001b). Dreyfus (1987) gives an example of this acting: a person is turning a doorknob to enter a room (Dias, 2007). In this everyday act, there is no conscious intention of the person directed towards the doorknob. There is a "seamless web of activity for the fulfilment of a purpose, in which both the person and the doorknob are participants" (ibid., p. 383). Hence, the doorknob 'disappear' for the user in the act of using it. In fully engaged or immersed acting exists no perceived distinction between subject or object as such. In other words, objects in the world are not perceived as isolated objects (Dias, 2006).

Heidegger uses the hammering of a nail to show the lack of perceived subject-object distinctions. He argues that human beings "do not require conscious reflective knowledge about the physical properties of a hammer and the physics of hammering. The tool is ready-to-hand and we simply hammer the nail into the wall" (Turk, 2001b, p. 87). So, a tool is 'ready-to-hand' when "it is being used in a practical, absorbed and even *invisible* (italics by the author) way, without any need for reflection" (Riemer & Johnston, 2011, p. 8). In Heidegger's example, the nature of a hammer "cannot be understood without the knowledge of everything that makes a hammer into what it is in a certain equipmental context … equipment needs to be appropriate (angemessen), which means it has to have its place in the totality of other equipment, shared practices, as well as user skills and social orthodoxies" (Riemer & Johnston, 2011, p. 7).

Through training, people acquire skilled routines in which the human body and technologies function as a single assembled unity (Dorrestijn, 2012b). Within such an 'equipmental whole' equipment is 'ready-to-hand' and employed towards a purpose (Dias, 2006). There is no perceived subject-object distinction. This 'equipmental whole' can also be found in design, construction or maintenance projects that almost invariably use equipment of one sort or other (whether computers, cranes or other tools) in this way.

Such equipment becomes 'present-at-hand' and the object of attention in the event of a breakdown, such as, for example, a broken hammer. Equipment is also present-at-hand when someone does not know anything about it and hence fails to use it skilfully. In these cases, there is a clear perception of subject-object distinctions. An 'in-between' state is when equipment is 'unready-to-hand'. Unreadiness-to-hand occurs when a user is learning and acquiring the skills necessary to be involved with equipment in an absorbed way (Riemer & Johnston, 2011).

A temporary breakdown provides an opportunity to understand the otherwise unreflective acting in regular practices (Chughtai & Myers, 2014). In most cases, human beings are able to deal with a temporary breakdown using 'thematic deliberation' (Dreyfus, 1987). They pay deliberate attention to what they do and how equipment helps them to do so (Hartmann, 2012). If the breakdown becomes too severe, users enter the mode of theoretical detachment. They then try to understand the abstract properties of the situation at hand by detaching from the actions they were unconsciously involved with (ibid., 2012).

### *Ways of engagement in construction research*

Heidegger's tool analysis and his concepts of ready-to-hand, present-at-hand and unready-to-hand make it possible to understand different ways of human engagement with the world. The three concepts have been used in ethnographic studies to explore if and how information technology shapes working practices (Hartswood et al., 2008). Voordijk (2019) looked at the implementation of Building Information Modelling (BIM) using Heidegger's concepts. As was the case with BIM, this technology first conflicted with existing work practices in a ready-to-hand work environment. Using BIM started with being instructed when the technology was still unready-to-hand. This situation changed when users acquired skills and expertise necessary to deal with BIM in an unreflective way (Dreyfus & Dreyfus, 1996). BIM became ready-to-hand as it gradually blended into 'the totality of shared practices' (Riemer & Johnston, 2014). Given certain problems, users sometimes reverted to the 'old' software because they could not work with the new BIM software. For these users, BIM remained present-at-hand.

Interpreting these temporary breakdowns in established practices makes it possible to understand different ways of human engagement in construction practice. An example of such a study is Hartmann (2012). Hartmann shows that Heidegger's modes of engagement are helpful to understand the dynamics during the implementation of decision-support systems by project teams. Hartmann distinguishes three modes of being when implementing such a system: thematic deliberation, immersed action and theoretical detachment. At the beginning of the implementation, the project team conducted regular meetings with the IT specialists to discuss the progress of creating 3D models in a mode of thematic deliberation. When meaningful 3D models became available, the project team started to use them to support ongoing project management tasks in an immersed way after some time. After temporary breakdowns of the system, members of the project team started to deliberate about the aspects of a 4D system for the current project. After a while, they detached themselves from the current project environment and developed new insights.

This type of studies shows the relevance of ethnographic approaches when a researcher investigates meaningful totalities of subjects and objects (Sandberg & Tsoukas, 2011) in construction organisations and projects.

## Critical ethnography

### *Struggle between lifeworld and the system*

The critical perspective recognises that a researcher and a research participant are part of a wider set of social, economic and political relationships (Myers, 1997). This leads us to critical ethnography. Critical research traditions in general acknowledge that truth claims are situated and implicated in power relations (Masta & Secules, 2021). As such, critical ethnography as a methodological approach addresses power relations in contexts. Alvesson and Deetz (2000) mention that "[c]ritical studies are inclined to pay attention to and interpret 'raw material' for advanced interpretations in terms of power and domination" (p. 150). According to Thomas (1993) "[c]ritical ethnographers describe, analyze, and open to scrutiny otherwise hidden agendas, power centers, and assumptions that inhibit, repress, and constrain. Critical scholarship requires that common sense assumptions are questioned" (pp. 2–3). Taken-for-granted assumptions, beliefs, ideologies and discourses are challenged (Richardson, 2005, p. 282).

Although critical ethnography is distinctive, it does not stand in opposition to other forms of ethnographic research (Myers, 1997). Rather, it is a type of reflection about the relationships among knowledge, culture, society and action (Thomas, 1993). Similar to other ethnographers, critical ethnographers tend to do "intensive empirical investigations of everyday, lived cultural reality" (Foley, 2002, p. 472). Masta and Secules (2021) state that where "an ethnographer would tend to treat the participant accounts and participant observation of that lived cultural reality as an accurate representation, critical ethnographers think about how power relations may impact the perspectives of each account" (p. 36). Critical ethnography challenges the status quo and "unsettles both neutrality and taken-for-granted assumptions by bringing to light underlying and obscure operations of power and control" (Madison, 2011, p. 5).

Several scholars used Habermas' concepts of system and lifeworld for critical ethnography (Cecez-Kecmanovic et al., 2002). Habermas distinguishes between the lifeworld, i.e., the sphere of everyday communicative interactions, and the system, i.e., powerful institutions such as the market and bureaucratic organisations. According to Habermas, social integration rests upon the concept of the lifeworld and on actions oriented towards achieving mutual communication or understanding (Habermas, 1985, p. 117). A shared background makes it possible for participants to reach a shared understanding. This mutual understanding is possible because of the lifeworld, but the lifeworld is also reproduced through processes of mutual understanding.

Habermas (1984) defines a system as a formally organised domain of action (Habermas, 1984). Within this system concept, actions activities and decisions are regulated through functional interconnections or 'steering media' such as markets (i.e., subsystems of the economy) and bureaucracies (i.e., administration subsystems). These steering media could become an alternative to mutual communicative understanding (Feenberg, 2000). If systems and subsystems become sufficiently complex, they could even become independent of the lifeworld. The system "escapes from the intuitive knowledge of everyday communicative practice" (Habermas, 1985, p. 173).

In other words, system integration (through the steering media) becomes decoupled from and replaces social integration (through mutual understanding).

Habermas describes this process as 'mechanization' or 'colonization of the lifeworld'. This critical conceptualisation of the impact of a steering system is similar to Heidegger's concept of enframing (Dorrestijn, 2012a). Heidegger's concept of 'enframing' is a core concept in studying the impact of technology on the 'pre-theoretical' lifeworld of human beings. According to Heidegger, working practices may be enframed when technology is implemented.

In Heidegger's critical analysis, modern technology means that through 'enframing' nature and humans are transformed to be 'standing reserves'; resources solely available for exploitation and consumption (Heidegger, 1977; Dias, 2003). This is because "everything is ordered to stand by, to be immediately at hand, indeed to stand there just so that is may be on call for a further ordering" (Heidegger, 1977, p. 322). Nature becomes 'Be-stand', thus a standing reserve of available resources to be exploited (Heidegger, 1994). In this process of instrumentalisation, methodical planning becomes dominant. Modern technology also requires human beings to view the world in a different way: it constructs a worldview that is *imposed* upon man and nature. Through enframing, the use of technology creates a new lifeworld (Heidegger, 1994). In this critical research on the impact of technological systems, Heidegger also, like Habermas, represents this impact as a struggle between two spheres in which the lifeworld must be protected against 'enframing' by the system (Dorrestijn, 2012b).

### *Critical ethnography in construction research*

Critical ethnography has been applied in construction research on the implementation and use of digital technologies in construction projects and organisations. This implementation and use has critically been viewed from a system and lifeworld point of view of Habermas and Heidegger (Voordijk, 2019). The concepts of Habermas were applied in critical ethnographic construction research because the use of interorganisational digital technologies in construction is sometimes forced upon organisations. This may have an important impact on the working practices of these organisations (Adriaanse et al., 2010). Other organisations can steer an organisation towards using these technologies in a way that is not in line with its lifeworld. This may result in user resistance and user rejection.

The lack of intrinsic motivation and the limited shared background knowledge can be important constraints on the 'rationalization' of the lifeworlds through mutual understanding (Voordijk, 2019). One of the reasons for the reluctance to use BIM is the perceived uneven distribution of the benefits among the different parties involved in a construction project. The way interorganisational digital technologies, such as BIM, are introduced may put constraints on the mechanisms influencing interorganisational communication. In construction, BIM entails standardising interfaces, protocols and various hardware and software components. The reliance on standards is also one of the key aspects in the Heideggerian notion of enframing (Ciborra & Hanseth, 1998). Standardisation of communication through BIM can therefore be interpreted as a process of enframing.

Drawing on Foucault-inspired studies of information systems use and surveillance, Davies and Harty (2012) analysed whether BIM itself can be regarded as a tool for surveillance and control. Starting point were critical studies using Foucault to understand management practices as part of an attempt to increase control of workers, through digital and other means of surveillance and through the internalisation of norms associated with an efficient worker (Zuboff, 1988). Based on ethnographic observations, evidence from case studies did not support the existence of a totalising 'panoptical' surveillance or an all-powerful controller that could force the staff of a construction project into specific ways of working. Evidence, however, also showed that negotiations over who has the power to enforce or resist technology are fundamental in shaping its implementation.

## Conclusions

This study discussed three major philosophical sources of ethnographic studies: phenomenology, hermeneutics and critical research.

Studying the 'lifeworlds' or lived experiences of human beings is an important part of ethnographic research. Phenomenology, the study of these experiences, is therefore presented as first philosophical basis for ethnographic research. In this context, the interpretive phenomenology of Heidegger is discussed. This philosophy is concerned with understanding the meaning of lived experiences and individuals' realities given the world in which persons live. Studies on lived experiences have also been fruitfully conducted within construction research. These studies, for example, examine the lived experience of those working in project delivery teams, construction supply chains, decision-making processes or how a certain concept appears to the consciousness of construction experts or practitioners.

As an interpretive methodology, ethnographic research recognises the multiplicity of meanings of participants of a research project (Sharp, 2019). Capturing these meanings through interpretive ethnography is based on research participants interpreting their reality and researchers interpreting these interpretations. Hermeneutics is therefore presented as the second philosophical basis for ethnographic research. Basic concepts of the founders of contemporary hermeneutics, Heidegger and Gadamer, are discussed: Heidegger's 'being-in-the-world' and 'thrownness' and Gadamer's 'hermeneutics circle'. 'Pre-theoretical' shared agreements or practices are the basis for everyday action when 'thrown' into a situation. This hermeneutic approach can also be applied to the way in which we understand construction projects and organisations. It is shown that Heidegger's 'thrownness' is relevant for ethnographic research on the unique contexts of construction projects that require context-sensitive action of the engineers and other practitioners involved.

Subsequently, based on Heidegger's notion of 'being-in-the-world', it is discussed how ethnography helps to investigate different ways of human engagement in general and in construction organisations and projects. Heidegger's tool analysis and his concepts of ready-to-hand, present-at-hand and unready-to-hand makes it possible to understand these ways of engagement. These concepts are used to explore if and how digital technologies shape working practices in construction projects. Based on Sandberg and Tsoukas (2011), it is shown how the concept of a temporary breakdown provides an opportunity to understand otherwise immersed or unreflective acting in regular practices. Hartmann (2012) distinguishes several modes of engagement during the implementation and breakdown of decision-support systems by project teams. Ethnographic approaches are in particular relevant when a researcher investigates meaningful 'totalities of subjects and objects' (Sandberg & Tsoukas, 2011) in construction organisations and projects.

Besides phenomenology and hermeneutics, critical research is a third important philosophical basis for ethnographic research. Critical researchers try to challenge taken-for-granted assumptions, beliefs, ideologies and discourses (Richardson, 2005). Several scholars used Habermas' concepts of system and lifeworld for critical ethnography. In critical ethnography, the impact of the system is represented as a struggle between two spheres in which the lifeworld must be protected against 'colonization' (Habermas) or 'enframing' (Heidegger) by the system. Critical ethnography has been applied in construction research on the implementation and use of digital technologies in construction projects and organisations. The use of interorganisational digital technologies in construction is sometimes forced upon organisations. The subsequent tension between system and the lifeworld may result in user resistance and user rejection. Implementation of digital technologies can also be interpreted as a process of enframing or an attempt to increase control of construction workers.

The three major philosophical foundations of ethnographic studies in construction research are relevant in the following ways: phenomenology underlies that examination of lived experience of those working in construction practice by describing how concepts appear to the consciousness of construction experts or practitioners. Hermeneutics is relevant for ethnographic research as it sheds light on how unique contexts of construction projects require context-sensitive action of practitioners involved. The hermeneutics circle shows how understanding is built up from whole-part analysis, and investigation of how humans engage differently in construction projects. Finally, critical ethnographic studies in construction represent structures and may particularly help conceptualise studies of digital technology as struggles between the lifeworld, the system and attempts to control construction workers.

## References


Aagaard, J. (2017). Introducing postphenomenological research: A brief and selective sketch of phenomenological research methods. *International Journal of Qualitative Studies in Education*, 30(6), 519–533.

Adriaanse, A., Voordijk, H., & Dewulf, G. (2010). The use of interorganisational ICT in construction projects: A critical perspective. *Construction Innovation*, 10(2), 223–237.

Akmam Syed Zakaria, S., Gajendran, T., Skitmore, M., & Brewer, G. (2018). Key factors influencing the decision to adopt industrialised building systems technology in the Malaysian construction industry: An inter-project perspective. *Architectural Engineering and Design Management*, 14(1–2), 27–45.

Alvesson, M., & Deetz, S. (2000). *Doing Critical Research Management*. London: Sage.

Berger, P., & Luckmann. T. (2003) [1966]. *The Social Construction of Reality: A Rreatise in the Sociology of Knowledge*. Garden City, NY: First Anchor.

Cecez-Kecmanovic, D., Janson, M., & Brown, A. (2002). The rationality framework for a critical study of information systems. *Journal of Information Technology*, 17(4), 215–227.

Chari, D. (2014). What is nanoscience? – A hermeneutic phenomenological study of nanoscience researchers' experiences. Doctoral dissertation, Dublin Institute of Technology.

Chughtai, H., & Myers, M. D. (2014). *The entwinement logic of practices: insights from an ethnography of young IT professionals*. Paper presented at the Information Systems and Global Assemblages.(Re) Configuring Actors, Artefacts, Organizations: IFIP WG 8.2 Working Conference on Information Systems and Organizations, IS&O 2014, Auckland, New Zealand, December 11–12, 2014. Proceedings.

Ciborra, C. U., & Hanseth, O. (1998). From tool to Gestell: Agendas for managing the information infrastructure. *Information Technology & People*, 11(4), 305–327.

Davies, R., & Harty, C. (2012). Control, surveillance and the 'dark side' of BIM. In S.D. Smith (Ed.), *Paper presented at the Proceedings of the 28th Annual ARCOM Conference*, Edinburgh, UK. Association of Researchers in Construction Management, London, UK.

Dias, W. (2003). Heidegger's relevance for engineering: Questioning technology. *Science and Engineering Ethics*, 9(3), 389–396.

Dias, W. (2006). Heidegger's resonance with engineering: The primacy of practice. *Science and Engineering Ethics*, 12(3), 523–532.
Dias, W. (2007). Philosophical grounding and computational formalization for practice based engineering knowledge. *Knowledge-Based Systems*, 20(4), 382–387.
Dorrestijn, S. (2012a). *The Design of Our Own Lives: Technical Mediation and Subjectivation after Foucault*. Enschede: Universiteit Twente.
Dorrestijn, S. (2012b). Theories and figures of technical mediation. In J. Donovan & W. Gunn (Eds.), *Design and Anthropology* (pp. 219–230). London: Routledge.
Dreyfus, H. (1987). Husserl, Heidegger and modern existentialism. In B. Magee (Ed.), *The Great Philosophers: An Introduction to Western Philosophy* (pp. 252–277). Oxford/New York: Oxford University Press.
Dreyfus, H. L., & Dreyfus, S. E. (1996). The relationship of theory and practice in the acquisition of skill. In P. Benner, C. Tanner, & C. Chesla (Eds.), *Expertise in Nursing Practice: Caring, Clinical Judgment, and Ethics* (pp. 1–24). New York: Springer.
Feenberg, A. (2000). From essentialism to constructivism: Philosophy of technology at the crossroads. In E. Higgs, A. Light, & S. Strong (Eds.), *Technology and the Good Life* (pp. 294–315). Chicago/London: University of Chicago Press.
Foley, D. E. (2002). Critical ethnography: The reflexive turn. *International Journal of Qualitative Studies in Education*, 15(4), 469–490.
Gadamer, H.-G. (1976). The historicity of understanding. In P. Connerton (Ed.), *Critical Sociology, Selected Readings* (pp. 117–133). Harmondsworth: Penguin Books.
Gill, M. J. (2014). The possibilities of phenomenology for organizational research. *Organizational Research Methods*, 17(2), 118–137.
Habermas, J. (1984). *The Theory of Communicative Action* (Vol. I). Boston: Beacon.
Habermas, J. (1985). *The Theory of Communicative Action* (Vol. 2). Boston: Beacon.
Hartmann, T. (2012). Triggers for the critical engagement with decision support systems. Paper presented at the Engineering Project Organization Conference, EPOC 2012.
Hartswood, M., Procter, R., Slack, R., Voß, A., Büscher, M., Rouncefield, M., & Rouchy, P. (2008). Co-realization: Toward a principled synthesis of ethnomethodology and participatory design. In M. S. Ackerman, C. A. Halverson, T. Erickson, & W. A. Kellogg, *Resources, Co-evolution and Artifacts* (pp. 59–94). London: Springer.
Heidegger, M. (1977). *Sein und Zeit (GA 2)*. Frankfurt a. M: V. Klostermann.
Heidegger, M. (1994). *Basic Questions of Philosophy: Selected "Problems" of "Logic"*. Bloomington: Indiana University Press.
Heidegger, M. (2008). *Ontology—The hermeneutics of facticity*. Bloomington: Indiana University Press.
Howarth, T. (1997). *A gadamerian hermeneutic approach to ethnographic construction research*. Paper presented at the 13th Annual ARCOM Conference.
Hudson, L. A., & Ozanne, J. L. (1988). Alternative ways of seeking knowledge in consumer research. *Journal of Consumer Research*, 14(4), 508–521.
Husserl, E. (1970). *The Crisis of European Sciences and Transcendental Phenomenology: An Introduction to Phenomenological Philosophy*. Evanston: Northwestern University Press.
Katz, J., & Csordas, T. J. (2003). Phenomenological ethnography in sociology and anthropology. *Ethnography*, 4(3), 275–288.
Kearney, R., & Rainwater, M. (1996). *The Continental Philosophy Reader*. London: Psychology Press.
Laverty, S. M. (2003). Hermeneutic phenomenology and phenomenology: A comparison of historical and methodological considerations. *International Journal of Qualitative Methods*, 2(3), 21–35.
Lopez, K. A., & Willis, D. G. (2004). Descriptive versus interpretive phenomenology: Their contributions to nursing knowledge. *Qualitative Health Research*, 14(5), 726–735.
Madison, D. S. (2011). *Critical Ethnography: Method, Ethics, and Performance*. Thousand Oaks: Sage.
Masta, S., & Secules, S. (2021). When critical ethnography leaves the field and enters the engineering classroom: A scoping review. *Studies in Engineering Education*, 2(1), 35–52.
Mitcham, C. (2003). *Philosophy of Information Technology*. Oxford: Blackwell.

Moran, D. (2002). *Introduction to Phenomenology*. Abingdon: Routledge.

Myers, M. D. (1997). Critical ethnography in information systems. In A. S. Lee, J. Liebenau, & J. I. DeGross, *Information Systems and Qualitative Research* (pp. 276–300). Berlin: Springer.

Neubauer, B. E., Witkop, C. T., & Varpio, L. (2019). How phenomenology can help us learn from the experiences of others. *Perspectives on Medical Education*, 8(2), 90–97.

O'Reilly, K. (2009). *Key Concepts in Ethnography*. Thousand Oaks: Sage.

Richardson, H. (2005). Consuming passions in the 'global knowledge economy'. In D. Howcroft & E. M. Trauth, *Handbook of Critical Information Systems Research* (pp. 272–298). Cheltenham/Northampton: Edward Elgar.

Riemer, K., & Johnston, R. (2011). Artifact or equipment? Rethinking the core of IS using Heidegger's ways of being. In *Proc 32nd International Conference on Information Systems*, 4–7 December, Shanghai, China (pp. 1–18).

Riemer, K., & Johnston, R. B. (2014). Rethinking the place of the artefact in IS using Heidegger's analysis of equipment. *European Journal of Information Systems*, 23(3), 273–288.

Samuel, C. E. (2020). *Leading Project Delivery Teams to Improve Collaboration and Performance: A Phenomenological Study*. Phoenix: University of Phoenix.

Sandberg, J., & Tsoukas, H. (2011). Grasping the logic of practice: Theorizing through practical rationality. *Academy of Management Review*, 36(2), 338–360.

Sharp, R. (2019). *Thinking with Heidegger: An Ethnography of the Place of Leadership*. Salford: University of Salford.

Thomas, J. (1993). *Doing Critical Ethnography* (Vol. 26). Thousand Oaks: Sage.

Turk, Ž (2001a). Multimedia: Providing students with real world experiences. *Automation in Construction*, 10(2), 247–255.

Turk, Ž (2001b). Phenomenologial foundations of conceptual product modelling in architecture, engineering and construction. *Artificial Intelligence in Engineering*, 15(2), 83–92.

Van Manen, M. (1997). *Researching Lived Experience*, 2nd edn. Ontario: The Althouse.

Venselaar, M., & Gruis, V. (2016). Studying intra-organizational dynamics in implementing supply chain partnering: A case study about work floor experiences in a Dutch housing association. *Construction Management and Economics*, 34(2), 98–109.

Voordijk, H. (2019). Building information modeling and its impact on users in the lifeworld: A mediation perspective. *Frontiers of Engineering Management*, 6(2), 193–206.

Voordijk, H., ter Huurne, R., & Olde Scholtenhuis, L. (2022). Adoption and use of ontologies in the utility sector: A technological mediation perspective. *Journal of Management in Engineering*, 38(1), 05021016.

Wimalaratne, P., & Kulatunga, U. (2022). A methodology to study the complexity of buildability in construction projects: Phenomenological research perspective. In Y.G. Sandanayake, S. Gunatilake, & K. G. A. S. Waidyasekara (Eds), *Proc. 10th World Construction Symposium*, 24–26 June 2022, Sri Lanka (pp. 14–26).

Zuboff, S. (1988). *In the Age of the Smart Machine: The Future of Work and Power*. New York: Basic Books.

# 8

# HOW ETHNOGRAPHIC METHODOLOGIES CONTRIBUTE TO CONSTRUCTION PRACTICE

*David Oswald and Helen Lingard*

*RMIT University, Melbourne, Australia*

**Key takeaways**

- This chapter demonstrates how ethnographic research can provide deep understanding of how work is done in practice.
- Through establishing rapport and participatory approaches, ethnographic researchers access new avenues to knowledge that could not be revealed from more distant construction research approaches.
- Ethnographic approaches are useful for workers to show how they work, explain challenges and reflect on important topics, such as their safety at work. The participation and observations from ethnographic encounters can thereafter lead to practical changes that improve workplaces.
- Future ethnographers should consider how their research design can promote rapport-building and rich encounters with construction employees.

## Introduction

Ethnographic research offers new routes to knowledge that would not necessarily be captured by more distant research approaches (such as those rooted in positivism). Ethnography, in its different forms, can complement other research methods to provide a more comprehensive understanding of the complex phenomena that characterise the construction industry context. In this chapter we demonstrate how new knowledge is acquired through ethnography and how it can be put to practical use in the construction industry. While ethnography provides valuable opportunities for the advancement of theory (see Chapter 20), ethnographic studies embedded in industry contexts also produce practical outcomes and improvements. Therefore, we propose that ethnography should be considered by construction professionals and consultants as a useful approach to inform work design and decision-making.

This chapter presents two exemplar cases to demonstrate the practical industry outcomes that ethnography can enable. These cases focus on construction workers' safety and, in particular, the way that safety is shaped by local contextual factors and social relationships in a workplace.

DOI: 10.1201/9781003379584-10

Before we focus in on the illustrative cases, we discuss relevant safety-related literature to provide context to the case findings.

Previous research into safety rules has widely accepted there is a gap between 'work as imagined' and how work is actually performed at a worksite (Borys, 2009). The way that safety is prescribed and documented in formal procedures and rules is not always practical or followed. Hence, Dekker (2005) suggested monitoring the gap between 'work as imagined' and 'work as done,' and to try and understand why it occurs. Knudsen (2009) more explicitly urged researchers to 'stop bitching' about the fact that this gap exists and instead try to close it. The causes of the gap and ways to resolve it remain areas of great interest to researchers and practitioners (Lofquist et al., 2017). However, traditional research methods (e.g., surveys and interviews) are arguably limited in advancing knowledge about the complexities and context that shape work and contribute to the (imagined vs work-done) gap. Firstly, because they may be more susceptible to social desirability bias, which is the phenomenon that people will tell researchers what they think they want to hear (e.g., agreeing in a survey that they 'always work safely'; when in reality they know they occasionally do dangerous work). Secondly, traditional methods also tend to collect data at a snapshot in time and do not allow researchers to thoroughly understand what happens within a workplace as new people join, social relationships evolve and events unfold. These dynamics particularly occur in the constantly changing environment of a construction project. Ethnographic methods are helpful in this industry to dealing with shortcomings of 'mainstream methods,' since they can provide insights on complex and dynamic topics deserving of a deeper understanding, such as safety rule-breaking.

Safety rules are designed to keep people safe, yet rules are commonly broken in organisations (Alper & Karsh, 2009). These deviations from the rules can be categorised as deliberate violations or unintended errors. While often viewed as being unsafe, it is important to note that deviations from the planned ways of working can also be positive, as they provide information that can improve systems, for instance, in exposing poorly devised rules that is continuously violated (and should be changed and improved) or latent errors which can be removed.[1]

Rules can be viewed within two different paradigms: Model 1 and Model 2 (Dekker, 2005). These describe how the development and use of work rules are managed and perceived (see Hale & Borys, 2013a). Model 1 is a classical top-down rational approach in which rules are set without freedom of choice and are imposed on operators at the frontline. It has strengths in that rule-making is explicit, easy to audit and the consequences for rule violation are clear (Hale & Borys, 2013a). This model can be particularly useful for novices or apprentices (ibid). However, there are also drawbacks in that violations are perceived as a negative behaviours, where operators can be likened to robots that need to be controlled (ibid). There is also a tendency (in Model 1) to move towards bureaucratic approaches to safety, which has undesirable secondary effects (Dekker, 2014). Hallowell and Gambatese (2009) demonstrated that interventions associated with bureaucracy are deemed the least worthwhile; and the Health and Safety Executive (HSE) stated that is imperative that workers and working groups disassociate 'safety' from 'bureaucracy' (HSE, 2003).

Model 2 is a bottom-up constructivist view of rules as dynamic, local, situated constructions where operators are the experts, and competence is perceived as the ability to adapt rules to the diversity of real-work situations (Hale & Borys, 2013a). Model 2 actions can be somewhat hidden, either deliberately by the frontline to retain their autonomy, or because management and supervision turn a blind eye to it (ibid). In terms of strengths, this approach recognises social processes as key to rule use, and the importance of managing exceptions and their link to violations (ibid). However, the rule-making and modification process lacks transparency for auditing and can hide differences of interpretation and competence.

The following ethnographic research explored safety-relevant work practices including if, when and why rules were or were not followed. The first case example describes a six-month project in which participant observation was used to understand the way that safety-related rules were interpreted and applied in a large construction organisation. The second case describes a video-based ethnographic study that shed light on the workability of safety rules and the way that aspects of the physical and social context of the worksite influence workers' behaviour in relation to safety rules. The first case is reflected upon in first person, and the second case is presented in third person.

## Case 1: How ethnographic work can capture profound knowledge on construction practice

Over a six-month period, I spent approximately 100 hours undertaking participant observation at the four construction sites. This involved spending time observing frontline leaders and having short informal conversations with people at different levels within the organisation. I used organised safety walks as a structured way to access all areas of the site, as well as undertaking more informal safety walkarounds. I used my mobile phone to type in initial fieldnotes at appropriate moments during the safety walks. Phones are often used in construction and are a normal part of the natural setting, so it was not a weird, or unusual, behaviour within this context. I then subsequently wrote up more extensive fieldnotes as soon as possible after leaving the site.

During months of site visits, I got to know the participants and had established rapport with many key gatekeepers and informants. They trusted me enough to explain what they really thought about safety on the construction site. In some cases, they opened-up with comments to me, that they may not have felt comfortable saying to their supervisors or senior leaders. Establishing such rapport can be particularly powerful on sensitive topics, such as bending safety rules; and it allowed me to understand when they would break safety rules and why. I started to notice through observation that health and safety (H&S) rules were sometimes broken on site and were not always challenged by supervisors or H&S advisors. This was because there were occasions when rules did not make workers any more safe, so intervening to insist on a rule being followed would only potentially damage relationships. One of my fieldnotes stated:

> I began to realise that keeping a strong relationship with the subcontractors was held in very high regard, even at the expense of safety rules. It was local-level management of safety, where rules weren't black and white, but instead applied most of the time, and relaxed at other appropriate moments. This kept strong relationships between the individuals from different organisations, which have different organisations expectations, rules and practices. The gap between rules and work actually performed was acknowledged, and had to be resolved at the frontline.

The above ethnographic fieldnote is an example of the local management of safety rules incorporating a Model 2 approach. I consistently observed that rules were not always followed 'by the book' (Model 1) and sometimes interventions were made by principal contractor employees, but sometimes they were not. Locally, the rules were viewed as a resource, which were required to be referred to at times when they were appropriate; but adapted at times when they did not reduce direct safety risks. The use of language and wording that was not authoritarian helped to deliver respectful communication, and by not challenging all situations, subcontractors could be kept 'on-side' when prescribed rules did not necessarily reduce the safety risks within the situated social context. These insights were showing how safety was actually being practised on the construction site through participant observations.

Near the end of the fieldwork, the participant observation approach was complemented with focus groups. As I had already built relationships with the participants, there is no need to 'cold call' for data collection, with the gatekeepers instead acting as facilitators for the research by booking meeting rooms and helping to recruit the desired number of participants for the focus groups. I had the time and opportunity to build trust with the interviewees prior to the focus groups, so believe they may have felt more comfortable to share their experiences and views.

The focus group members included foremen, supervisors, leading hands, site engineers, H&S advisors, shop stewards and workers at each of the four sites. Their explanations and reflections were explored in one-hour audio-recorded sessions. I found that the prior participant observation had not only provided benefits for the recruiting process, but also allowed me to ask more poignant questions. Having spent vast amounts of time in the setting, I could more clearly visualise interviewee answers (which can be very important in a complex and dynamic setting such as construction). The following two examples highlight depth of data capture when rapport has been established and participants demonstrate openness and sharing of their perspectives, even on sensitive topics.

**BOX 1: Ethnographic Data Extract Example: H&S Advisors Not Intervening with H&S**

The following extract presents why sometimes H&S advisors admitted they did not intervene with H&S rules being broken:

*H&S advisor:* 'Oh yeah, today would have been a classic. You might have been working underneath a soffit there's no danger there [of an object falling on your head]. Just if you get 32, 33 degrees the frigging helmet retains a lot of heat in your head. It's very refreshing [to take it off]. I'm not saying that's right, I just understand why it's come off.'

*Interviewer:* 'And if you saw that, I mean I guess by the book you should say something, but would that …'

*H&S advisor:* 'I should, and if I saw it on a day like today I perhaps wouldn't police it as hard. I might say to them I understand why you've done it, just make sure if you move into an area where there's a risk, you remember to put it back on.'

*Interviewer:* 'And that keeps that relationship?'

*H&S advisor:* 'Yeah.'

*Interviewer:* 'And guys understand?'

| | |
|---|---|
| *H&S advisor:* | 'They don't necessarily think I'm being an a***hole. Like glasses can be very difficult.' |
| *H&S advisor 2:* | 'Muggy days. Fog up.' |
| *H&S advisor:* | 'Oh on muggy days, your glasses were fogging up. A lot of our management don't understand that anti fog just doesn't work unless the subcontractors, and unless we, are providing glasses worth hundreds of dollars each, well …' |
| *H&S advisor 2:* | 'It's more dangerous to have them on, than have them off.' |
| *H&S advisor:* | 'And all of a sudden I've got a steel fixture that's walking around on steel carrying heavy frigging bars that can't see. What's the biggest risk? …' |
| *Foreman:* | 'It's just that practicality coming into play, really.' |
| Foreman 2: | 'Responsibility and practicality.' |

The discussions continued and presented further open dialogue on how safety was managed in practice, and why it not always aligns to the rules.

**BOX 2: Ethnographic Data Extract Example 2: The Disconnect between Rules and Practice**

The reflections below show how workgroups felt uncomfortable about bending rules, but still believed that in some circumstances it was the most appropriate action.

| | |
|---|---|
| *Interviewer:* | 'So I guess one question I have is do you feel at any point uncomfortable, say when you make those slight adaptions to the rules for what you're saying I think is common sense, but I guess to some people it might be seen as bending them. Do you feel uncomfortable that …' |
| *Supervisor:* | '*Yeah.*' |
| *H&S advisor:* | 'Oh we definitely do yeah, cos we get that sort of …' |
| *Foreman 2:* | 'But that's cos of the pressure we get from above, you know?' |
| *H&S advisor:* | 'We've got to assess the common good.' |
| *H&S advisor 2:* | 'And what are your priorities? I mean one of our leaders is, quite regularly says, "never seen a bloke get killed for not wearing gloves", you know, look at the big issues. You don't want to be walking around being PPE police and letting the big issues, fallen materials, live edges, electrical issues … you've got to work out where you're spending your time too, and we haven't got 50 hours a week to be patrolling the whole job. We've got to make sure if we're going out there, we're picking the big-ticket items, and they're the important, the high risk, activities. Not going out there on a gloves and glasses blitz for two hours, it's just a waste of everybody's time …' |
| *Foreman:* | 'It all comes to common sense.' |

| | |
|---|---|
| *Supervisor:* | '... as an industry.' |
| *Interviewer:* | 'So when we were talking about the pressure from above that you feel, that makes you feel uncomfortable when you're adapting the rules ...' |
| *H&S advisor:* | 'Yeah, I think, yeah.' |
| *Interviewer:* | 'What's *that, like site manager, is it above that? Or is it ...'* |
| *Foreman 1:* | 'Yeah it's senior manager.' |
| *Foreman 2:* | 'It's the company as a whole.' |
| *Foreman 1:* | 'They're all after you.' |
| *H&S advisor:* | 'You feel almost guilty. If you're doing a walk with management and yeah. And three blokes are over in the corner not wearing a hard hat, I mean there's nothing on top of them [that could fall on them]... but you go "oh...f**k" because you feel like you've got to intervene with them.' |
| *Foreman 1:* | 'And that's where they're so far out of it that they just have no practicality about it. They just think here's the rule book, you've got to follow it.' |

The above ethnographic data reveals perspectives of construction professionals that are unlikely to be openly shared without first developing trust and rapport with a researcher. H&S professionals admitting they did not always intervene with workplace safety issues, and feeling uncomfortable when senior management visit site as the rules and local management of safety did not always align. The foreman and supervisors shared their view that senior management did not fully understand the practicality of safety management at the site-level, instead assuming that safety rules can and will be enforced at all times. We contend that it was only because of the rapport and trust that was established between research participants and the researcher that these findings came to light. The ethnographic approach substantially increased the ability to establish such rapport and trust as significant time was spent in a workplace 'getting to know' the participants and their daily routines.

The results of the research revealed that frontline leaders believed acknowledging the gap, managing it appropriately and focusing on the high-risk areas instead was the approach that should be taken. This represents a practice-based approach where safe behaviours are not based on rule compliance in the local context, but on managing the gap between rules and the most appropriate situational action. These findings were contributing to understanding the social construction of safety at the construction (see Lingard & Oswald, 2020, for further discussion on this work). The ethnographic insights within this project provided new avenues to knowledge by explaining how social relationships, localised management and negotiation shaped safety on construction sites. A paper written from this project won the 2020 *Journal of Construction Engineering and Management* best paper award, with the Editor-in-Chief stating:

> The authors documented an ethnographic study that examines ways that actors at the prime contractor/subcontractor interface understand and respond to safety; it illuminates how social relationships shape safely practices on a construction site. The methodology relies on over 100 h of social embedment at four construction sites over a 6-month period with the researcher accompanying construction site leaders for both prime and subcontractors in a variety of formal and informal settings, especially safety walks. This research found that safety

> culture is developed by prime and subcontractors though a process of negotiation that is influenced greatly by requirements for productivity, and the stakeholders' personal preferences and previous experience in maintaining safety, in addition to prescriptive rules developed by the prime contractor. This contrasts with other theories that emphasize the development of worksite safety though the enforcement of universal rules with little context and flexibility. The results highlight that a desirable culture of safety develops on a worksite though effective social interaction and a measure of flexibility in addition to relying on rules enforced in a top-down manner.
>
> *(Garza, 2021)*

Ethnographic research has the potential to find new routes to knowledge and provide practical insights that would not be captured by more traditional research methods. This is further expanded on in the following visual ethnographic case example.

### Case 2: Using visual ethnography to understand and change construction practice

Visual ethnographic methods are sometimes used to understand workplace safety from the perspectives of people performing the work. For example, Tutt et al. (2013) used photography during fieldwork on construction sites, inviting participants to use the camera to 'show' aspects of their experiences of safe working, as well as photographing significant aspects of sites.

Video is also used to understand behaviours and practices in complex and high-risk settings (van Nieuw-Amerongen et al., 2011). For example, Pink et al. (2014) used video re-enactment to understand hand gel use in healthcare settings. Mackenzie et al. (2004) analysed video recordings of medical emergency situations to explore how aspects of the physical environment and pressure exerted by other team members contribute to the omission of steps in the procedure of patient airway management. Reiman et al. (2014) show how the analysis of video of truck drivers performing routine delivery tasks stimulated discussion about work-related risks and possibilities for improvement between workers and ergonomists.

Another video-based ethnographic study in construction explored how construction workers stay safe in the social and material context of the construction site (Pink et al., 2017). The researcher spent time at three sites at which workers were making instructional safety videos using a participatory video (PV) approach. This 'group-based activity [...] develops participants' abilities by involving them in using video equipment creatively to record themselves and the world around them and to produce their own videos' (Shaw & Robertson, 1997, p. 1). Unlike other forms of video development, PV is reflexive, and participants are both subjects and film makers. PV also enables people to 'shape issues according to their own sense of what is important' and decide how they want to be portrayed (Lunch & Lunch, 2006, p. 10). PV is more than a data collection approach and is, in part, an intervention. The aim is not simply to produce video materials, but to use the process of video production to generate critical thinking and 'empower people with the confidence, skills and information they need to tackle their own issues' (Shaw & Robertson, 1997, p. 26).

In the current study, researchers spent time at sites during which time they filmed and interviewed workers who, in turn, were themselves engaged in making videos about the safety aspects of their work. On completion, the worker-made films would be shared with other workers as safety training resources. The PV process was facilitated by an external consultant who engaged workers in brainstorming content, developing a story board and filming their work.

Data collected during the observation and filming of the PV process was analysed to reveal insights about safety in construction site contexts and also about the effect of the PV intervention on construction workers' safety-related attitudes and behaviours. These insights would have been very difficult to uncover using alternative data collection approaches, such as interviews or surveys. The video-based data collection approach captured habitual and taken-for-granted behaviours that differed from the way that work processes were documented in company operating procedures. These deviations between documented and actual work practices were not spoken about or known to managers within the company, as illustrated by the following example.

At one site visit, workers were making an instructional video about how to access ceiling spaces safely to install insulation materials. Ceiling spaces are typically accessed via a manhole at a ceiling height of between 2.4 and 2.7 metres from the floor. The company standard operating procedure requires a straight ladder to be placed at an angle of 45 degrees and extend 900 mm beyond the 'step off' point. The worker involved in making this video was observed to be angry and explained that it is practically impossible to follow the standard operating procedure because the small size of manholes does not allow adequate entry for the ladder, the worker and the pack of insulation to be installed. Moreover, the extension of 900 mm is difficult due to conduits, cables, beams and other obstructions. The company H&S Manager who was at the site on the day of filming observed: 'shooting it and viewing it through the camera's eye, we had to stop… the camera doesn't lie…To place a straight ladder at the 1:4 ratio just doesn't work, you can't get a body in there as well because it blocks off the access and you have to contort yourself to actually get in [to the ceiling space].' Instead, workers routinely used an 'A-frame' ladder to access ceiling spaces, 'which they do because they can't use a straight ladder' but, in doing so, they have to step off the top rung of the A-frame to get their bodies up into the ceiling space.

The practical difficulties associated with following the safe operating procedure became apparent during the PV process and required that modifications be made to the company's standard operating procedure because what was considered to be theoretically possible could not be practically implemented in the physical/material context of work. The H&S Manager observed how: 'During the process, they said "oh actually what we do in practice is not what the document says. Actually we do this." So it actually resulted in improvements to written safe work method statements to better reflect what crews were doing.'

Importantly, when workers were asked to describe how they work and, in particular, how they stay safe in the site environment, they explained the difficulty describing how they work because showing how, is easier than describing what is to be done: 'It's a lot easier to show someone what we're trying to say. We could just sit here and verbally speak about it but if you put your verbal words into a video, people are going to sit back and go "now I know what he's actually trying to say".'

It was also apparent that much of construction workers' knowledge about how to work safely is tacit, i.e., it is difficult to transfer to another person by means of writing it down or verbalising it (Polanyi, 1958). For example, knowing how to use a complex piece of equipment or perform a complicated work task safely are forms of tacit knowledge. In many cases people who possess tacit knowledge are not aware of their knowledge and therefore cannot easily communicate it to others, highlighting the strength of visual ethnography, and video-based methods, for capturing accurate information about in-situ work practices.

Construction workers involved in the PV process also identified social characteristics of workplaces that influence the way they work. For example, workers described being asked and expected to violate safe operating procedures by the builders (principal contractors) who engage their employer as a subcontractor. In some instances, this led them to take serious safety risks. For example, a team of young insulation workers was developing an instructional video about the practice of isolating and 'locking out' the electricity supply before commencing installation work. The workers described how they often receive no information about the location of cables in walls and ceilings or about whether any cables are live: 'if we're shooting into concrete with live wire, then you would think someone would say something, or show you a plan.'

The workers also described facing resistance from principal contractors if they request that the power be isolated before the undertake their work. They explained that these principal contractors do not want to isolate the electricity supply, even temporarily, because it would disrupt the work of other trades at a worksite. The insulation firm's H&S Manager described how her company was even threatened with termination of a contract when one principal contractor was made aware of the company's electrical isolation procedure. She explained: '*a situation had developed out on site where it had become a gentleman's agreement between the trades and ourselves "okay you need power so we won't isolate because of the inconvenience that it would cause out on job sites". Rather than challenge the status quo, that situation had become expected.*'

Data collected during the study also shows how participating in the PV process was experienced as empowering for workers, in some cases reducing their willingness to take risks. One worker became emotional when describing a situation in which he refused to access a roof space from the outside of a residential building without an appropriate scaffold or access system. The worker explained the correct procedure would be to access the roof space from an internal manhole inside the building. However, the builder did not want him to enter the nearly completed house – so as not to dirty the newly laid carpet inside. The worker refused: 'They wanted me to go in through the ceiling tiles to do some installation, and I just went…given what we've done [referring to his involvement in the participatory video process] … I know that's not the right way to do it, so I refused to do it. I said, "Nah, I can't do it guys. Unless you can provide me with a handrail on the side of the house there, I am not going to do it." [They said] "Oh but every other installer does it that way …!" [I said] "Well get one of the every other installers to do it mate, because I'm not doing it. It's not the safest way to do it, it's not the easiest way to do it…let's be fair dinkum about it …"' At this point in the interview the worker walked over to the PV facilitator to hug them, physically demonstrating positive reaction to the PV process. The video-based ethnographic methods were able to capture the strength of this worker's powerful emotional response to participating in the PV process, which was evidenced by what he did, not just by what he said (Figures 8.1 and 8.2).

The disciplines of construction and work H&S tend to privilege written and verbal forms of communication over visual communication. However, the video-based ethnographic study described in Case 2 enabled researchers and participants to develop new insights into the factors contributing to the gap between work as imagined and work as done and think creatively about ways to improve workplace safety practices. Visual thinking has been described as an

FIGURE 8.1 Participatory video process showing participants and researchers.

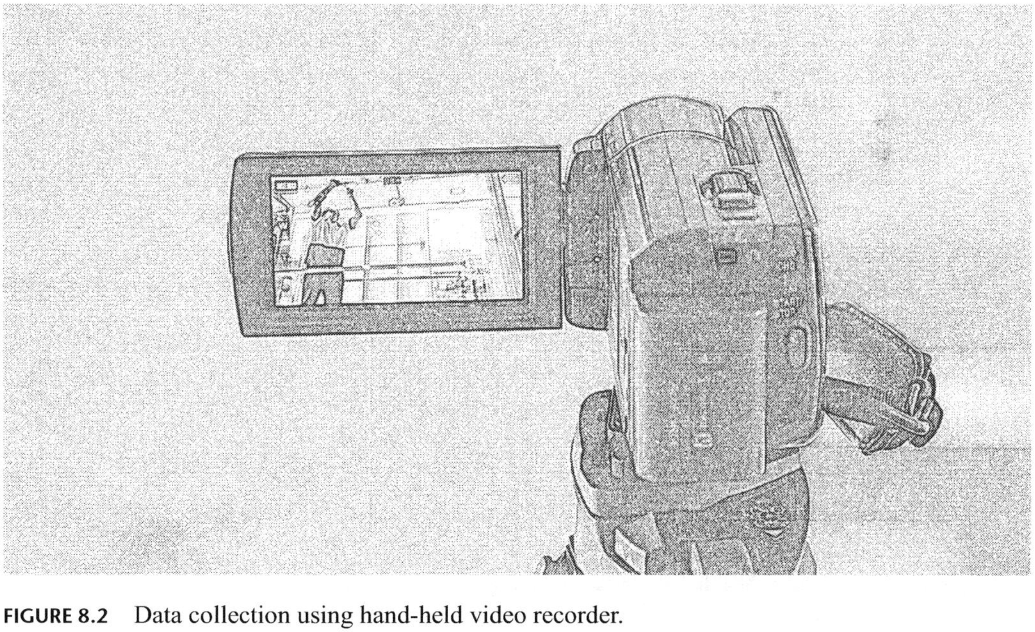

FIGURE 8.2 Data collection using hand-held video recorder.

active problem-solving process, in which familiar objects and processes are seen in new ways from different perspectives (Brumberger, 2007). The case study also demonstrates how PV is simultaneously a research method and an intervention.

This case shows how PV enabled workers and managers to understand how rule violations in the installation of insulation are everyday occurrences that are routinely undertaken (and sometimes expected). PV brought these hitherto unseen violations to light and allowed rule-makers to gain an understanding of what really goes on in the workplace and, importantly, why rules are broken.

The insulation installation workers also expressed a strong preference for using video-based methods to help them to understand the practical application of safety rules. One commented: 'I just think it's a lot easier to visually show that you're not to shoot a pin into this area, as to 500 words or something to describe the same thing.'

## Conclusions

Ethnographic approaches can provide deep insight into what actually happens on a construction site. Without immersion into the setting, it would have been difficult to observe and understand when, why and how work is undertaken. This depth of detail into how work is performed is essential for understanding reality and how construction practice could be improved.

Future researchers show consider on how to encourage rapport-building and rich encounters with participants for deep insights. In the first case study, the researcher began with weekly safety walks to 'get to know' the workers, with focus group discussions recorded at the end of the project after rapport was built. The establishing of rapport with participants meant that there was greater openness and sharing of perspectives, even on sensitive issues, such as when H&S rules were broken and when H&S advisors chose not to challenge rule-breakers. In the second case study, the intense encounter through use of video-based methods meant workers were able to show and explain (rather than only verbally describe) the challenges they had complying with safety rules. The participatory nature this ethnographic approach meant powerful practical contributions were possible, since following the participatory experience workers chose to refuse to do work that they believed was unsafe, and were emotional and thankful to be involved.

Future ethnographers should consider how to build rapport and encourage rich encounters when designing their research and should continually reflect on this during the study. Through ethnographic approaches, where immersion, participation and observation are core, there is clear opportunity to reveal a deep understanding of the context, and make positive changes into construction practice, that are unlikely to have been possible from more distant research methods, such as surveys.

## Note

1 A latent error is a hidden organisational weakness that lies dormant. They can go largely unnoticed for long periods of time until they contribute to an incident.

## References

Alper, S., & Karsh, B. (2009). A systematic review of safety violations in industry. *Accident Analysis & Prevention*, 41(4, July 2009), 739–754.

Borys, D. (2009). Exploring risk-awareness as a cultural approach to safety: Exposing the gap between work as imagined and work as actually performed. *Safety Science Monitor*, 13(2), Article 3.

Brumberger, E. R. (2007). Making the strange familiar: A pedagogical exploration of visual thinking. *Journal of Business and Technical Communication*, 21, 376–401.

de la Garza, J. M. (2021). 2020 Best paper award. *Journal of Construction, Engineering and Management*, 147(6), 01221001–1.

Dekker, S. (2005). *Ten Questions about Human Error: A New View of Human Factors and System Safety.* Mahwah: Lawrence Erlbaum

Dekker, S. (2014). The bureaucratization of safety. *Safety Science*, 70(December), 348–357.

Hale, A. R., & Borys, D. (2013a). Working to rule or working safely? Part 1: A state of the art review. *Safety Science*, 55(June), 222–231.

Hallowell, M., & Gambatese, J. (2009). Construction safety risk mitigation. *Journal of Construction Engineering and Management*, 135(12), 1316–1323.

HSE. (2003). Casual Factors in Construction Accidents: Report No. CRR 156. Prepared by Loughborough University and UMIST for the Health and Safety Executive. London: Health and Safety Executive. https://www.hse.gov.uk/Research/rrpdf/rr156.pdf

Knudsen, F. (2009). Paperwork at the service of safety? Workers' reluctance against written procedures exemplified by the concept of 'seamanship'. *Safety Science*, 47(2), 285–303.

Lingard, H., & Oswald, D. (2020). Safety at the front line: Social negotiation of work and safety at the principal contractor–subcontractor interface. *Journal of Construction Engineering and Management*, 146(4), 04020024.

Lofquist, E., Dyson, P., & Trønnes, S. (2017). Mind the gap: A qualitative approach to assessing why different sub-cultures within high-risk industries interpret safety rule gaps in different ways, *Safety Science*, 92 (February), 241–256.

Lunch, N., & Lunch, C. (2006). *Insights into Participatory Video: A Handbook for the Field*. InsightShare.

Mackenzie, C. F., Xiao, Y., & Horst, R. (2004). Video task analysis in high performance teams. *Cognition, Technology & Work*, 6, 139–147.

Pink, S., Lingard, H., & Harley, J. (2017). Refiguring creativity in virtual work: The digital-material construction site. *New Technology, Work and Employment*, 32(1), 12–27.

Pink, S., Morgan, J., & Dainty, A. (2014). The safe hand: Gels, water, gloves and the materiality of tactile knowing. *Journal of Material Culture*, 19(4), 425–442.

Polanyi, M. (1958). *Personal Knowledge: Towards a Post-Critical Philosophy*. Chicago: University of Chicago Press.

Reiman, A., Pekkala, J., Vayrynen, S., Putkonen, A., & Forsman, M. (2014). PV-assisted evaluation of truck drivers' work outside cab: Deliveries in two types of transport. *International Journal of Occupational Safety and Ergonomics*, 20(3), 477–489.

Shaw, J., & Robertson, C. (1997). *Participatory Video: A Practical Guide to Using Video Creatively in Group Development Work*. London: Routledge

Tutt, D., Pink, S., Dainty, A. R., & Gibb, A. (2013). 'In the air' and below the horizon: Migrant workers in UK construction and the practice-based nature of learning and communicating OHS. *Construction Management and Economics*, 31(6), 515–527.

van Nieuw-Amerongen, M. E., Kremers, S. P. J., De Vries, N. K., & Kok, G. (2011). The use of prompts, increased accessibility, visibility, and aesthetics of the stairwell to promote stair use in a university building. *Environment and Behavior*, 43(1), 131–139.

# 9

# ETHNOGRAPHY FOR CONSTRUCTION 5.0

*Sarah Pink*

*Emerging Technologies Research Lab, Monash University, Melbourne, Australia*

**Key takeaways**

- Ethnographic research challenges dominant narratives concerning the future of automation and robotics in the construction industry, to offer new and more realistic visions.
- Examining real-life examples of successful technology development and implementation in the construction industry enables new modes of optimism for the future.
- Innovative digital and futures-focused ethnographic methods equip us to research in and towards the construction industry 5.0 as it evolves.

## Introduction

Automated and robotic technologies increasingly participate in the present and imagined future of the construction industry. This is set to grow with the emerging focus on industry 5.0 (defined below) and interest in future human-machine collaboration. Thus, there is an urgent need for researchers to investigate, analyse and engage with existing imaginaries for automated and roboticised possible futures, and to understand their implications for organisations, workers and training in the industry. This research will deliver new insight and foresight for planning in industry and government and should generate new, likely, plausible and realistic visions of plausible human and environment-focused ethnographic futures. However, new methods are required to approach this new research field, to encounter the sites where it is emerging and to research the possible futures it implies. In this chapter I outline this new research context and demonstrate a series of innovative steps for doing ethnography in the shifting present and possible futures as automation and robotics become part of the industry's digital transformation. In doing so, I draw on my experience and expertise in digital, in-person, on-site and interview-based ethnographic research in the Australian construction and robotics industries.

## Ethnography for an industry in digital transformation

The digital transformation of the construction industry has been characterised as slow, in comparison with other industries (e.g., Melenbrink et al. 2020). However interest is growing, for instance, a recent review by Liang et al. (2021) offers a taxonomy of developments in human-robot collaboration for the industry. Dominant narratives concerning the digital transformation

DOI: 10.1201/9781003379584-11

of the construction industry have until recently focused on Industry 4.0, as a technological solution whereby the introduction of automated and robotic systems and technologies has been seen as benefitting the construction sector by "reducing injury rates, handling repetitive tasks, and helping to enable construction in settings not currently feasible" (Melenbrink et al. 2020: 1). Academic work, driven by business and engineering approaches, envisions that technology disciplines involving "architects, civil engineers, mechanical engineers, computer scientists and roboticists" (Melenbrink et al. 2020: 14–17) could deliver this change.

Alongside this, academics and business consultants have advanced the opinion that the participation of workers would be focused on their being re-skilled in response to technological innovation (McKinsey & Co 2019; Melenbrink et al. 2020).

In contrast, worker-focused research has rightly critiqued the techno-optimist narratives of Industry 4.0 (Sherratt et al. 2020: 1), urging for attention to the social and ethical consequences of Construction 4.0, Sherratt and colleagues argued for accountability of how "technique will reshape the industry to suit the incoming technologies rather than adopt and adapt technologies to enhance the industry as is". This is in a context where as they note, engineers have been glorified and workers who perform the dangerous and dirty work have been forgotten (2020: 4).

However, industry visions have now overtaken this critique, with a new narrative which equally invokes social and ethical questions. The new focus on industry 5.0 has shifted the agenda, through its emphasis on the integration of industry and technological transitions with societal needs and environmental sustainability. This is embodied in the European Commission (2021) report *Industry 5.0 – Towards a sustainable, human-centric and resilient European industry.* Advisory and consultancy groups are now leaping onto the construction industry 5.0 narrative.[1] These organisations depict a future where collaboration is key, as different stakeholders, human workers and automated and robotic technologies and systems need to work together to create quality outcomes, and worker safety and physical and mental health would be improved.

Therefore, the social sciences generally – and new ethnographic insights and foresights in particular – are needed more than ever to generate understandings of how human-machine futures will develop, to understand the career trajectories of new tech-construction professionals, to inform education and training policies and to create realistic and ethical narratives to guide industry 5.0. As I argue in this chapter, it is crucial for social scientists to investigate ground up, by identifying and learning from examples of how and where successful integration of automated and robotic systems and technologies are already coming about in the construction industry. The ethnographic methods discussed in this chapter are precisely designed for that purpose.

## The ethnographic field in industry 5.0

Investigating an anticipated, nascent and still emergent shift towards a *possible* Industry 5.0 requires an eye both to predicted futures, and to generating realistic visions of futures as they come about, and as they are likely to be influenced by real people. In industry 5.0 visions people are acknowledged, but this doesn't mean the visions will play out as expected.

What and where, as we approach an industry which is envisioned to be undergoing a "digital transformation" towards Construction Industry 5.0, is the ethnographic field? To understand how industry 5.0 impacts the construction industry, conventional fieldsites for ethnographic research in the construction industry described in earlier work (Pink et al. 2013; Oswald & Dainty 2020) have to shift, to account for not just how people work on site and in offices, but also for companies developing new systems, technologies and services. While site visits and in-person encounters remain important, ethnographic research methods must be tailored to a digital-material and multi-sited field and to cross into future temporalities that ethnographers are not accustomed to working in (Pink, Fors, et al. 2022). The example discussed here contributes to a collaborative project – AUTOWORK – which investigates the future of work in relation to automation and robotics in Norway and Australia. I focus on my research in one organisation – FBR (originally Fast Brick Robotics), based in Australia. FBR, and other start-ups developing automated technology for construction, which we encountered in AUTOWORK, had originated from the visions of their founders who remain centrally involved and have developed new technologies in tandem with developing and training new workforces.

Technology (tech) companies necessarily have a strong online presence in which they narrate their own stories and possible futures. Tech companies involved in the construction industry are no exception. As exemplified below, this offers an excellent starting point for any ethnographic analysis of the way in which the construction industry is, through the use of automated and robotic technology and systems, moving forward into possible futures. Tech company founders tend to have prominent profiles and are often visible online, in podcasts, interviews, tech talks and other contexts. Thus, the online presence of companies, people and also the technologies and services that they develop constitutes part of the ethnographic field.

Therefore, a *Digital Ethnography* (Pink, Horst, et al. 2016) approach which understands the digital and the material/physical to be part of the same environment and sets of relations is integral to understanding how construction industry futures are constituted and imagined. Digital ethnography should not however be confused with online ethnography – that is, ethnography that is done solely online (see, for example, Hine 2015; Kozinets 2020), for instance by participating in, and analysis of, social media or other online activities.

Technology companies are also continually designing, developing, testing and employing technologies, in a field that brings together the digital and material in the experiential everyday worlds of workers and those who come into contact with their work. When investigating this dimension of the ethnographic field, *Visual* (Pink 2021) and *Sensory Ethnography* (Pink 2015) approaches enable researchers to participate in these everyday situations, using the camera as a mode of probing, exploring or discussing experiences, locations and practices with workers. Such methods invite participants to show us the realities of everyday workplaces, and for memories and examples to be invoked, performed and recounted during interviews or video-recorded tours. These research encounters allow researchers to imagine and seek to empathise with the experiences of participants in research, the kinds of *being there* in the environments where participants usually work can generate situations of sharing the same foods, drinks or material objects, in such ways that bring researchers closer to the lives of workers even when they cannot (for safety and other reasons) participate in the same activities as them.

The digital and material fieldsites mentioned above are in the first instance encountered in a present which continually rolls over into the past. They are invoked through sharing and viewing video. But they are also the everyday places where people may imagine futures – including their own career futures, the future iterations of the technologies that they are involved in working with or the future of their organisations. That is to say, future imaginaries are part of the way people live in the present. For example, as discussed elsewhere (Pink, Lingard, & Harley 2016), one of the most powerful moments in a past project was when our team viewed a fieldwork video in which a participant spoke of the anticipatory sentiments of construction work invoking an everyday imaginary of what could happen next. Because the industry is so dangerous, the notion of getting everyone home safe each day was central, meaning that the anticipation of risk is a way of life in evaluating work to be done.

Thus ethnographic research in the contemporary construction industry is shifting to a digital-material site, which, as demonstrated in the example discussed below, is likely to interweave throughout research processes.

## Ethnographic methods for construction in digital transition

In this section I situate the ethnographic approach discussed later through a reflexive account. I then propose and demonstrate a series of practices to guide the way ethnography in the digitally transforming construction industry is designed and undertaken.

### *Doing ethnographic research in the construction industry*

Ethnographic practice is reflexive (Pink 2015, 2021) and acknowledges how the trajectories of researchers, fieldwork sites and participants interweave and together shape the findings. In my ethnographic practice methods evolve, both within and through projects; methods developed in one project cannot be directly transferred to another. Rather methods developed earlier become templates to inspire the tailoring of new methods to new research questions and challenges. The design and application of ethnographic methods is thus a reflexive and ethical process, which involves understanding one's own situatedness; ethnography involves discovering one's own positionality and passions as much as understanding those of the participants in research.

The first time I stepped onto a construction site was in 2006, invited by Phil Bust, who was then a research fellow in construction management at Loughborough University. Phil had been videoing on the site and we connected over my work in video ethnography. As we interviewed a worker, his story was to remain imprinted in my memory as one of the reasons why I immediately felt committed to continue to work in this field. As a subcontractor, the sooner he completed the job, the sooner he would be paid and could move onto the next, so rather than losing over 30 min walking over to the other side of the site to collect his protective goggles he would be quite prepared to take the risk of getting a steel splinter in his eye, finish the job and then get to emergency to have it removed. I was hooked on wanting to understand what safety really meant to people working in this complex industry environment, and how ethnographic research could help make one of the most dangerous industries in the world safer. Our first paper, led by Phil, with Alistair Gibb (Bust et al. 2008) was published in *Safety Science*, and we subsequently collaborated in a series of funded projects, alongside colleagues including Andy Dainty, Dylan Tutt and Jennie Morgan. Our ethnographic research inspired me because we were bringing new methods to research in a field that anthropologists had conventionally neglected. I also felt increasingly strongly about the need for social scientists to orient themselves towards what we might see as life-and-death industries, where apart from their concerning levels of fatalities, workers are very frequently injured. While anthropologists do attend to many important applied questions, there was little attention to construction (Pink et al. 2013).

Our work also increasingly directed our attention to technology along with the evolution of uses of the mobile phone on construction sites (Tutt et al. 2013; Pink et al. 2014). We found emerging technologies were increasingly becoming part of safer work practices, shaped and often innovated by workers themselves, as further evidenced when from 2012, I collaborated with Helen Lingard, and James Harley in Australia. Lingard, Harley and I encountered an increasingly technological field, through our collaboration with CodeSafe Solutions, whose platform-based video training had changed the lives and safety of workers as a new "digital pedagogy" within the industry (Pink, Lingard, & Harley 2016). My later work within AUTOWORK, mentioned above, focuses primarily on the Australian construction industry, in collaboration with Ben Lyall, and involves engaging with start-ups, the unions, traditional large construction companies, tool libraries and people ranging from Chief Executive Officer (CEO), Chief Technology Officer (CTO), Chief Research Officer (CRO) to new tech workers and trades people. Below, to exemplify how I have developed ethnographic methods for an industry in a process of digital transformation, I draw on my experiences of researching with a robotics start-up FBR (formerly Fast Brick Robotics), which based on the technology innovations of CTO Mark Pivac has developed Hadrian, the world's first robotic bricklayer.

My ethnographic practice has evolved through a series of theoretical, practical and empirical dialogues, which bring interdisciplinary theory and practice to real-world questions in *Visual Ethnography* (Pink 2021, 4th edition),[2] *Sensory Ethnography* (Pink 2015, 2nd edition),[3] *Digital Ethnography* (Pink, Horst, et al. 2016),[4] *Design Ethnography* (Pink, Fors, et al. 2022)[5] and futures ethnography practices.[6] Collectively these approaches frame an ethnographic practice, which uses and analyses video and photographic technologies and media, is attuned to how people, bodies and minds sense and feel, account for digital technologies, data and content, learns from the experimental nature of design and are/is always attentive to as yet unknown and uncertain futures. Applying this approach to emerging technologies – considered to be on the cusp of becoming ubiquitous, entering a stage of mass production and playing a significant role in our futures – enables us to explore that moment "at the edge of the future" (Pink 2023a). Therefore, I investigate the everyday life sites where new and emerging technologies are designed and developed and where they are used or fail. I seek to pre-empt the possible futures represented in dominant techno-determinist narratives about the predicted impact of technologies on society.

### *Methods for researching automation and robotics in the construction industry*

In this subsection I draw on my own experiences to propose and demonstrate a series of methods tailored to a digitally transitioning construction industry.

#### *Online ethnographic sites*

There is a growing literature on online ethnography (Hine 2015), remote ethnography, and social media ethnography (Postill & Pink 2012). This literature highlights different but related possibilities for ethnographic practice in environments characterised by the digital materialities of a world which software and hardware participate in, and where digitally mediated elements of our environments are equally important to investigate as are in-person and material sites. Starting fieldwork online therefore was a familiar first step. I outline the process of initiating

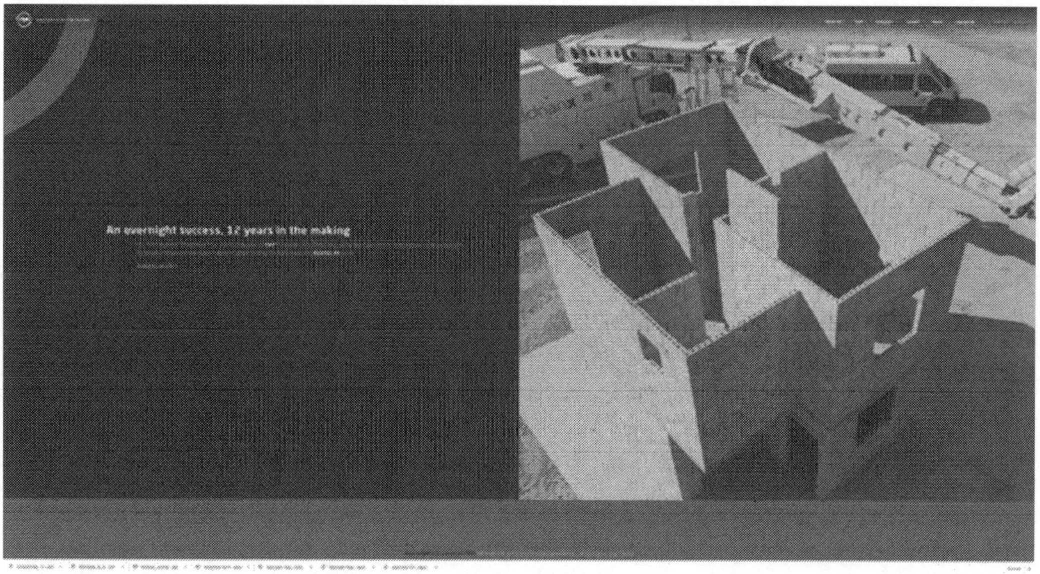


**FIGURE 9.1** Screenshot from FBR's website giving an idea of the scale and work of Hadrian.

*Source:* https://www.fbr.com.au/

online research as it developed in my own work, before reflecting on the implications of this for researchers contemplating new projects in a digitally transitioning construction industry (Figure 9.1).

I first learned about FBR when commencing the AUTOWORK project, Google searching for robotics in the construction industry, I was excited to learn that a world leading robotic bricklayer had been developed in Australia. The FBR website and its growing collection of videos provided an insight into what was possible in the industry. At that stage, during the Covid-19 pandemic, it was impossible to do any in-person fieldwork, but I followed FBR through their website, and have continued to as their work evolves and expands.

I next encountered another online resource – a robotics podcast series hosted by Melbourne-based Nicki Rossouw, who I later met in person due to our shared interest in robotics. Nicki had interviewed Mark, FBR's Chief Technology Officer (CTO), giving me the opportunity to listen to his and FBR's story online. The interview was fascinating in terms of what I learned about the technology development process, and along with the company's website and videos they had posted, it became part of my online ethnographic fieldsite. I made notes and transcribed significant quotes from Mark's interview with Nicki as it unfolded, recognising similarities between the long process of technology development that Mark had undertaken and the experiences of other engineers I had interviewed about their inventions in the past (e.g. in Fors et al. 2019; Pink 2023a), and wrote up a short piece for publication (in Pink 2023a).

Emerging technologies are, by definition, in a process of coming about, they might be in technological development, in the process of finding their markets (the right moment) or waiting for the new engineering innovations (like smaller batteries or new materials) to come about. FBR's robotic bricklayer Hadrian was a fascinating example of the pathway a technology may take, framed by the particular characteristics of the construction industry. For me as a researcher, it also offered an opportunity to consider the interface between two areas I had been researching independently – the construction industry and automated and robotic technologies. The podcast and website were freely available online, but my anthropological ethics are rooted in collaboration and consultation and the next step for me was to contact Mark to ask him to look over what I had written about him and FBR and to ensure that he was happy with how both were represented.

As these examples demonstrate, in a digitally transforming industry there are several aspects that ethnographers benefit from accounting for. This is for two reasons. First online resources – such as company websites, their self-documentation and the materials that they develop for the benefit of others, news and media about the organisations reported in tech or industry, mainstream media or social media items and discussions, podcasts and other materials – provide access to elements of the evolving context and experiences in situations where in-person fieldwork is impossible due to distance, cost or pandemic circumstances. Second, these materials are not separate from the digital transition that the industry as a whole is experiencing, or from the processes of tech development, marketing and engagement that companies are engaging in. In this context, online ethnography is not an add-on, but rather a starting point, and an ongoing commitment while investigating the experiences and developments of a digitally transforming industry. Researchers moreover participate in the same online environments as people involved in the industries we study. For instance, the Emerging Technologies Research Lab, where my work is based, publishes a series of industry accessible reports (and will do so based on the AUTOWORK research), I announce my publications, have videos of my talks online and share these materials on online platforms where participants in research are also active. Such participation extends beyond fieldwork *in situ* to become part of a broader professional practice.

### *Online ethnographic interviews as fieldwork encounters*

Later as my work with AUTOWORK developed, I had new questions about Hadrian and FBR, but at this stage of the Covid-19 pandemic, Western Australia, where FBR is located, was still closed to the rest of the world. Mark agreed to an online recorded, and subsequently transcribed, video interview, which lasted about an hour. I had prepared a set of questions, which also guided my online interviews with other construction industry experts. The one-to-one interviews were important in enabling me to learn in a very focused way, the details of Mark's and FBR's story, which to give readers a taste, started near the moment where, he said:

> I had always thought that there would be a really good application for a very large, computer-controlled machine that was stabilised by optical measurement methods, and until about 2005, all of the ideas that I had for that, there was a very, very limited market for the machinery. And then, in 2005, there was a shortage of bricklayers in WA [Western Australia], because there was a building boom on, and that's when it finally clicked. One of my customers at the time said, "Mark, you should develop an automated bricklaying machine."

I soon realised that in addition to his engineering and robotics expertise, Mark had a deep knowledge of the construction industry and its workforce. Hadrian's contribution was important not only due to the shortage of workers, but much of the heavy work Hadrian performs is precisely the kind of work that leads to the long-term injuries and fatalities that shaped the context for much of my earlier collaborative safety research, in the industry, and which make working in construction unappealing to young people.

Our discussion included how FBR and Hadrian worked with building companies and their future plans to expand, the new workforce that the company is developing and other bricklaying technologies that are being developed around the world. For example, Mark described how at that stage:

> Our typical team on site involves the actual Hadrian operators, and then we have construction technicians, and they're responsible for driving the telehandler and loading the machine, and doing any other construction-related tasks around the building site. So that can include quality control work, installation of brick ties, installation of lintels, things which the machine doesn't do. And there's a range of different guys there. Some are guys who don't have a formal qualification but have worked in the construction industry before. One's a registered builder. There's one who was a roof tiler. Another one was a plasterer. So guys who have got that construction background and understand the industry and know what a building site looks like. They end up in the non-operator roles which are on site.

This interview, along with the podcast and FBR's growing video archive on their website, helped me to understand their story and the ways in which the company was not just delivering and new technology, but a service to the industry, and was creating a new differently skilled workforce around the development, production and operation of Hadrian. We ended the interview in agreement that I should visit once the State borders opened again in Australia.

Perth, where FBR is located, is a 5–6 hour journey, including a flight, from Melbourne where I'm based, and in fact "There's no city of comparable size anywhere in the world that's so remote".[7] Therefore, even had we not been in pandemic circumstances, I would have most likely contacted Mark for a first meeting online before travelling there. In part, this is due to the time-sensitive nature of researching digital transformation processes, because this very much entails witnessing an industry in transformation as well as the technology development and testing processes of a robotics start-up, the opportunity to witness these processes of change, through ongoing research over time, is important. Therefore, the podcast, followed later by my online interview with Mark, and the later in-person interview discussed next, enabled me to create three touch points with the work of FBR over a two-year period. Indeed, one of the characteristics of ethnographic research is that interviews are never undertaken in isolation or analysed simply for what they contain. Rather the ethnographer is always thinking in dialogue with other fieldwork encounters in the same and other projects.

As mentioned above, the processes through which Hadrian came about reminded me of the findings of other ethnographic projects with engineers and in start-ups, whereby technology development and testing is a long and winding path (e.g. Lanzeni & Pink 2021). This finding brings further insight for fieldwork in the digitally transitioning construction industry: ethnographic research always enables us to notice that thing that is deeply relevant to understanding the context, but which is not necessarily directly obvious.

In the case of FBR, my interview with Mark demonstrated that while on one level Hadrian provided an excellent specific example of how robotics was coming about in the industry, this was coupled with the insight that similarly successful automated or robotic developments might similarly remain hidden, invisible, latent or even unanticipated, until the right moment came about. This means that it is very difficult to predict what technologies will feature in the future of the industry, partly because they may grow ground-up out of the work of start-ups which don't even necessarily yet exist.

### *Construction site ethnography*

Almost a year later, I arrived in Perth, having stayed up to date by viewing the growing collection of videos and news items on FBR's website. I was incredibly grateful to Mark and his colleagues who had set up a construction site visit for me on my first day. I drove out to the site in the rain to meet Kiel (Chief Commercial Officer) and Josh (Site Manager). Hadrian had been laying bricks that morning, and while Hadrian can work in the rain when human bricklayers cannot, if the rain is blown by the wind and wets some of the optics, Hadrian sometimes has to stop. I was allowed to film on the site and although my camera screen stopped working in the rain and I had to switch over to my iPhone, the chance to view the machinery invoked a series of questions about how the technology worked and the ways in which FBR operators and engineers worked with it, already informed by Mark's interview, months back, the range of expertise and training they had.

The project I visited, Kiel said, was one of a series of projects each of which presented new learnings and challenges, telling me that "we specifically pick our projects for that purpose, all those learnings go into the next iteration of the machine". I gained a sense of what a practical research and innovation project like FBR's looked and felt like in practice. He described the work being done and the human roles that were performed as part of it as we looked on to the machinery, while anticipating the new iterations of the technologies, and how they would operate in their next projects (Figure 9.2).

Construction sites are complex spaces, and there are many ways to immerse oneself in or learn about construction work on site ethnographically (Pink et al. 2013). I did not have permission to go onto the site itself, as this would have necessitated a safety induction, and

**FIGURE 9.2** Looking to the construction site from the street. Photograph by Sarah Pink

moreover I was only making one visit, to learn about the work of Hadrian on site through research with Kiel and Josh. The most suited method for this was the video tour (Pink 2021), a method which I have developed over many years, to learn about the sensory, material, social and emotional environments people live and work in. Video tours can also be developed as a mode of "short term" ethnography (Pink & Morgan 2013); they create an immersive experience and a situation in which the researcher has the possibility to explore a site with participants in the research and to ask focused questions often inspired by the environment itself. Likewise, being in a specific environment also assists participants in research to point out and demonstrate the experiential and visual dimensions of that place, in ways that would be impossible without actually being there. As we toured some of the houses that Hadrian was collaborating in the building of, Josh pointed out to me the difference between the wall that Hadrian had built and the external finished wall next to it, which was being built by a brickie (human bricklayer) using traditional methods and mortar whereas Hadrian uses a kind of glue to secure the bricks. I was able to see the difference and gain an understanding of what machines and humans working collaboratively on the same construction site might entail. I learned how the amount of time it would take them to complete Hadrian's work on this particular project also depends on the trades, because the homes being built had two storeys. At the time I visited, they were waiting for the trades to finish their work so that they could proceed and build the second floor of the houses. As Keil expressed it: "We're just fitting in with all the other trades that are required to bring something like this to life".

### *Organisational ethnography*

I visited FBR's offices the next day, starting in their offices where we started with a video interview with Mark (CTO) and Mike (CEO), in their boardroom. This was to be the most intense day of the fieldwork. I was able to learn much more about the company, its development, international ambitions, and importantly for my research also about the technology and new roles that were being created in the industry through FBR's work. The senior colleagues I interviewed were carving new and, in some cases, unique roles specific to the trajectory they collectively envisaged, of introducing Hadrian to a large international market. Being there also helped to consolidate my understanding of FBR as a tech start-up, as much as a company that was implicated in the construction industry. For the first part of the interview Aaron FBR's video maker – who was also responsible for the videos I had been viewing on the website – filmed me doing the interview. This is not the first time I've been filmed and photographed by people or organisations who I've been doing fieldwork with. For example, during my PhD about women bullfighters in Southern Spain, when I was featured in local media, when I was researching Slow Cities in the UK, a member of the movement documented me researching their projects. Ethnographers need to be prepared to expose themselves to the same scrutiny that participants in research open themselves up to, and to be willing to make our research processes open to view (Figure 9.3).

After these interviews, Mark took me over to their massive workshop a few 100 metres up the road. We went into the offices, where I met some of the staff working on software development, and the many other departments, and then we went out into the expansive research and innovation workshop where Hadrian machines are developed, made and tested, building whole houses inside (Figure 9.4). The company does almost everything in-house, since, Mark explained, in Perth outsourcing can be time consuming because it is dependent on contractors who

FIGURE 9.3 The spray cabin, which is big enough to fit Hadrian into, inside the massive workshop where Hadrian is developed, gives a sense of the scale of the space. Photograph by Sarah Pink

FIGURE 9.4 A selection of the brick technologies on display at FBR. Photograph by Sarah Pink

themselves need to source materials and labour. Hadrian is created and tested on site, they have built and disassembled 12 houses in their workshop. Mark took me on a tour of the space, which was too extensive to outline in full, but here I emphasise the way in which this tour, during which I could only take limited photographs, enabled me to understand the atmosphere, materiality and diverse sets of skill and knowledge involved in the research and innovation work being undertaken. For example, as we toured Mark showed me a series of samples of bricks which had been developed specifically for Hadrian's use, giving me a sense of their density, touch and weight (Figure 9.4). I was also able to see how making robotic hardware both involves collaborations between workers and some automated technologies, and to gain a sense of the materialities involved, the new iteration of Hadrian still being worked on, stood in the middle of the building. Steel shavings, created during the process, lay beautifully in a skip.

The day I visited FBR it was the company's seventh birthday, which was celebrated with a delicious fish and chip lunch. The food serving tables were set up in a row in front of the nearly complete latest iteration of Hadrian, and Natalie, the HR director, who had organised the event later, told me that she had counted nearly all the staff lining up for lunch. As she put it, they had "something special", which had also become increasingly palpable to me through the day. I met and interviewed a flow of people who were passionate and committed to the company's vision, which could potentially change the construction industry and improve worker safety in a situation where there is a global skills and housing shortage. FBR's employee numbers were in the early 1990s when I visited, soon to reach about 100 once the next positions they were recruiting for were filled. She emphasised that the people who worked for FBR "are here because they want to be", given that there was a very competitive recruitment market at the time, as she put it "every CV you see, there's dreams behind that CV". Some of the people who work at FBR have come from the traditional construction industry, but they are diverse ranging from physicists, computer scientists, electricians, a legal team and including a full-time video maker. Before and after the birthday lunch I immersed myself in interviewing people working for the organisation in a room to the side of the workshop. I was keen to understand their roles in the company, and how they anticipated and imagined the future of work in an automated and robotic construction industry. The research findings are discussed elsewhere, however here I want to emphasise that by being there in person I gained a sense of a group of people working towards shared goals in an environment which represents what a successful vision of industry 5.0 might be. Ethnographic research can precisely gain such understandings of how such companies, grown from an idea, with a vision which promotes safety and seeks to benefit society in a dangerous industry, are successful. FBR has created this environment because it is doing something inspiring and that matters to people and like my analysis of CodeSafe Solutions (Pink, Lingard & Harley 2016; Pink 2023a), mentioned above, offers an example of how digital transition is successfully happening.

## Ethnographic futures: Methods for construction industry 5.0

Industry 5.0 is likewise a future-focused concept. Above I have described how consultancy and advisory groups have begun to chart or hype an industry 5.0 future for the construction industry, which maps out the benefits that adoption of this model could have. The notion of construction industry 5.0 is likely to expand in the near future, offering a model for successful digital transition. At the time of my ethnographic research industry 5.0 remained an emergent concept which frames a possible future, or emergent and contingent field of activity. By this I mean that ethnographic research about industry 5.0 or about a future automated and robotic construction industry cannot be located in one single fieldwork site, but rather the various fieldwork sites in which we research give us the opportunity to glimpse possible futures as they are anticipated on the ground and as they are in processes of coming about. Ethnographic research of the kind I have described above works differently to models of industry or construction 5.0, because it detects and analyses processes as they come about, in ways that already work, rather than seeking to model and impose them from above. In this sense, doing ethnography in this industry in digital transition is itself a critical move, because it highlights the need to attend to how modes of working alongside automated technologies in the industry are already happening, to learn from these examples and to engage experts who are experienced in these realities. As I have outlined above, the expertise involved in creating and successful robotic bricklaying machine is immense, both in its depth and its breadth, and it is moreover not a fixed or static model that can be captured. The research, innovation and testing cycles of technology start-ups are ongoing; they are concerned with building and marketing emerging technologies.

There is nothing new in the idea that ethnographic fieldsites are always in a state of flux and change. Indeed, as I have emphasised elsewhere, ethnographic research in the construction industry demonstrates this point explicitly as construction sites themselves are very obvious places of tangible material and social change, as people, things and processes continually reconfigure. Construction 5.0 offers a different perspective on a changing fieldsite, since it is located at what I call "the edge of the future"; it appears poised to come about, or it is a moment in the making, a process in progress. It thus constitutes a particular set of possibilities into which we might start to imagine and interweave other elements of possible futures and to explore these with participants in research. But to encounter and develop a thorough academic and applied understanding of its possibilities we need to seek out the actual examples of where elements of it are already coming about, and to understand the possibilities they demonstrate.

## Conclusion

As I left FBR, Natalie drove me back over to the offices and made up a corporate gift bag for me, with a water bottle, a towel, keyring, pins and a pen. But it was not simply the material gift that I left with but the gift of being allowed to become part, although for a very short time, of the life of an organisation, community or group of people. I have been doing ethnography for over 30 years, since I first arrived in Southern Spain to do fieldwork about women in the world of bullfighting. Over those years, the generosity of participants in ethnographic research, to show researchers what matters to them, to talk of their experiences, their plans, anticipations, dreams and possible futures, has always made me feel immensely privileged. Doing ethnography enables us to sense the intensities of other people's lives, not necessarily in exactly the ways that they feel them, but enough to ensure that we can gain an understanding of what matters. As Natalie so perfectly expressed it: "you would feel it today, every single person out in that workshop, yeah, has got a passion for Hadrian, they've actually even got a passion before they walk in our door". Other ethnographers will know this experience when they feel it. It gives us both amazing insights into life as lived and anticipated, but it also entails a deep ethical commitment, to treat the knowledge we have been offered about what matters to others responsibly and respectfully. Ethnography does not end when we leave the fieldwork site, but the commitments it demands extend right through to where, how and when we represent our findings. This is part of what I elsewhere call the "ethnographic place" (Pink 2015), which involves not simply the fieldsite in which we do research but the full spectrum of further academic and public engagements that surround the work.

The construction industry is one of the most compelling places to do ethnographic research, and as its "slow" digital transition comes about, we need to be attentive to both the wider narratives of change touted by the consultancies and taken up by government, and the real processes of technology innovation and possibility that are being driven by tech start-ups and that are already having impact in the industry. FBR offered me the perfect example through which to explore how automated and robotic technologies could really be used in the industry. The ethnographic methods I engaged to do so were tailored to the case. I never recommend that other researchers "use" my methods (see, for instance, Pink 2021), but rather invite ethnographers to draw inspiration from my practice, to adapt my methods and to tailor the way they are used to their own cases, while researching ethically and responsibly. It is in this spirit that I present the above account of my research experiences. Each subsection of the previous section of this chapter describes a particular method or set of methods that evolved as suitable for the different stages and modes of investigation I undertook and connects them to existing literature in which they are discussed in further detail.

Much existing research in the social sciences and humanities is both pessimistic and critical of the role automated and robotic technologies might play in our futures, and like conventional tech-solutionist narratives tend to assume that the future will happen to people, done to them by institutional powers and automated technologies. My colleagues and I have critiqued such approaches elsewhere (Pink, Ruckenstien, et al. 2022) on the basis of sociological and anthropological research into what really happens as people engage with emerging technologies and shape them to their lives and workplaces. Thus, I have proposed the practice of *anthropology forward*, which seeks to generate imaginaries for and constitute possible futures that are optimistic (Pink 2023b). This involves a reorientation of the ethnographic agenda, to ask how, on the basis of what we might be able to know in the present, might we be optimistic about what can "go right" (Willow 2023) in possible futures. Undertaking ethnographic studies that can help us to identify how things can go right is part of the journey of doing anthropology forward.

I invite and encourage construction industry ethnographers to take seriously the emerging relationship between automation and the future of work and technology in the industry. This is, as discussed elsewhere, building on the comments of industry actors, a complex transition, given the fragmented nature of the industry, its unpredictable terrains and short-termist investment cycles. Yet it is happening and therefore digital and emerging technologies and all that comes with them, are part of the fieldsites of future ethnographies in this field. These technologies need to be accounted for and the ways in which we can successfully move forward with them into the future need to be considered.

## Acknowledgements

My first thanks are to all the people who generously participated in the research discussed in this chapter, for sharing their experience and knowledge with me. In particular, I thank Mark Pivac for hosting and introducing me at FBR and for reading and correcting the first draft of this chapter. The project "Workers in transition through automation, digitalization and robotization of work" (AUTOWORK) is an international research project between the Norwegian University of Science and Technology, Norway and Monash University, Australia, funded by a grant from the Norwegian Research Council's Welfare, Work, and Migration program, Project no. 301088. I thank my AUTOWORK colleagues for our shared discussions and interest in the future of work.

## Notes

1 Such as: Trillium Advisory Group, https://www.trilliumgroup.io/post/industry-5-0-in-engineering-and-construction; ARC Advisory Group https://www.arcweb.com/industry-best-practices/industry-50-vision-future-industry; and Tech Republic https://www.techrepublic.com/article/industry-5-0-human-machine-interface/.
2 Using video and photographic methods and media in ethnographic research and representation.
3 Being attentive to sensory experiences and categories throughout the ethnographic research process.
4 Acknowledging that digital technologies and media are inseparable from both the research process, as technologies of research, and from the everyday life situations that ethnographers investigate.
5 Bringing together design and ethnographic practice to create new interdisciplinary methods and approaches to research and intervention.
6 Ethnographic practice that focuses on investigating possible futures, rather than focusing on the present and past.
7 https://www.theguardian.com/cities/2015/aug/19/where-worlds-most-remote-city.

## References


Bust, P., Gibb, A. & Pink, S. (2008). Managing construction health and safety: Migrant workers and communicating safety messages. *Safety Science*, 46(4), 585–602. https://doi.org/10.1016/j.ssci.2007.06.026

European Commission, Directorate-General for Research and Innovation. **(2021, 5 January)**. Industry 5.0 – Towards a sustainable, human-centric and resilient European industry. https://research-and-innovation.ec.europa.eu/knowledge-publications-tools-and-data/publications/all-publications/industry-50-towards-sustainable-human-centric-and-resilient-european-industry_en

Fors, V., Pink, S., Berg, M. & O'Dell, T. (2019). *Imagining Personal Data*. London: Bloomsbury.

Hine, C. (2015). *Ethnography for the Internet: Embedded, Embodied and Everyday London*. London: Bloomsbury.

Kozinets, R. (2020). *Netnography: The Essential Guide to Qualitative Social Media Research*. London: Sage.

Lanzeni, D. & Pink, S. (2021) Digital material value: Designing emerging technologies. *New Media & Society*, 23(4), 766–779. https://doi.org/10.1177/1461444820954193

Liang, C.-J., Wang, X., Kamat, V. R. & Menassa, C. C. (2021). Human–robot collaboration in construction: Classification and research trends. *Journal of Construction Engineering and Management*, 147(10), 03121006. https://doi.org/10.1061/(ASCE)CO.1943-7862.0002154

McKinsey & Company. (2019, December). The impact and opportunities of automation in construction. https://www.mckinsey.com/business-functions/operations/our-insights/the-impact-and-opportunities-of-automation-in-construction

Melenbrink, N., Werfel, J. & Menges, A. (2020). On-site autonomous construction robots: Towards unsupervised building. *Automation in Construction*, 119, 103312. https://doi.org/10.1016/j.autcon.2020.103312

Oswald, D. & Dainty, A. (2020). Ethnographic research in the construction industry: A critical review. *Journal of Construction Engineering and Management*, 146(10), 1–13. https://doi.org/10.1061/(ASCE)CO.1943-7862.0001917

Pink, S. (2015). *Doing Sensory Ethnography*. 2nd Edition. London: Sage.

Pink, S. (2021) *Doing Visual Ethnography*. 4th Edition. London: Sage.

Pink, S. (2023a). *Emerging Technologies/Life at the Edge of the Future*. Oxford: Routledge.

Pink, S. (2023b) 'Doing anthropology forward: Emerging technologies and possible futures' in A. Willow (ed) *Anthropological Optimism: Engaging the Power of What Could Go Right*. Oxford: Routledge. pp177–191.

Pink, S., Fors, V., Sumartojo, S., Duque, M., Lanzeni, D. & Strengers, Y. (2022). *Design Ethnography: Research, Responsibility and Futures*. Oxford: Routledge.

Pink, S., Horst, S. H., Postill, J., Hjorth, L., Lewis, T. & Tacchi, J. (2016). *Digital Ethnography: Principles and Practice*. London: Sage.

Pink, S., Lingard, H. & Harley, J. (2016). Digital pedagogy for safety: The construction site as a collaborative learning environment. *Video Journal of Education and Pedagogy*, 1(1), 1–15.

Pink, S. & Morgan, J. (2013). Short term ethnography: intense routes to knowing. *Symbolic Interaction*, 36(3), 351–361

Pink, S., Morgan, J. & Dainty, A. (2014). Safety in movement: Mobile workers, mobile media. *Mobile Media & Communication*, 2(3), 335–351. https://doi.org/10.1177/2050157914537486

Pink, S., Ruckenstein, M., Lupton, D. & Berg, M. (2022) 'Everyday Automation: Setting a Research Agenda' in S. Pink, D. Lupton, M. Berg & M. Ruckenstein (eds) *Everyday Automation*. London: Routledge.

Pink, S., Tutt, D. & Dainty, A. (eds) (2013) 'Introducing ethnographic research in the construction industry' in *Ethnographic Research in the Construction Industry*. London: Taylor & Francis, pp 1–22.

Postill, J. & Pink, S. (2012). Social media ethnography: The digital researcher in a messy web. *Media International Australia*, 145(1), 123–134.

Sherratt, F., Dowsett, R. & Sherratt, S. (2020). Construction 4.0 and its potential impact on people working in the construction industry. *Proceedings of the Institution of Civil Engineers – Management, Procurement and Law*, 173(4), 145–152. https://doi.org/10.1680/jmapl.19.00053

Tutt, D., Pink, S., Dainty, A. & Gibb, A. (2013) 'Our own language' in Pink, S., A, Dainty & D. Tutt (eds) *Ethnographic Research in the Construction Industry*. London: Taylor & Francis, pp 40–57.

Willow, A. (Ed.) (2023). *What Could Go Right: Anthropological Engagements with Optimism*. Oxford: Routledge.

# SECTION III

# Doing ethnography in the construction industry

# 10

# DOING SHORT-TERM ETHNOGRAPHY

*Richard Brett*

*Anglia Ruskin University, Chelmsford, United Kingdom*

**Key takeaways**

- Building rapport with research participants is essential for effective fieldwork.
- Using video in short-term ethnography is highly recommended as an accompaniment to traditional field notes, but not as a replacement.
- A researcher must be willing to commit much energy to undertaking short-term ethnography to gain close access to construction work.
- Short-term ethnography is useful for generating rich data about tightly defined aspects of construction culture.

## Introduction

The motivation behind the fieldwork that forms the basis of this chapter was doctoral research to explore the way everyday behaviours of construction workers reveal the presence or absence of craft traits. A new housing site and a building conservation site were selected for comparison. I carried out a period of short-term ethnographic fieldwork on both. The former was building new houses for sale on the open market. The latter was a building conservation project where extensive refurbishment was being undertaken to a listed building of significant historical interest. I spent ten consecutive working days on the new housing site. The second spell of fieldwork, on the conservation site, lasted only six days. This was shorter than intended because the data collection was halted by the Covid-19 pandemic of spring 2020.

Short-term ethnography is so-called because less time is spent in the field compared with traditional ethnography where, typically, the fieldwork takes place over a period of many months or years. Although we are using the label of short-term ethnography (Pink & Morgan, 2013), there are other terms in use describing something approximately similar. For example, LeCompte (2010, p. 104) talks of "rapid or compressed research". They argue that this cut-down version is suitable where ethnography is the most appropriate research design, but "time money and staff do not permit a full-fledged ethnography".

Choosing short-term ethnography is not just a pragmatic question of limited researcher resources. There are social settings, especially those based in the workplace, where prolonged stays by the researcher are not welcome or desirable. Furthermore, the phenomena of interest may be practised in specific and fleeting events. For example, a typical house in the UK is built by lots of

DOI: 10.1201/9781003379584-13

different construction workers doing their own specialist role as part of a sequence of activities. Once they have completed their specific task they move onto another house or site. As a result, some of our research participants were only on site for a day or two in the entire duration of the project. The depth of engagement that occurs while both researcher and participant are present is what matters. Staying longer in some contexts can be of limited value (Löwstedt, 2015).

While the duration of time spent in the field is reduced, the intensity of the data collection is increased. To achieve this, Pink and Morgan (2013) suggest a "deliberate and interventional approach" on behalf of the researcher. In my fieldwork experience, this intensity manifests as tireless activity from the researcher. There is no time to waste when the fieldwork opportunity is short. Every moment is filled with some form of data collection activity. It takes effort – both physical and mental – to get close to workers, to follow them around, to study their activity intently and to ask what they are doing. It is this that provides depth to the data because of what is revealed when one gets close. Standing back and observing passively will cause much to be missed.

A researcher's familiarity with the field setting is a key component of success in short-term ethnography (LeCompte, 2010). Ethnographers reflect on their emic or etic perspectives. This is the degree to which they have an insider's knowledge of the social setting they are studying. Emic is the insider's perspective, held by those within a group. The etic perspective is that held by outsiders and is the explanatory language used to describe that group to others. Rather than being two opposite positions, these perspectives lie on a continuum (Fetterman, 1989). This is particularly true in construction research which has a strong applied element, and many researchers have connections to, or a work history in, the construction world.

The way I navigated the fieldwork was constantly informed by my prior experience of working in the construction industry. In fact, it was that work experience which generated the idea for the research from the outset, long before the actual fieldwork began. This gave me more than just knowledge of the field. I had an insider perspective which I harnessed to make decisions about where to go and which events might be generative of rich data about craft. In this way, the brief stay on site was very fruitful.

In this chapter I reflect on my experience of undertaking two spells of short-term ethnographic fieldwork. First, I consider how I developed a rapport with potential participants. This starts with seeking out a suitable site, negotiating access and proving my competency on a construction site. The next step is making the first contact with participants in a positive way so a fruitful relationship can be built up. I then reflect on my take on participant observation and the challenge of obtaining informed consent from construction site workers. In the time-bound world of short-term ethnography, one only has a few hours to build this all-important rapport, not weeks or months as enjoyed by traditional ethnographers.

Second, I reflect on the lack of control the ethnographer has over events. Construction sites are very dynamic. The physical form and the social group are constantly changing as the building project progresses. A short-term ethnographer is well positioned to cope with this by being active in the field and retaining a flexible approach to take advantage of events as they occur.

Third, I emphasise how important the use of video cameras as a means of data collection was to me. The cameras help compensate for a researcher's lack of experience in conducting ethnographic fieldwork. Away from site, re-watching and re-analysing the video files can give confidence in the empirical data upon which findings are made. Most importantly, re-engaging with video footage away from site enriches the data that was captured from the short period of fieldwork.

## Entering the field

Prior to the beginning of data collection, the researcher will already have selected an appropriate fieldwork site and obtained the relevant permissions which can be a lengthy process. Of more critical importance for short-term ethnography is the need to align the researcher's own availability with the site programme. When the researcher only has a few weeks to spend on a project, choosing when to deploy needs careful thought. If there are particular activities the researcher wishes to incorporate in their research, then they need to timetable their fieldwork accordingly. This is easier than it sounds given the almost inevitable delays and changes that beset construction projects. Arriving on site too early or too late in the building programme may mean missing preferred activities. In real-world research, particularly ethnography, the researcher is not in control of events and must adapt to them. The construction site management team will not alter their programme to suit the researcher. Site selection and timing are typical "pragmatic considerations" of ethnography (Hammersley & Atkinson, 2007, p. 30).

In my experience, gaining access to one site was very easy, while it was very difficult for the other. With the housebuilding site I used a connection I had with a senior person in the company developing the housing. He was supportive from the outset, and access was arranged with little more than a phone call and a visit to site to meet the management team.

Contrastingly, it took a year to secure approval for the conservation site. The building owner was a high-profile national charity which owns and operates a wide range of buildings of national and international heritage significance. It had concerns (which were unfounded) about what the fieldwork might show and how the practices of the organisation would be portrayed. These issues were overcome through explanatory dialogue with the charity's project management team. The purpose of the fieldwork was made clear by emphasising the focus of the research would be on the actions of workers, rather than the organisation itself. Eventually permission was granted with no conditions attached. The length of time taken was down to the slowness of dealing with a large organisation where there are a number of people who become involved in the decision-making. And of course, my request to spend time on site was not an enticing offer for the organisation. There was no great motivation to assist me, other than their own goodwill. The only course of action when seeking access is to allow time.

For my interest in craftwork there were certain building activities such as carpentry, brickwork, roofing and leadwork, that I felt would be more informative for my inquiry into craftwork than others. When I started negotiating access for the housebuilding site, the developer offered a range of projects. I chose the project where the site programme schedule for the superstructure phase of the houses coincided with my own availability. On the conservation site, the building project was on a much larger scale. The carpentry and roofing works were both scheduled to last many months, making it easier to schedule in the fieldwork. This was just as well, as events turned out. Having confirmed a start date for the fieldwork, it was delayed at the last moment because of structural problems with the building. I didn't return for three months while the architects and client explored options for remediation. Experiences such as this underline how there needs to be flexibility in short-term ethnography.

Once the permission had been obtained and the dates set, there was still one more thing to do, which was to prove to the site management team that I was competent to be there. On the housebuilding site there were no concerns as the management team had a very relaxed approach. I was free to go wherever I wished. It was quite different on the conservation site where much of the work was taking place a hundred feet above the ground. The work platform was an extensive scaffold. Towards the end of the first day on site, I had a conversation with the site manager, talking about how the day had gone. The site manager remarked that, having watched me in action on the scaffold platform, he was happy for me to return another day (neither site required me to provide any form of safety or competency certification). Unbeknown to me, the site manager had been watching to make sure I was competent enough to be on his construction site.

This exchange was a reminder of one of the features of ethnography: the researcher is an object of interest to the participants. The way the researcher acts will have an influence on the way he or she is accommodated on site. The last thing a site manager needs is to have a potential liability or disruptive presence on site. Displaying a modicum of site sense (Aboagye-Nimo & Raiden, 2016) and respect for one's surroundings is imperative for developing an important dimension of short-term ethnography: rapport.

## Building rapport

Ethnographic fieldwork is a very sociable undertaking. The researcher must push themselves out into another world. This can be foreboding, especially to a novice. The choice of ethnography as a research approach places the researcher onto a course of action in which, eventually, they will have to make contact, and build relationships, with people they do not know. The theoretical work of reading, writing and navigating the ethics process is all building to the first day on site. The initial contact with the first research participant is a seminal moment. It requires a degree of personal bravery. The researcher must put aside their doubts, ignore that nagging thought of why they did not just send out an online questionnaire and summon up the courage to talk with the site workers.

While these needs are not exclusive to short-term ethnography, the reduced time frame means the researcher does not have the luxury of dwelling on site, patiently developing a position within the social group. In our fieldwork, the time spent with any one participant varied from a few hours to three or four days. From the very outset of the fieldwork, it is imperative to make productive relationships on the site. Establishing rapport with participants needs to be quick and successful. It is recommended that there be a strategy to developing rapport (Hammersley & Atkinson, 2007), but in my case it emerged on site through practice, rather than something pre-meditated.

The first contact with a potential participant is an anxious moment. What do you do if they tell you to go away? Over the course of our fieldwork, I evolved a protocol for making contact and generating a fruitful relationship quickly. The first step was asking the site manager to let the workers know there was a researcher on site who might come and speak with them. It was made very clear there was no obligation to participate, although given many of the workers were subcontractors, this message was probably unnecessary.

The potential research subjects were engaged in their everyday work. This meant, in almost every case, climbing up a ladder to get to the scaffold platform where they were working. In itself, this is something of an event. Scaffolds are small and there is not much room to manoeuvre when there are more than a couple of people in the same space. My entrance could not go unnoticed. However, the forced proximity was a good thing. I needed to get physically close to the workers to hear what they were saying and to see what they were doing. It is also impossible to avoid conversation when you are squashed together and so some dialogue between researcher and participant was inevitable.

Once the (ungainly) assault on the scaffold has been made, there is nothing for it but to exude a confidence that might not really be there and make an introduction. This always started with a brief explanation of who I was, and what I was doing, and would they mind if I spent some time with them while they worked. My opening gambit was then to ask the workers to explain what they were doing. With that, I was up and running.

One of the strongest impressions from the fieldwork is that, to a man, the workers liked an interest being shown in them. They were all effusive about their work and open about sharing their thoughts and experiences. I did not knowingly put on any affectations to be more like the construction workers to help generate rapport. In fact, I downplayed the full extent of my construction knowledge for fear of this becoming a threat to the participants. While I did not seek the role of an "acceptable incompetent" (Brewer, 2000), I was mindful to remember the worker was the expert, not I. I was always conscious of being a guest in their world. It is difficult when reflecting after to appraise whether my construction background helped in forming relationships, but I suggest it was the way I conducted myself during the fieldwork, rather than my prior experience, which was the key. Once the courage is summoned for the initial approach, establishing rapport is no more complicated than being polite, respectful and interested in the workers. Nodding agreement, consenting to opinions (no matter how outlandish) and not criticising the work all aide in generating a positive and open atmosphere. Once the relationship has begun, the conversations take many different directions. The key is to make the initial approach a successful one. In my experience, the autonomy many construction workers enjoy in their everyday work is reflected in their confidence. They typically have a lot to say, both about the work they are doing and a multitude of other topics. The workers are a rich seam for a short-term ethnographer.

## Participant-Observation

As part of the research design, it is necessary to establish the extent to which the researcher is participating or observing. The phenomena under investigation will influence the way this is decided. There are plentiful examples among construction ethnographies of researchers who are on site for a long period of time (Thiel, 2012). It is possible for ethnographers to become very involved. This can mean taking on the role of an apprentice to gain first-hand experience of what it is like to be a construction worker (Baarts, 2009; Löwstedt, 2015).

In my case, I came to the research already armed with a "long and varied apprenticeship in building" (Brett et al., 2022). Of interest to me was what the workers themselves were doing, as the embodiment of craft traits, and what the narrative around craft was on site. This meant picking up on the verbal (and non-verbal) communication within the workers. Participation in the action was not necessary because I already had prior insight into the way of working. It was not so important for me to learn how or what they were doing (as my prior work experience went a long way to answering this), but instead to learn about why they were doing it, analysed through the theoretical lens of craft. Furthermore, as a short-term ethnography with extremely limited time on site, the apprenticeship role would hinder the freedom of mobility of the researcher to jump around from event to event.

I opted for a very overt role of researcher, best described as a "minimally participating observer" (Clark et al., 2021, p. 229). This involved much interaction with group members without participating in what they were doing. It also means sharing the work environment, including its hardships. A construction site in England in January is a cold place for both a worker and a researcher. Standing very close to the workers was essential to achieving the level of intensity in the data collection that we needed. For example, on the housebuilding site, I was interested to observe the bricklayers having some difficulty laying coping stones. The quest for quality is an important trait of craft, and so the moment when the bricklayers evaluated the quality of their own work promised to be revealing. When it came, a glance and a brief comment between them were the only visible sign they were happy with the work. Only by being right next to the workers, and waiting for that moment, was I able to witness their approval and then ask follow-up questions about it (Brett et al., 2019). This was critical in achieving "intensive excursions" (Pink & Morgan, 2013) into the everyday working lives of the participants.

## Ethics approval on the fly

Ethics approval for the fieldwork required each individual participant to give recorded, informed consent. It is impractical to suggest that ethical consent could have been obtained before any form of data collection commenced, in the way that it might with an interview or questionnaire because the fieldwork begins upon the moment of entering the construction site. With ethnography, the act of engaging with a site worker sets the analytical ball rolling for the researcher, but the very first dialogue cannot be a request to sign an ethics form. Conceivably, it would be possible for the site manager to collect all the workers together at the start of the day, the researcher could then explain themselves, hand out the information sheet and consent forms, and willing participants could be recruited this way. However, on both sites I visited this was not feasible. There were too many workers and, while the researcher's presence was welcome, it did not extend as far as stopping work on site and gathering all the workers together, which is a rare occurrence. Furthermore, with new workers arriving and old ones leaving daily, it would be far too intrusive to execute.

After much deliberation on how to meet the challenge of obtaining informed consent, I opted for the most straightforward approach. After introducing myself to a participant and an initial chat, the researcher-participant relationship was up and running. It was then a case of waiting for an opportune moment to broach the issue of ethics. The researcher explained why the university insisted on ethics approval and asked if the respondent was happy to participate, also telling them there was a participant information sheet to read. At this point, if the participant had said no, any data collected (either on video or in the notebook) could be erased with minimal research effort wasted.

As it turned out, every participant consented. They also all signed the participant consent form so ethical approval was obtained in paper form. Another option may have been to record verbal consent on the video cameras. On reflection, obtaining ethical approval was a clumsy process. Just as the researcher was developing nascent ties with worker based on informal dialogue, the need to formalise affairs with the informed consent is disruptive. However, with short-term ethnography, the need to obtain consent from participants early on is given extra imperative by the short timescales. A participant may not attend site the next day or the day after. Putting off obtaining recorded ethical approval runs the risk of never getting the approval. Given how important it is to achieve robust consent, the recommendation is to spend some time building a relationship with the participant and then introduce the need for informed consent at the first available opportunity. Once it is obtained, the issue is resolved, and the data collection can proceed.

## The researcher's lack of control

Upon entering a construction site, the researcher is a guest, whose presence is only allowed by the good grace of the site management team. While this provides the kind of access from which research findings can be generated, the researcher has no control. Workers will start and finish work as they please. They will take breaks according to their habits (during the fieldwork, I made a point *not* to spend break time with the workers so both parties had some time out). The work they do is not planned with the convenience of the researcher in mind. Of course, it is not even desirable for workers to change their behaviour if the ethnography has been proposed on the basis of seeing the workers in their natural setting. From undergraduate dissertation student to postgraduate researcher, one's own sense of importance in their work is strong, or at least it should be. The data collected and the ethnographic narrative that will flow from it will have some purpose, be that achieving a degree or publishing a paper. But its importance to the researcher is not shared by the construction workers. Even knowing and accepting this, it can be frustrating when the activity of the site is not in the researcher's favour. On the conservation site I was very keen to see some lead welding take place because this had already been observed on the housing site. However, obtaining the hot works permit was a very drawn-out affair for the site team and the welding nearly did not happen while I was on site. There was nothing I could do to affect the situation to bring the lead welding about. By going onto a construction site, the researcher is giving up a degree of control over the success of their project There is an inherent element of risk attached to doing fieldwork in this way.

In my experience, the site context also had a significant impact on the behaviour of the workers. On the housebuilding site, the workforce was young, and everyone was paid by the amount of work they did. This is a motivation to work hard and work fast. It is one of the basic principles upon which most houses are built in the UK. Time spent not laying bricks or fitting roof trusses is unpaid time. On our site, the workers were able to talk with the researcher and work at the same time, but work came before conversation.

On the conservation site, the workers were a lot older and were paid for their time, not how much work they did. For these workers, conversing with the researcher was just another opportunity to put their tools down and stop work. Some workers were so keen to speak about their work that they would stop for 10–20 minutes at a time to explain what they were doing. While this provided a great insight into their work world, it was their choice to be generous with their time.

The response to this loss of control was to retain flexibility in my approach and always be looking for opportunities. Rather than allowing events to wash over me, I took the lead in seeking moments of interest. I was on site to explore craft traits in construction workers so anytime a worker was doing something that had the potential to be revealing about craft, I would be there. Participant observation, carried out up close to participants, can deliver rich data. But on a construction site where much is happening, the researcher needs to be mobile and alive to opportunities. Across all our fieldwork, there were very few moments when I was observing the site or the workers as a whole. Instead, I was always observing a particular worker, engaged in a particular activity, whether this was carpenters fitting rafters, or leadworkers dressing lead sheets to form a parapet.

## Becoming an active ethnographer

The final reflection is on how active construction site ethnography can become. The time constraint of short-term ethnography adds an element of pressure to the fieldwork, especially for a novice. Knowing there are only a few days on site provides an imperative to always seek out events in the field. This can manifest itself physically. Construction sites accommodate many things happening at the same time in different places. To see what is happening, the ethnographer must be mobile.

On the housebuilding site, I was interested to spend time with the leadworker who came on to site to install some roof flashing details. His job is one of the few that has not been transformed by prefabricated components that only need assembling rather than making. The leadworker still has to shape materials into the building. It was only by chance that he was on site while the fieldwork was underway, and he was scheduled to be on site for just two days. This was an opportunity not to be missed, but there was only a small window of time available. To capture as much of the everyday work of the leadworker as possible, I had to climb up and down the roofs and across the scaffold as he quickly moved from roof detail to roof detail. This was physically quite demanding (and required some confidence with heights), but retaining physical closeness was essential to be able to watch the leadworker in action and embark upon a dialogue about what he was doing.

Intense and mobile engagement with workers was typical for me throughout the fieldwork. Informed by my prior insight of construction work and sensitisation through craft theory (Brett et al., 2022), the fieldwork quickly became a string of discrete events that started, finished and ended within a few hours (or sometimes much less). The physical and mental demands of this approach left me very tired at the end of each day.

## Using video

Visual ethnographic methods are varied. The video footage can be the output of the fieldwork, presented in movie form (Vannini & Vannini, 2019), or participants can be invited to comment on their own behaviours by watching them back on video (Jarrett & Liu, 2018). My use was rather conventional, and the decision to use video cameras was an easy one to make. Given the interest in the detail of the everyday lives of construction workers, the thought of being able to record them in action was instantly appealing.

Initially, I saw the main benefit of the cameras as the ability to fact check my memory and the notes taken in the field, or to provide a means of capturing construction site activity alongside, or instead of, the field notes. Across the two projects, I collected 50 hours of video footage. Just watching this through once is more than a week's work, so the data generation from the cameras is considerable. As it turns out, the real value was in the potential of the video to enable re-engagement with the construction site at a time and place far removed from the original fieldwork (Pink & Morgan, 2013).

The video footage was used in two ways. First, it was a record of what happened. The footage could be used to cross-reference the notebook to see if the description scribbled down on site, plus any contemporaneous analysis, was supported by the video. This performed the invaluable function of giving great confidence in the events presented in the ethnographic narratives: the video data is evidence that those events did occur. It did, however, pose an unanticipated problem. Just as the video footage gave confidence that a remembered event did take place, there were a few occasions where my memory and or the notebook have something that was not shown in the video footage. This may have been because the camera was not turned on, or it took place out of shot. The lack of footage then serves to undermines one's own memory and written record. This is not easily resolvable after the fact.

Experience, gained through more ethnographic fieldwork, is likely to increase the quality and confidence in the field notes so that the video footage and the notebook can be equals. Using the video cameras without the notebook would be feasible, but it has a flaw. Video cannot be created after the event – it only exists as a record of a snapshot of many moments, stitched together to make a digital record. There is a lot happening on construction sites and the cameras, with a narrower field of vision than a human, miss more than they capture. Should something happen out of shot, the researcher can capture it mentally and then record it in the notebook, just as traditional ethnographers have done. Over-reliance on the video camera could drastically limit the breadth of data collection.

The second way I used the video footage was to enable very close analysis of events. While out in the field, the typical arrangement was to have one camera in a fixed position and one in the researcher's hand for capturing specific moments up close. The site scaffold was the perfect place to attach a camera. I was able to watch a moment of interest as it happened, then record it in the field notes shortly afterwards and have it captured by one or two cameras. This provides multiple perspectives of one event for minimal expense of effort at the time. The effort comes later at the analysis stage. The video footage enables continual re-engagement with events. Even now, four years after the first fieldwork, the video footage is very evocative. The sounds of the construction site and the familiar faces recreate the construction site in the researcher's own workspace. It would still be possible to analyse the video footage today and make new findings. The ability to watch a video, and watch again or share with collaborators, is the source of the richness in the data that enables the production of thick description.

## A narrow focus

From before the fieldwork began, right through to the analysis and development of the ethnographic narratives, the research had a narrow focus. The aim was to concentrate on what the construction workers were doing as they carried out their jobs on site. Craft theory, used by way of sensitisation initially, provided the lens through which I viewed events on site. But the context of the site determines how the fieldwork plays out. The place of theory is continually revised and reconsidered, almost on an hourly basis, during the intensity of fieldwork. In ethnography, there is no separation between data collection and analysis. The conversations with workers, or events witnessed, constantly throw up new questions and new directions. It is typical for ethnographic studies to evolve over their lifetime, but the lifetime of short-term ethnography is, obviously, short. There is a danger the researcher might become distracted or diverted in a way that muddles the fieldwork, and from which there is no time to recover.

Any construction site process, practice or behaviour of interest to a researcher will be unstable. It will emerge, mature and then dissolve away. Long-term traditional ethnographic fieldwork could learn about these phenomena just by being there long enough and waiting for them to come and go. But short-term ethnographers cannot wait. They must be ready to seize opportunities as they arise. Using the video cameras helps retain the focus. In the field I used the cameras to record a specific event happening. The camera was held close to the worker to capture the detail of the act and the accompanying dialogue. Where I did use a camera to capture an overview of the site from a vantage point, the resulting footage lacked detail and was of no use. At the end of the fieldwork, I was left with a set of video files of discrete events shown in detail. Using these for the analysis obliged me to retain focus. For a short-term ethnographer, there may not be another chance to see a particular work activity again. But with the video cameras, a version of that event has been captured and retained.

While the fieldwork is brief, it is also the case that construction sites are ever-changing environments. The physical context changes with every hour of building work done. Almost as quickly, the people constituting the social organisation of the construction site also change. As the building work progresses, work gangs come and go (Sykes, 1969). Few construction workers get to see an entire project through from beginning to end. This makes construction sites interesting and lively places to be. There is always something happening to catch the eye. It is only remembering the research focus that will keep the fieldwork on track. On the inside cover of my field notebook, I wrote out the phenomenon I was there to explore. From time to time, I would turn to the inside cover and read what I had written. This was enough, amidst the hubbub of the site, to retain my focus on why I was there.

## Conclusion

My immediate reflection upon completing the fieldwork was about just how much the researcher invests into it. The construction site demands physicality from the researcher. Just as the workers are never static, so it is the same for the researcher. Unless one is observing the management team in the site accommodation, the researcher will never get to sit down, will always be on the move and will share all the environmental hardships of the workers. The researcher must always be on the lookout for those moments that are revealing, or the opportunity to ask a question before the moment is lost. Energy is expended talking with workers, and this kind of extroversion may not be in the nature of every researcher. But, in my experience, the effort is repaid with the depth of findings. The openness of the workers about their work, how they felt about it and what it meant to them was always revealing. Partly this may have been a natural outcome of showing great interest in the work of people who typically go about their work uncredited. But I would also like to think it was due to the effort put in to establishing rapport. I recommend anyone thinking of researching site workers, to consider how they will create productive relationships and to not be deterred by fears they may decline to participate.

Without the video cameras my fieldwork would have lost much of its impact. The time spent reviewing the video footage was necessarily long, but it was during this post-fieldwork phase that the majority of analytical insights were made. The researcher initially under-appreciated just how evocative the video footage would be. It does not just complement the observations made on site; it allows the generation of new findings. I would argue the video supercharges the fieldwork, enabling analysis and narrative to flow from very brief events that would not have been possible, at least in my case, if I had only used traditional field notes.

Finally, my purpose in undertaking the fieldwork remained constant throughout, and it was a simple one: I was looking for traits of craft. It is easy to be distracted by events. Construction sites are lively and interesting places, full of characters and stories. Trying to capture too much breadth in short-term ethnography is difficult. For me, short-term ethnography was a rapid-fire succession of event after event. Because of the detailed focus that is applied to the everyday lives of the research subjects, the researcher does not come away with a grand narrative of a construction site as a social organisation. When reporting my results, I introduce them as vignettes, to convey their brevity and evocative nature. Rather than producing ethnography that speaks of a community or a place, short-term ethnography has, for me, produced a highly detailed narrative of individuals navigating events within a social setting.

## References


Aboagye-Nimo, E., & Raiden, A. (2016). Introducing site sense: Comparing situated knowledge in construction to coalmining. In P. W. Chan & C. J. Neilson (Eds.), *Proceedings of the 32nd Annual ARCOM Conference, ARCOM 2016* (pp. 467–476), 5–7 September 2016, Association of Researchers in Construction Management, Manchester, UK.

Baarts, C. (2009). Collective individualism: The informal and emergent dynamics of practising safety in a high-risk work environment. *Construction Management and Economics*, 27(10), 949–957. https://doi.org/10.1080/01446190903147501

Brett, R, Thomson, D., & Dainty, A. (2019). Coping with stone: A short term ethnography of skilled work in construction. In C. Gorse & C. J. Neilson (Eds.), *Working Papers of the 35th Annual ARCOM Conference* (pp. 12–21), 2–4 September 2019, Association of Researchers in Construction Management, Leeds, UK.

Brett, R., Thomson, D., & Dainty, A. (2022). Exploring craft in construction with short-term ethnography: Reflections on a researcher's prior insight. *Construction Management and Economics*, 1–15. https://doi.org/10.1080/01446193.2022.2046827

Brewer, J. (2000). *Ethnography*. Milton Keynes: McGraw-Hill.

Clark, T., Foster, L., Sloan, L., & Bryman, A. (2021). *Bryman's Social Research Methods* (6th ed.). Oxford: Oxford University Press.

Fetterman, D. M. (1989). *Ethnography: Step by Step*. Newbury Park: Sage.

Hammersley, M., & Atkinson, P. (2007). *Ethnography: Principles in Practice*. Abingdon: Routledge. https://doi.org/10.2307/2070079

Jarrett, M., & Liu, F. (2018). "Zooming with": A participatory approach to the use of video ethnography in organizational studies. *Organizational Research Methods*, 21(2), 366–385. https://doi.org/10.1177/1094428116656238

LeCompte, M. D. (2010). *Designing & Conducting Ethnographic Research an Introduction. Designing & Conducting Ethnographic Research an Introduction* (2nd ed.). Lanham: AltaMira Press.

Löwstedt, M. (2015). 'Taking off my glasses in order to see': Exploring practice on a building site using self-reflexive ethnography. *Construction Management and Economics*, 33(5–6), 404–414. https://doi.org/10.1080/01446193.2015.1066021

Pink, S., & Morgan, J. (2013). Short- term ethnography: Intense routes to knowing. *Symbolic Interaction*, 36(3), 351–361. https://doi.org/10.1002/SYMB.66

Sykes, A. J. M. (1969). Navvies: Their work attitudes. *Sociology*, 3(1), 21–35. https://doi.org/10.1177/003803856900300102

Thiel, D. (2012). Builders, bodies and bifurcations: How London construction workers "learn to labour." *Ethnography*, 14(4), 412–430. https://doi.org/10.1177/1466138112463656

Vannini, P., & Vannini, A. S. (2019). Artisanal ethnography: Notes on the making of ethnographic craft. *Qualitative Inquiry*. https://doi.org/10.1177/1077800419863456

# 11

# DOING LONG-TERM ETHNOGRAPHY

*David Oswald, Fred Sherratt, and Simon Smith*

*RMIT University, Melbourne, Australia*

*University of Colorado, Boulder, CO, USA*

*University of Edinburgh, Edinburgh, United Kingdom*

**Key takeaways**

- Long-term ethnographic studies in construction have challenges associated with participants not fully understanding your research and why it takes significant time (e.g. years) to complete your work.
- Finding a role and fitting in to the studied context are essential for overcoming these challenges, establishing rapport and gathering quality ethnographic data.
- The nuances of how to fit in can be learnt through constant reflection of how the researcher is viewed within the context by the participants.

## Introduction

'Classic' forms of ethnography involve spending extended periods of time with the people being researched (Pink et al., 2010). These long periods of ethnographic fieldwork have been described as stressful by a number of researchers (e.g. Thiel, 2013). Pink et al. (2013, p. 7) acknowledge it involves 'anxiety, alienation and frustration as well as feelings of solidarity and understanding'. Long-term ethnographic approaches typically involve observation, participation in activities, writing extensive fieldnotes, interviewing and self-reflexivity for many months or even years. Since these approaches do involve significant time commitments, there has been a growing number of researchers opting for different forms of ethnographic approaches that are shorter-term (see Oswald & Dainty, 2020). However, long-term ethnographic work does provide the opportunity to gather significant quantities of rich, fine-grained and in-depth data. For instance, Phelps (2008, p. 58) participated in field study over a three-year period involving:

- '1000+ hours on site observing behaviors and interacting with the project team
- Attendance at 150+ key meetings
- Extensive notes regarding 65+ meetings
- Semi-structured interviews with 12 key project team members
- Numerous informal interviews (i.e. casual conversations) with project members
- Collection and review of 3000+ pages of meeting notes, plans, and other written artifacts'.

DOI: 10.1201/9781003379584-14

This benefit alone means long-term ethnographic approaches should be considered when time commitments can be navigated within the research project design.

This chapter reflects on a three-year research study to highlight key considerations for the success of a long-term ethnographic project. The research study (a PhD project) aimed to explore unsafe acts on a large multinational construction project in the United Kingdom (UK), on which four different principal contractors were working together in a joint-venture consortium. Adopting an interpretivist paradigm, this study gathered qualitative data through ethnographic methods. A moderate participant observer approach was implemented, where the researcher adopted a role as a PhD 'student' member of the health and safety (H&S) department and frequented the research setting approximately twice a week for almost three years. This role included attending site walk-arounds and meetings, where it was possible to observe and query H&S activities; undertaking on site ad-hoc discussions with workers; and being present at accident and incident responses.

The focus of this chapter is on two key practical considerations that contribute to the success of a long-term ethnographic study. These include 'finding a role' and 'fitting in' to the research setting. It is essential to find a role within the research setting that feels natural and is socially accepted by those being researched. For example, a researcher adopting the non-participatory role of standing observing construction workers while smartly dressed, with sunglasses on, and a clipboard out, will only raise suspicions amongst workers that they are being scrutinised in some way, especially if no prior rapport has been established. When the researcher's role feels unnatural there is a risk that the observations being undertaken are not going to be representative of the natural setting. This clearly could lead to poor quality fieldnotes and less desirable results.

In conjunction with the focus on the researcher's role, there are reflections within this chapter on 'fitting in' with the research setting. The world of construction and the world of academia are very different. Academics may concentrate on words such as ontology, epistemology, methodology, reliability and validity; while on the UK construction site (in this case), words such as boom, jib, snotter, big red, rubber duck and the more colourful expletives were used. While these worlds are very different, an immersed long-term ethnographic researcher in the construction industry will have to negotiate the various difficult and daunting challenges that can emerge when trying to 'fit in' to the field.

This chapter seeks to discuss such challenges faced with undertaking a long-term ethnographic project within a construction research setting. This will provide insights for future researchers that are considering adopting an ethnographic approach in similar industries or contexts.

### *Finding your role, and fitting in, for the long-term*

The following sections discuss two key considerations for doing long-term ethnographies including finding your role and fitting in. These practical considerations are of real significance for long-term projects, as unsuccessful integration into the research setting could result in extensive periods observing with only limited and poor-quality data being captured. In shorter more intensive forms of ethnography, the researcher may have to be more intrusive as there is less time to collect data. There may also be a narrower, more focused research question that is needed to be answered quickly. While short-term ethnographers may have a 'small window of time available' (see the previous chapter) to gather data and connect with participants, longer-term ethnographers have the luxury of more time to establish rapport, explore the problem and refine their research approach. While there is perhaps greater opportunity for long-term researchers, time still needs to be used effectively to capture quality data. To do this, the researcher needs to 'find a role' and 'fit in' to the research setting – both of which are discussed in first person below.

The construction site setting is male dominated, culturally, as well as numerically (Fielden et al., 2010). This was the case on the project in focus, and as a male I was amongst the vast majority (in terms of gender) on the project. At the time, I was in my mid-20s, and I found myself adopting a PhD 'student' role. Considering I also spent most of the working week at the University this was an identity that appeared acceptable to participants within the research setting. The only other PhD student on the project was in his late 30s, was a full-time employee, and had adopted the identity of a 'researcher' when engaging with participants, highlighting that employment status, maturity and age, was an important consideration in seeking out an appropriate and acceptable identity. However, adopting this student role brought with it many presumptions from the construction workforce, and also singled me out as 'different', as I did not have a trade or construction profession like everyone else on the site.

As an ethnographic researcher you are different, which can also mean that you become an 'easy target' for the punchline of jokes. What could be considered as offensive, or even bullying in some contexts, can simply be accepted, and even expected within the industry. For example, being able to swear in front of you, or at you, make jokes about you, or just saying anything they want in front of you, is often a sign of being accepted by others on the site: the workers feel they are able to 'be themselves' with you, in what is a very masculine and macho industry (see, for instance, Lacuone, 2005). As a male, I was expected to be able to cope with these jokes and fit into the macho culture, which Lacoune (2005) categorised by horseplay and aggressiveness. For women, these macho attitudes have also previously been highlighted as an obstacle that women must overcome (Gale, 1994). The women on site were treated differently: swearing in front of them would often be followed by the common site-phrase 'excuse my French', as the perpetrator acknowledged their words as inappropriate for those within hearing; and comments about 'girl-watching' and jokes made about women were avoided within their presence. Conway et al. (2018) revealed that related objectification of women needs to be addressed and altered in order to aid the industry in encouraging, welcoming and retaining more women. Previous female ethnographic researchers have reported the need to be able to 'cope with the crude and at times sexist jargon' (Baarts, 2009, p. 951), and although my experience inevitably differed from theirs, I was socially expected to accept such comments, rather than challenge them as inappropriate.

I also had to cope with jokes around my student status, which drew on many presumptions and stereotypes from amongst the participants. Often jokes were made about the arrival of my next student loan, drinking beer and tough early mornings. One construction worker asked: 'The other day my laddy [his son and college student] came back from a night out, was sick all over the kitchen and then fell asleep in the dog kennel … are you like that?' The moderate participant observation approach adopted also lent itself to light-hearted comments about being a 'part-timer' as I was typically only at the research setting twice a week. I would be asked questions like: 'presume the student union must be closed today if you are in?' Shipton (2013) found it 'a bit annoying' after a while but 'played along with the lazy student jokes to keep things light-hearted'. She also got the 'impression that people were not even listening when I talked about my studies', suggesting that they weren't very interested or didn't understand. A PhD isn't a likely path for any of the construction workers to have taken, as many left school as soon as they could, and most did not go onto university. The challenges involved doing a PhD were not readily appreciated, or even appreciated at all on the construction site – this was a stark contrast to the university setting.

It can be challenging to accept such comments around 'laziness', 'being a part-timer' or 'get a real job' considering I was working very hard on achieving my PhD; but I felt I had to shrug this off as banter. Participant relationships are paramount for the research success, and so on balance I felt continued relationships were more important than my feelings of self-validation. Indeed, the jokes I was subject to were typically light-hearted, and certainly not what I considered bullying; nevertheless, staying mentally resilient was still very important in order to help both achieve and develop continuing relationships.

The relationships with the H&S team provided a natural way to access all the different areas on the construction project, as each H&S advisor was assigned a particular location across the site. Thus, rather than just randomly showing up without a clear purpose, I could travel with the H&S advisors to their areas, and attend the meetings, incident responses or walkarounds they were carrying out as part of their daily work activities. This was crucial in my acceptance within the different social settings across the large project. Finding activities that suited my identity as a student member of the H&S team, such as shadowing a H&S advisor on a safety walk, were therefore expected, non-threatening, natural behaviours for the social setting. The H&S advisors, acting as gatekeepers, helped ease the passage of entry to the field by introducing me to other people on the project, such as foremen, site supervisors and workers, essentially developing a snowballing-sampling strategy. Ad-hoc discussions with these different informants across the project made the surroundings and contexts more visible and understandable. While these short excursions into the lives of the participants were often intense and very valuable, this close researcher-researched interaction also created challenges that are discussed in further detail below.

Indeed, early in the project I found that one of the main challenges was realising how I could collect high-quality, fine-grained ethnographic data, where I could get an in-depth understanding of what was really going on. When doing ethnography for a long-term basis there is potential to gather vast amounts of data, so it is worth reflecting on what is considered 'action' and what is not (see Goffman, 2006). For example, within construction safety research is the ethnographic 'action' capturing safe, eventless, normal days of work to understand how safety is achieved, or is the action the rare significant unsafe events that cause harm? The ethnographic research will not be able to observe, capture and write fieldnotes on all types of action over a long period of time, so this is a key consideration for those doing long-term ethnographic research. Establishing rapport and creating strong relationships was an essential foundation to capture such data (see Oswald et al., 2014).

Considering an ethnographic researcher's main tool is typically participant observation, it was natural that I would ask questions, share my views, politely point out acts or conditions that I thought could be unsafe for discussion. If there was a condition that appeared dangerous, I stopped short of intervening with participants on safety issues, instead informing H&S advisors so they could intervene. This kept in line with both my role, as a student, and the H&S advisor's role, for whom it was natural to intervene on H&S issues as part of their profession. It also avoided the risk of ruining relationships with participants, who may not have appreciated being told what they were doing was unsafe by a young student who had no clear authority within the site environment.

Ethnographic researchers are often advised to 'find a role' for themselves in the research context. These roles can range from non-participation to complete participation (see Spradley, 1980) or peripheral membership to full membership (see Adler & Adler, 1987). As discussed above, my role involved both observation in the research setting and participation in some H&S activities; but not all, as I would, for example, usually not participate in H&S interventions. My role was what Spradley (1980) would term 'moderate participation'.

The ethnographic data was primarily collected through both informal discussions and observations in the research setting. Following ad-hoc and dynamic discussions with participants, initial fieldnotes were taken, which were typically typed onto my mobile phone at appropriate moments. Hammersley and Atkinson (2007) explained a common joke about ethnographers relates to frequent toilet trips to hastily write up notes in private. Taking notes with pen and paper would be more obtrusive in the natural setting and could generate distraction or distrust amongst participants (Hammersley & Atkinson, 2007), as they would act as a reminder to people that they are being studied (O'Reilly, 2012). The natural setting dictates how these initial notes would be taken. While typing into a mobile phone is natural on the construction site, doing this in a meeting could be perceived as rude, so notes were taken on the meeting minutes, again in keeping with the social norm for this context. In the office setting, notes would be directly typed into my laptop, also in keeping with familiar behaviours and norms of that environment.

While these approaches helped with collecting data in a natural way, there were still many challenging intricacies around 'fitting in'. Ultimately, as an ethnographic researcher, you are different. Everyone else in the setting typically already has a clear purpose, role or responsibility. Indeed, ethnographic researchers can simply be advised to just 'rock up' and see what happens. This was an approach that I took, but one that initially wasn't particularly successful. While the majority of informants were very welcoming, some did struggle to understand the purpose for me being there. They would ask 'what is it you want to achieve?', 'what are you trying to get out of this?' and 'is there any way we could help?' As ethnographers often turn up and see what happens, having objectives and clear answer to these questions was difficult, and I didn't exactly know what I was looking for. Indeed, within long-term ethnographic approaches, there is an opportunity to be less direct, and less structured, in the research design with broader, more open-ended research questions, to allow for themes to slowly emerge throughout the study period. For example, in Thiel's (2012, p. 3) long-term ethnographic study of builders on construction site, he stated: 'I had held a vague notion that I would investigate gender and informal economic activity but, otherwise, I would simply work on and hang around a building site and wait and see "what came up"'. This is not always the research design within ethnography, as there can be more intense and targeted ethnographic approaches, where there may be a focus to answer more specific research questions within a shorter time-frame.

Within the long-term project discussed in this chapter, there was a vague notion of investigating unsafe behaviours in construction. I was not exactly sure what to observe or who to speak to but, like Thiel (2012), I would see what emerged and go from there. However, this brought about one of the first challenges of doing a long-term ethnographic project around 'fitting in'. The following fieldnote is from the first month of my long-term (PhD) study. It provides an indication of the types of challenges ethnographic researchers can be confronted with when having a 'vague notion' of study.

> Establishing rapport with the health and safety department was difficult to begin with, mainly due to the high levels of staff turnover and inter-department conflicts. I knew one member didn't really want a PhD student in the team, and he was very confrontational and autocratic. On bad days he would ask me 'what the f**k are you doing' and made comments such as

> 'all you seem to do is just chat and observe'. It appeared to me that he seemed to think I offered nothing to the team and that when I was at 'work' I seemed to just watch and talk; a key part of my ethnographic research but perhaps a misconception of being lazy or unfocused when within the work environment. I found this initial period quite stressful, especially as it came right at the start of my research and I was still 'finding my feet'. I wanted to give clear answers to his questions about my purpose; but as is often the case with PhD research investigations, and especially bottom-up ethnographic studies, this does not often become clear until later into the project… However, some of those around me appeared to want clear and concrete answers, which I found difficult to give at the beginning of the research. I was also just beginning to learn and develop understanding of ethnography as a research methodology. I was questioning, whether I 'was doing it right?', 'what data should I collect?' and 'were these challenges normal?' I decided to take some time away from the research setting to try to answer some of these methodological questions myself, and to develop a better understanding of my research area in the hope that I could feel, to some extent, worthwhile to the business.

My role involved activities including observing and communicating with participants, so I was able to understand the social context that the participants were in. There were days I had been very productive in achieving these activities, yet to others in the environment, these activities could be misconstrued as doing nothing (gathering observation data) and chatting (gathering data by communicating with participants). In an industry that is under significant pressures for production (Oswald et al., 2019), and in which individuals are usually always very busy, the perception of me doing 'nothing' and 'chatting' was a significant risk to fitting in. For example, the comments made included: 'all you do is just chat and observe', 'why don't you pick up a shovel?' or more explicitly and directly 'what the f**k are you doing here?' Such questions came early in my research and made me realise the importance of finding a clear role that is understood by others in the setting. I didn't want to be seen as the lazy student that offered no value – so I needed to adapt my role.

Following these comments, and self-reflection, I would make sure I would spend a significant amount of time at the research setting undertaking 'computer work' on my laptop when in the office, as well as just 'safety walk work' on site. This felt more natural than only doing participant observation work on the site, as H&S advisors would move between site and office frequently, and this way, I could do something similar. Hence, rather than always writing up my fieldnotes out of the research setting, which can be typical for other ethnographers, I began writing them in the setting which gave me a recognised task; albeit not one directly contributing to the construction project itself, but one that was commonplace within the work environment and therefore acceptable to others. This altered the data collection process, as rather than my observations being solely on the front-line of the construction site, I was now also recording observations in the H&S office setting, of H&S activities, actions and incidents, all while I was typing fieldnotes into my laptop either from an earlier site-visit or from an observation in the office. Both settings ultimately provided valuable insights into my research, enabling data to be collected from both the site and H&S office contexts. This approach helped me to find a role that was not out of place within the setting and significantly reduced the perceptions of 'laziness' that were all too easily associated with the initial approaches made to participant observation.

I also removed the flexibility I started with, to avoid the perception that I could just turn up and leave whenever I wanted, and had similar commitments of time to others working on the project. Thus, I kept a consistent week structure, where I would dedicate days I would be at the research setting, typically Tuesdays and Thursdays (8 am to 5 pm) every week. They expected to see me, and I expected to be there. I gave this structure priority in my week and worked my other responsibilities around these days. I would re-schedule meetings at university that were on Tuesdays and Thursdays, avoid teaching opportunities on these days, and other research commitments where possible. I also helped the department with tasks related to the safety climate surveys, accident frequency rates or other documentation that needed attention as this furthered my development; could be interesting for my research; made me feel like I was contributing; and also improved relations within the H&S team, as these were often unwelcomed administrative tasks that the H&S advisors did not themselves enjoy undertaking.

The start of my ethnographic fieldwork was no doubt the most challenging and this is why many ethnographers often delay the start of the fieldwork, due to a lack of confidence or as a consequence of apprehension (Murchison, 2010). It did help to be a member of the H&S department, which meant I had a common interest with my closest informants (as my PhD was within the field of H&S). However, integration into the team was initially challenging, perhaps because the majority of the department did not come from a university background, most of the advisors had either construction trade experience, known as coming 'off the tools', or came from the military. I listened to them and their experiences, providing insights when I was asked for my input; but mostly adopted an 'apprentice-like' role asking questions on their perceptions, which was beneficial for unpacking their views of the social context, and for establishing rapport with my closest informants and gatekeepers. The vast majority of the team were from the UK, and so, culturally, I fit in, although there were times I was identified as an outsider. As I was studying a PhD in unsafe behaviours, occasionally they would mock behavioural theories, with statements such as 'It is all behaviours, isn't it Dave? Just change it with a big group hug and behavioural-based safety, then we will all be safe'. The academic and construction worlds were so different, I knew there would be some derisory perspectives on PhDs, research and theory, but I would just shrug this off. However, once rapport with the H&S team became more established potential informants began assisting and engaging with the research project more regularly. In my case, the H&S advisors could provide opportunities for observing events on interest. For example, a H&S advisor may ask me: 'I have just had a call about a safety issue on the road networks, do you want to come along?'.

The levels of rapport within these established relationships did not however stand still; further successful engagements were found to increase these levels, while non-engagement resulted in these levels dissipating with time. The levels of rapport established were difficult to quantify, but there are several indicators that can be reflected upon. For example, if participants:

- have prolonged conversations with you;
- go 'out their way' from their normal day tasks to help you with your research;
- invite you to meetings and events that they think might be of interest to you;
- let you know of incidents that have occurred while you were not at the research setting, or somewhere else on the research setting;
- allow you access to (or actively send you) photos, emails or project documentation of interest to your research; and
- make you feel like an insider by inviting you to company events (e.g. the Christmas party) or celebrate a birthday in a similar way to others in the office (e.g. lunchtime pizza or cake).

Essentially, the quality of the data collected is highly dependent on the relationships that have been developed within the community under study. When strong relationships are developed, participants are more likely to open up and explain detailed accounts of the goings-on, as they trust the researcher. I found it more important to keep participants 'onside', even if the researcher is then subjected to jokes from time to time. Having such mental resilience and being able to rationalise this throughout the fieldwork experience was therefore key for success.

While posing such challenges, it is worth noting there are many benefits to undertaking long-term ethnographic research. Indeed, proponents of ethnography argue that ethnography has several benefits associated with generating a rich understanding of social action and subtleties within different contexts (Reeves et al., 2008). Benefits can also be in forms of validity, as they are conducted in natural settings over long periods of time, which reflect the reality of life experiences of participants more accurately than contrived settings (LeCompte & Goetz, 1982). They can come in the form of data quality as, for example, they can provide insights that go beyond the participants' impression management, which can be problematic with faster methods such as interviewing (Delamont, 2004). Furthermore, if ethnographic approaches also include interviews, the quality of the interview can benefit, as the interviewer has spent vast amounts of time in the setting, so can clearly visualise interviewee answers; can ask more relevant questions based on their observations as an insider; and has had the opportunity to build trust with the interviewees so they may feel more comfortable to open up to the researcher. By building relationships with key gatekeepers, there is no need to 'cold call' for data collection, with the gatekeepers instead acting as facilitators for the researchers needs.

## Conclusions

The challenges of 'fitting in' and 'finding a role' have been discussed within the setting of a UK construction site to help prepare future researchers and supervisors for long-term ethnographic research.

Mental resilience is specifically relevant to the macho and masculine construction industry, where being made fun of can be a sign of acceptance. Having the awareness that a PhD isn't valued to the same extent in the construction site context is important. The enormous effort and energy involved in a PhD won't necessarily be understood or even appreciated by those you are spending considerable time with. Resilience had two forms: an ability to 'shrug off' the jokes on site and being able to continue to work long and hard hours on the PhD despite the seeming lack of interest in and even respect for the PhD life choice.

Ethnographic texts recommend finding a role, from non-participation to complete participation. While this can help the researcher to identify their scope, it merely acts as a foundation for the long-term study. Establishing the finer details to help achieve being socially accepted in a defined role requires self-reflection and adjustment during the early stages of the ethnographic study. In this study, the role initially adopted created the perception of a lazy part-timer, who attended sporadically and just watched and chatted. This was poorly received by those in what was a very busy and stressful work environment.

Instead, there was greater success after setting out a clear weekly structure with hours that aligned with the participants, consciously spending time doing similar workplace activities as the others to fit in; conducting more subtle observations, rather than scheduled observations; and undertaking the informal conversations, which are paramount to ethnographic research, to help build rapport and gather data (that improved in quality as relationships were built). While it may not take long to identify a role for the researcher, refining the nuances of the role to 'fit in' with the context is a key consideration for success of ethnographic research projects.

## References

Adler, P. A., & Adler, P. (1987). The past and the future of ethnography. *Journal of Contemporary Ethnography*, 16(1), 4–24.

Baarts, C. (2009). Collective individualism: The informal and emergent dynamics of practising safety in a high-risk work environment. *Construction Management and Economics*, 27(10), 949–957.

Conway, Z., Wade, F., & Smith, S. D. (2018). A participant observation study of gender dynamics on construction sites. In C. Gorse & C. J. Neilson (Eds.) *Proceeding of the 34th Annual ARCOM Conference* (pp. 291–300), 3–5 September 2018, Belfast, UK, Association of Researchers in Construction Management.

Delamont, S. (2004). Qualitative research practice. In C. Seale, G. Gobo, J. F. Gubrium, & D. Silverman (Eds.), *Ethnography and Participant Observation* (pp. 217–229). London: Sage.

Fielden, S., Davidson, M., Gale, A., & Davey, C. (2010). Women in construction: The untapped resource. *Construction Management & Economics*, 18(1), 113–121.

Gale, A. W. (1994). Women in non-traditional occupations: The construction industry. *Women in Management Review*, 9, 3–14.

Goffman, E. (2006). Where the action is. In *The Sociology of Risk and Gambling Reader* (pp. 222–251). Abingdon: Routledge.

Hammersley, M., & Atkinson, P. (2007). *Ethnography: Principles in Practice*, Third Edition. Abingdon: Routledge

Lacoune, D. (2005). "Real men are tough guys": Hegemonic masculinity and safety in the construction industry. *The Journal of Mens Studies*, 13(2), 247–266.

LeCompte, M. D., & Goetz, J. (1982). Problems of reliability and validity in ethnographic research. *Review of Educational Research*, 52(1), 31–60.

Murchison, J. (2010). *Ethnography Essentials: Designing, Conducting, and Presenting Your Research*. Hoboken: John Wiley & Sons.

O'Reilly, K. (2012). *Ethnographic Methods*. Oxon: Routledge

Oswald, D., & Dainty, A. (2020). Ethnographic research in the construction industry: A critical review. *Journal of Construction Engineering & Management*, 146(10), 03120003.

Oswald, D., Sherratt, F., & Smith, S. (2014). Handling The Hawthorne Effect: The challenges surrounding a participant observer. *Review of Social Studies*, 1(1), 53–73.

Oswald, D., Sherratt, F., & Smith, S. (2019). Managing production pressures through dangerous informality: A case study. *Engineering, Construction and Architectural Management,* 26(11), 2581–2596.

Phelps, A. F. (2008). Lean information flow in complex healthcare projects and the role of the "integrator" as information flow manager. PhD Thesis. The Pennsylvania State University.

Pink, S., Dainty, A., & Tutt, D. (2013). *Ethnographic Research in the Construction Industry*. Oxon: Routledge.

Pink, S., Tutt, D., Dainty, A., & Gibb, A. (2010). Ethnographic methodologies for construction research: Knowing, practice and interventions. *Building Research & Information*, 38(6), 647–659.

Pink, S., Tutt, D., Dainty, A., & Gibb, A. (2013). *Ethnographic Research in the Construction Industry*. Oxon: Routledge.

Reeves, S., Kuper, A., & Hodges, B. (2008). Qualitative research methodologies: ethnography. *BMJ*, 337, 512–514.

Shipton, C. (2013). *Making changes: an ethnography of client changes on a construction project*. PhD Thesis. University of Reading.

Spradley, J. P. (1980). *Participant Observation*. New York: Holt, Rinehart and Winston.

Thiel, D. (2012). Builders, bodies and bifurcations: How London construction workers "learn to labour". *Ethnography*, 14(4), 412–430.

Thiel, D. (2013). Ethnography and flux: Identity and epistemology in construction fieldwork. In S. Pink, D. Tutt, & A. Dainty (Eds.), *Ethnographic Research in the Construction Industry* (pp. 73–90). Abingdon: Routledge.

# 12

# DOING MULTI-SITE ETHNOGRAPHY

*Andrea Yunyan Jia*

*Hong Kong Polytechnic University, Hong Kong, China*

**Key takeaways**

- Almost all ethnographic studies are multi-sited by nature.
- The multi-sites can be nodes and interfaces of a project network, different levels in a hierarchy, different channels of communication, different time points of a project or distinctive discourses and subcultures in an organisation.
- *Empathy* is the anchoring logic of ethnography. The ethnographer's empathetic self is the instrument to draw out data from the field.
- The key challenge to multi-site fieldwork is to enact empathy through sensible *positioning* of self in diverse and shifting situations.
- The enactment of empathy is a Discourse work. It's a coordination between the ethnographer's natural identity, the situational elements and the performing practice of the ethnographer.
- The principle of asking questions in the field is to constantly watch out how far you are legitimated to go by the community of the field. Question follows legitimacy.
- Data collected from one of the communication channels must be interpreted as part of an integral whole. Interpretation by data from a single channel could lead to seriously wrong conclusions.

## Introduction

> Counsel in the heart of man is like deep water,
> but a man of understanding will draw it out.
>
> *(Proverbs 20:5)*

What is the very essential feature that distinguishes ethnography from non-ethnography? Atkinson et al. (2014: 8) summarise the commonality of different traditions of ethnography as 'grounded in a commitment to the first-hand experience and exploration of a particular social or cultural setting on the basis of participant observation'. Falzon (2009: 1) describes ethnography 'privileges an engaged, contextually rich and nuanced' kind of research with data of 'fine grained daily interactions'. O'Reilly (2012) suggests the essence of ethnography is to draw out people's inner motives by being 'here and now' with the participants in their daily practice.

DOI: 10.1201/9781003379584-15

However, there are ethnographies without ethnographer's personal participation in the field. Diane Vaughan (1996), for example, wrote an ethnography on NASA's organisational culture based on inquiry transcripts on the Space Shuttle Challenger disaster; and she defined this approach as 'historical ethnography' (Vaughan, 1996: 61). Similarly, Prof. Andrew Hopkins published many ethnographies on major accidents as outcomes of organisational cultures, drawing data from records of accident inquiries. He calls his approach 'desk ethnography' and defines ethnography as 'immerse myself in the world of the organisation, without ever setting foot on its premise' (Hopkins, 2016: 94). These exceptions suggest that 'being physically in the field' is not necessarily the essential feature of ethnography (and therefore doing multi-site ethnography is not about being physically in multiple sites). One can be present in the field to do a non-ethnographic study, or work out an ethnography from second-hand data. There is something deeper that features ethnography and therefore the skills needed in doing multi-site ethnography.

A definition of ethnography in Alvehus and Crevani (2022: 237) is 'understanding the meaning system of the Other'. The word 'Other' with a capital O is a philosophical term that signifies the existence of beings that are fundamentally different from the world of 'self' assumed after one's own likeness (Honderich, 2005). This definition is a reminder that 'us', the readers and the ethnographers, are not standing on a value-free universal ground, but are embedded in 'another' meaning system which is different from, yet of equal weight with, that of the field we are trying to research. In other words, ethnography is not a scientific study of a 'village of the savages' by 'the civilised society', nor the investigation of a set of objects and symbols by 'the humanity'. It is the negotiation and communication between two meaning systems of equal intelligence and sophistication. As such, the essential feature of ethnography as both a product and a process is *empathy*, the ability to resonate the Other's feeling and rationality enacted in material practice (Oliveira-Silva & Gonçalves, 2011). On this basis, we may define ethnography as *the establishment of a discourse between two meaning systems: one of the audience, the other of the field; in-between is the agency of the ethnographer who perceive, comprehend and interpret the field, and communicate it to the audience.*

Building on this ground, multi-site ethnography is one that recognises the 'field-end' of the discourse is an arena of multi-realities that co-exist in one or more physical spaces. These can be physically dispersed places in a space, or different temporal locations in a process, or subcultures in an institutional hierarchy, or nodes and links in a social network, or different channels of communication, or multiple discourses in the same organisation. In this sense, almost all ethnographic studies are multi-sited by nature. Seeing the existence of plural realities is the pre-condition for new dialogues that produce new insights and knowledge. Traditionally, multi-site ethnography is conducted by a single ethnographer over a prolonged period of time. This requires the ethnographer to be highly sensitive and skilful to resonate the diverse rhythms of the multiple sites. In recent years, multi-site ethnography is increasingly conducted by multiple ethnographers. This takes systematic planning of research teamwork, a level of standardisation in the research process, and exercise of project management and data management. This chapter addresses these features of multi-site ethnography by providing a practical guide on research strategies and principles for multi-site fieldwork in construction-related fields. It also provides the vocabulary and frames for construction researchers to communicate their ethnographic research in publications.

The rest of the chapter is organised in this way: firstly, we introduce the key concept, empathy, as a spiritual anchor to set a ground for the rest of the themes that follow. Next, we introduce strategies for research design, logics for sampling the sites, and management of ethnographic teamwork and multi-site data. The next section introduces more specific strategies of doing multi-site fieldwork. Then another section is devoted to introducing the art of 'positioning', the ethnographer's key activity across diverse sites. The chapter concludes with recommendations and implications.

## Empathy: The anchoring logic of multi-site ethnography

*Empathy*, the spirit of resonating and sharing the feeling and rationality of the Other, is the essence of ethnography as both a product and a process. In fieldwork, ethnographic data is not 'collected', but 'co-produced' through the thick interaction between the ethnographer and the field; the 'person' of the ethnographer is the major instrument for data collection (Goffman, 1989; Falzon, 2009). Empathy is the key attribute of an ethnographer that helps draw out valid data from the field. It involves the ability to put aside one's own emotion, rationality and interest to embed oneself in those of the observed (Oliveira-Silva & Gonçalves, 2011). Empathy is more than a feeling, an attitude or a mind status; it has to be embodied in language and practice. In multi-site fieldwork, empathy is the anchoring logic that underlies all research activities across diverse contexts, starting from a respect for and recognition of the co-existence of multi-realities that different communities are embedded in and living by. It is enacted by a broader resonance to all the sites which are often in conflicting interests and views with each other, through sensible positioning of self in diverse situations. Throughout the research process, empathy should be embodied in all the situational activities of an ethnographer across sites, the data analysis process and the writing of the final narrative or theory.

Sensitised to plural realities in a field, the ethnographer can enact empathy by either resonating or complementing people in their respective cultures. In studies of highly professionalised context – such as a design project team of architects and engineers, technical expertise is an integral part of the culture. The ethnographer needs to have a level of compatible expertise to be able to resonate the field culture of the professional teams. Such a compatibility in professional expertise constitutes part of the ethnographer's capacity of empathy. One of such examples is Abdirad's (2022) autoethnography in an engineering firm that produced rich technical details and deep insights on how project routine was disrupted by and realigned with the use of Cloud-BIM technologies. In the study, his own professional expertise played a key role in legitimating him as one of the professional community in the data co-production process. In cases that the ethnographer does not have the compatible expertise to resonate the field, they can assume a complementary role, i.e., position themselves as a student of the professionals, or stay passive (Hartmann et al., 2009). To assume a complementary position is another type of empathy (see an in-depth discussion on empathy and ethics in Goldman, 1993). Before applying any research strategies or fieldwork techniques, ethnographers must sincerely and honestly position themselves as an equal or a complement (never a superior) to the field people, give full respect to the culture of the community and enact an 'empathetic self' to enable a data co-production process.

## Strategies for research planning and design

Research practitioners find that ethnography as a methodology is mix-method and iterative-inductive in nature (Pink et al., 2010) and needs to hybrid frameworks and concepts across disciplines to address the research question (Cury et al., 2019). As such, ethnographic research is essentially data driven, which generates new theories.

In fieldwork, a common challenge for multi-site fieldwork is the ethnographer cannot be everywhere at the same time (Marcus, 1995). For time-constrained research projects, this dilemma has to be solved by multiplying the number of ethnographers or shortening the time of immersion in each site, or a combination of both. Such research projects will need to plan and manage ethnographic teamwork.

This section will first introduce a conceptual lens to help researchers see and define the 'sites' in their research design, and then introduces strategies for site sampling. This is followed by strategies for planning ethnographic team work; and managing the ethnographic data from multi-sites.

### *Where are the sites?*

The first questions that one may ask in designing a multi-site ethnography are: Where are the sites? And, what are the sites? Although not explicitly acknowledged, the traditional ethnographic fieldwork 'already crosses many potentially related sites of work' (Marcus, 1995: 100). The ethnographic study is multi-sited when the focal phenomenon, people, narrative or structure is mobile or spatially dispersed, or involves connections or conflicts that span over domains (Falzon, 2009). Here the 'domains' can be multiple physical spaces that are located at distinctive geographical places or temporal locations; they can also be defined by invisible boundaries that differentiate between cultures, norms, communication channels or discourses. Gille and Ó Riain (2002) suggest multi-site ethnography is particularly suitable for the study of globalisation to understand how the global processes play out in multiple contexts which extend in time and space. In the construction field, the study of migrant workers is potentially such a global process which takes globally multi-site ethnography to understand, extending beyond the construction site of their immediate workplace (e.g., Tutt et al., 2013; Hamid & Tutt, 2019). On place-based research, the multi-site can also be identified through analysis of project documents, that is, analysing project archives based on where the project activities happened. Other than physical spaces, the multi-sites can also be social spaces that exist in the same physical space, for example, different groups of subcultures in one organisation (Marcus, 1995; Hannerz, 2003). In the study of strategic issues of projects, the sites can be multiple organisations, their networks and interfaces. Below we discuss four types of sites that are not often in the awareness of ethnographers in the construction field.

#### *Sites defined by organisational hierarchy*

The multiple sites can be identified by the hierarchical levels in a firm or project organisation. Each of the level is a distinctive site that has its own boundary, rhythm, language and sub-culture. When reporting the case study of a project, researchers hence need to be explicit about the reality presented in their study represents the reality at which level of the project. Construction researchers often collect data for a case study primarily from project meetings and documents. Analysis of a project based on these data sources alone will produce a reality of the project's official activities, which in general represents the reality at the middle management level. This, however, is often decoupled from its actual practice at ground level and the business decision-making at top level. To triangulate the official reality, for example, sites at the operational levels need to be sampled, observed and understood. Practically, the findings can also inform management decisions at senior level of the organisation with a more realistic understanding of what their own team can do. On another extreme, traditional ethnographic studies often over-focused on individual factors to the extent of losing sight of macro-level structures (Sydow, 2022). This is illustrative of earlier ethnographic research in the construction field, most of which were single-level studies focusing on the site workers, who were studied as a cluster of individuals as if their work context is an indigenous community. These studies overlooked an essential ground that modern construction work activities occur in a business organisational context of hierarchical structures and institutionalised processes, with in-built role-based accountabilities, financial incentives and performance measures. If the assumed context is accounted for better, sites at multi-levels of the project organisation need to be studied as the necessary context of workers' actions and decisions.

### *Sites defined by time slices*

The ancient Greek philosopher Heraclitus' well-known statement, 'One cannot step twice into the same river' enlightens us that people, things, structures and organisations could be so different at another time such that they can be seen as a different substance. This temporal perspective is more explicit in the 'projects as temporary organising practices' literature, which suggests the existence of project organisation is enacted moment-by-moment by the practice (e.g., Grabher, 2002; Grabher and Ibert, 2011). Visibly on a construction site, a project organisation at the commencement stage is dominated by subcontractors specialised in foundation work; towards the completion stage, the project organisation is a very different set of actors: the foundation subcontractors has been replaced by the glazing or finishing subcontractors. The people, the work and the organisation are so different at these two stages such that they can be seen as two different organisations. As such, the multi-sites can be defined and sampled by different time slices, be it clock time or project time (Rämö, 2002; Prasad & Shadnam, 2023). For example, Jia et al (2017)'s ethnographic studies with a sample of 29 construction sites included two that were the same construction sites at two different moments, each with a completely different cohort of workers on different trades in different work situations. Vanni and Crosby (2023) conducted an ethnographic study to understand user experiences of the 'place' of a public square by walking through the focal place in different weathers and seasons over 12 months' period of time, noting down the variances in visual, audial, smell and touch senses. In these cases, the same 'place' was sampled as multi-sites defined by 'moments'.

### *Sites defined by channels of communication and the digital sites*

Multiple channels of communication co-exist in organisational process and inter-organisational relations, including, e.g., official letters of notice, emails, phone, social media or public media; each can be studied as a site of its own (Hannerz, 2003). The digital space is an increasingly major site of ethnographer's work and impact (Akemu & Abdelnour, 2020). When planning research and interpreting data from the digital channel, it is important to ensure data from one channel is treated as 'part' of the ethnography, instead of the whole. This is because people's activities in and outside of the digital space are part of a continuum where these elements contextualise and shape each other (Brooker, 2022: 4). Interpreting people's digital activities in isolation could lead to seriously wrong conclusions.

Apart from the channels of communication, a plethora of sensors and digital platforms are emerging to multiply the channels for capturing the 'big data' (Bjerre-Nielsen & Glavind, 2022), e.g., environmental conditions continuously recorded on a construction site, or real-time recording on people's health conditions and material characteristics through use of in-built sensors. The data captured in these digital channels are high in temporal resolution (that is, the data are normally captured at a frequency of every one minute or second,) but low in situational details. They need to be synthesised, grounded, calibrated or enriched by the 'thick data' from ethnographers' situated observations (Bornakke & Due, 2018).

In this case, concurrent research design is needed; sampling should be stratified in a way that ensures meaningful connection between data from the different channels. Sensor-captured datasets can go through an initial analysis to map out key situations, anchoring time points and typical scenarios to guide ethnographic data collection (Bjerre-Nielsen & Glavind, 2022). For example, in our heat stress research on construction sites, each of the participating workers was wearing a heart rate monitor and a skin temperature sensor, doing their daily work as usual; while a heat stress monitor was mounted in their work environment to capture data of temperature, humidity, solar radiant heat and wind speed. All these sensors were set to produce a data point every one minute. Two working days monitoring for nearly 300 workers produced a huge database. Such big data cannot be properly interpreted and understood without concurrent ethnographic data of workers' activities on site. This was done through researchers' on-site observations and ethnographic interviews with each worker at the end of the work day to identify the time range of lunch, breaks and recognisable activities. The ethnographic data thus can be used to *label* or annotate the sensor-captured data for further analysis.

#### *Site defined by distinctive discourses*

The multi-sites can also be distinctive discourses or subcultures in an organisation. In the construction-related field, one of the creators of such sites is the horizontal fragmentation in a project organisation, i.e., conflicting interests and incompatible values between professions and firms. Project organisations of different contractors' workforces mingled in one construction site, each having its own report lines. Even in the same design team, each of the professions has its own language and grammar. In megaproject sites in Australia featured by strong union presence, for example, at least three distinctive discourses are observable on the focal issue of work health and safety: the unions see safety issues as opportunities to bargain more welfare for their members; the employers' safety practice at organisational level involves an important task to 'quarantine' the union representatives to minimise potential interruption on productivity; and the safety advisors work to improve safety through complementation and innovation (cf., Jia et al., 2019). The ethnographer thus found herself at a time mingling with the workers, at another time discussing with the managers; a time shadowing the safety advisor, another time talking to the engineers; a time chatting around on construction sites, another time sitting in the project office; a time speaking the union's language, another time speaking the administrator's language; so on and so forth. The ethnographer needs to be sensitive to the subtle shift of discourses and be aware of the invisible boundaries when travelling out of a discourse 'site' entering into another.

### *Sampling the sites*

The overall sampling strategy of multi-site ethnography is theoretical sampling. Theoretical sampling means the researcher's decision on what will be the next sample is iteratively guided by the emerging theoretical insights from the progressive analysis of existing sample (Glaser & Strauss, 1967; Glaser, 1978). Even when the sites are approached by convenient sampling, the ethnographer has a large enough agency to decide in the field what to look at, what to follow, what to resonate and what data to draw from the field. Or, in cases that the data have been collected at a massive scale, the ethnographer still has much discretion in the data analysis process on what kind of data to focus on for understanding the meaning systems of the communities in the field. As such, the research design for multi-site ethnography is *formative* throughout the research process, guided by the progressively formulated theoretical insights. Thus, rather than starting with a detailed plan of data collection, a sampling logic specific to the research inquiry is more practical for guiding prioritisation of attention during fieldwork.

### *Sampling logics*

At research design stage, a sampling logic may be set for systemising prioritisation decisions during the research process; this sampling logic is an extension of the research inquiry. If the research inquiry is to construct a narrative or a full picture, then the sampling logic is to follow the people, the focal phenomenon, the event, the inquiry or the narrative (Marcus, 1995; Falzon, 2009). For example, ethnography sited at nodes and connections of a project network can render rich meanings to the quantitative results of social network analysis (e.g., Berthod et al., 2017; Sydow, 2022). If the research aims to improve an existing policy by reviewing its impact on practice, then multiple sites need to be sampled to cover sufficient variety of spatial context, temporal context, layers of social context or personal psychosocial context. If the research aims to fully understand a group of people's motivations and logics of action, then it needs to study enough variety of situations to compare how they respond and react. Finally, if the research inquiry is to generate a formal theory, or develop a digital product, then the sampling logic is 'constant comparison', and the sites are selected to maximise the variety of contexts in which practice, action or application takes place (Glaser & Strauss, 1967; Glaser, 1978).

The major sampling logic should be superseded by a criterion for exception, that is, when certain critical event emerges in situations of routinised activities, the ethnographer should follow the event. Once on a megaproject site, when I was shadowing a contractor's safety advisor to observe the daily work activities, the client's consultant came to complain about the safety issues that he found in a newly completed set of scaffolds, which would need a massive rework. This appeared to be a significant event that would involve all the key stakeholders, through which organisational networks, procedures, cultures and response patterns would become observable. On this emergent event, I left behind the 'site' of daily work routines to follow through the investigation process of this incident.

### *Criteria for adequacy of sample size*

There are two viable logics in determining the sample size. One is by *saturation*, that is, when the joining of new data does not generate any new categories to the existing findings, the sample size is considered as adequate. For example, Zerjav (2015) closed his data collection when he found 'significant new critical issues were no longer being reported' (p. 1773).

Another logic for judging the adequacy of sample size is by *satisficing* (Falzon, 2009:11, who borrowed from Herbert Simon, 1957), which means when the data collection progresses to situations that 'both satisfy and suffice'; or in other words are 'good enough' for the purpose. Falzon (2009) suggests two reference criteria for satisficing: by the test of common sense; or by the researcher's knowledge of literature which represents a mass of observations by precedent researchers who have studied compatible phenomena. In both criteria, it is not the decision of an isolated ethnographer, 'but the whole methodological and epistemological complex which they are part of' (Falzon, 2009: 12).

Under time and resource constraints, decisions on site sampling often involve a trade-off between the number of sites and the time of immersion in each site. Thus, the fieldwork on each site is often done as *rapid ethnography* (Millen, 2000; Loosemore et al, 2015) or, in another term, *micro ethnography* (Pink & Morgan, 2013), characterised by short intensive fieldwork, guided by a clear research problem for a more focused inquiry (e.g., Alvehus & Crevani, 2022, study of project leadership in emerging situations). For instance, I once organised a study on heat stress in construction sites (cf., Jia et al., 2016, 2017), where we had more sites willing to participate than we could study. This was because the research topic, climatic heat, has a natural constraint on the time available for fieldwork, which must be completed within a Hong Kong summer, during which a construction is paused. This period ranges from late May to early September, excluding weekends and days of typhoon or heavy rain. Thus, to maximise the variety of construction projects, the eventual research design decided on a sample of 26 sites, two days on each site. A pilot study was conducted in the year before on five sites, four days on each site.

### *Managing multi-ethnographers' teamwork*

When you have a team of ethnographers in a research project, a level of standardisation is necessary. The research phases and procedures need to be standardised to a level so that the work can be passed from one researcher to another, from one site to another, and can be revised and compared from time to time, with sufficient flexibility. A set of routines may be set up to provide a frame for team collaboration to ensure everyone is constantly inspired for continuous writing, for example, a briefing session before and a debriefing session after each site visit; completing fieldnotes three days after an on-site observation; routine team meetings for discussing fieldnotes, insights and coding results, etc. (Alvehus & Crevani, 2022).Before the fieldwork, a set of tools and protocols may be prepared for guiding research activities on site; a well-structured data sheet can help set up a work routine for the team. Questionnaire or interview templates can be used to frame data collection or stimulate conversations in the field. However, bear in mind most of the instruments are progressively formulated and situationally adapted (see also Bresnen, 1988).

The division and assigning of tasks among the research team members should be based on each member's personal characteristic. Before commencement of the fieldwork, it is necessary to set aside a time for team building to enable personalised coordination in the field. It is also necessary to run a training session to sensitise the research team members on the use of data collection tools, awareness of site politics, generative questions, activities and their boundary conditions (see Glaser, 1978 for more details on 'theoretical sensitivity'), and other fieldwork techniques as will be introduced in the following sections.

Before data analysis, a training session on coding and checking of codes is needed for the research team (cf., Pitsis et al., 2003). This is particularly important when the team members come from incompatible backgrounds. Cury et al. (2019) suggest to arrange a period of time for an interdisciplinary team to fully immerse into one another's disciplinary world to understand others' perspectives, values and priorities and in, assigning fieldwork, send interdisciplinary researchers in pairs to enable observations of complementary perspectives and aspects. Professor Steve Rowlinson used to assign his PhD students Martin Tuuli and

Tas Yong Koh to go in pair for stakeholder interviews, which proved to be fluent and fruitful (e.g., Tuuli et al. 2010). I did this in several studies, some successful (e.g., Jia et al, 2016, 2017, 2019); some a bit more challenging (e.g., Chan & Jia, 2019). In our heat stress research in Hong Kong, researchers did fieldwork in teams of three, with myself as the anchoring member, which turned to be highly efficient and productive. My megaproject study in Australia arranged fieldwork in two years: in the first year, I went alone on site; the HSES Manager and a safety advisor acted as my co-ethnographers, which produced excellent data. In the second year, I invented an idea at last minute to invite a colleague from the UK to do a joint fieldwork with me so that we can triangulate observations. This experiment was not very successful in the sense of executing the planned research but generated many new ideas from our experiences of surprises. Some of the lessons learned will be introduced in the following sections. What is missing in this case is the time for teambuilding, without which it would be unrealistic to expect that personal and disciplinary paradigms can be aligned and coordinated for an ethnographic teamwork. More importantly, the training and teambuilding session should aim to raise team members' awareness of one's own limitation and therefore remain open to other researchers' observations and interpretations. Even on completion of a well-researched ethnography, ethnographers should remain open that it could be a very different story by the experience of another insider, even it might sound untrue to our own experience. The following conversation between two philosophers in *Autumn Flood* (Chuang-Tzu, 369BC-286BC) provides a good metaphor:

Master Chuang and Master Hui walked about on the bridge over the Liang River.

| | |
|---|---|
| ***Master Chuang*:** | Look at these fish flitting around so cheerfully.<br>This is the enjoyment of being a fish! |
| ***Master Hui*:** | You are not a fish. How do you know the fish enjoy it? |
| ***Master Chuang*:** | You are not me, how do you know that I don't know the fish enjoy it? |

### *Managing ethnographic data from multi-sites*

It is necessary to spend some time to culminate a database for the ethnographic data and keep it over time. To do so, the data need to be structured to a level to set a common ground for consistency, which enables comparison across sites, between multi-ethnographers, or connection to other datasets. For example, raw transcripts and fieldnotes can go through initial coding to reach a level of structure (Dohan & Sánchez-Jankowski, 1998). A well-structured database will also enable comparison of results and interpretations over time or future re-analysis with new methods for different research objectives. For example, Mahalingam's (2022) recent interpretation of comingling of institutional logics in a project is based on re-analysis of a dataset collected two decades ago.

Research ethics used to expect people-related data be discarded six months after the completion of the research project. This convention has changed in recent years. Funding agencies such as Australian Research Council now ask for a data management plan in every funding application. Instead of being discarded, the data is anonymised and set in a structure, then either stored in a safe place such as a locker in a Lab, or a server provided by the university, or shared to an open data platform (e.g., the Australian Research Data Commons) for others' use.

## Doing multi-site fieldwork

This section introduces practical strategies for doing multi-site fieldwork, including how to access the site, key activities in the field, and arrangement of deprogramming sessions to prevent a status of 'going native'.

### *Access the multi-site field, gatekeeper and rapport building*

Any field has more or less a boundary and a level of exclusion to distinguish the out-group from the in-group (Hofstede, 1980). Ethnographers firstly need to find the gate to cross this boundary to access the field. For the sites to be available for your study, researchers need to strategically build up a network and cultivate partnerships with industry partners. People in the network are potentially the *gatekeepers* who can bring you into the site. Alternatively, the ethnographer can approach the gate through institutional network, for example, if your institution has established partnership with some industry partners, or if the research projects you are doing is advised by an industry panel, the site access can be introduced through their referral. In the case of Van Marrewijk (2018), the lead researcher himself was a respected expert in the project organisation being studied, and this identity legitimated him to bring in five student assistants to conduct the ethnographic research. In the ethnographic study of Ventrella et al. (2020) on developing customised design solutions of clean energy cooking system in least developed countries, the lead researcher was located in the US while the field-ethnographers and the participants were in Guatemala, Uganda and Honduras. Site access was granted through the research institute's long-term partnership with local Non-Governmental Organisations (NGOs). Important is these strategies must be built on empathy, i.e., the researcher first of all needs to be interested in and sincere to the field that he or she is to study.

Once entering the site, ethnographers need to be aware that the field is a set of multi-sites, and the gatekeeper is located in but one of them. The ethnographer should be vigilant in discerning the invisible walls, subcultures and ongoing discourses within a visible site and build personal *rapport* with different communities to 'earn' access to other sites. Lalonde et al. (2012) provide an example on how an ethnographer, after being admitted to the field of a large architecture firm, built trust with the participants and gained access to further sites in the organisation. Appointed as an official member of the project team, the ethnographer 'arrived at project meetings long before the scheduled time and was usually the last to leave the site' (p. 422). He used these extra times to initiate informal chats with the field people and observe informal interactions among the project team members, which helped build up trust with them and enabled him to be introduced into three more committees.

Particularly, the gatekeeper might have a different or conflicting interest with the participants that you are going to talk to in the field, as defined by their role in the organisation and their identity. The ethnographer needs to think through how the gatekeeper's identity may influence, shadow and be influenced by your relationship with the participants in the field. For example, workers in a megaproject might want to talk about their unions to you, but if a manager lead you to them and stand on side, they would shut off this area of information.

### *Key activities in the multi-site field*

In the field, the ethnographer co-produces data with the participants through three key activities: participant observation, ethnographic interview and ethnovention. Throughout the process, the ethnographer needs to sensibly watch over appropriate timing to request and collect project artefacts as they become available.

### *Participant observation*

*Participant observation* is the practice of an actively involved observer in contrast to that of a detached observer or a non-reflective participant in the field (Goffman, 1989). An effective participant observer has an identify that can be legitimated as one of the community in the field. This identity of an empathetic self needs to be enacted in the multi-sites by constantly positioning oneself into the shifting context. The ethnographer needs to be sensible about one's level of legitimacy endorsed by different sites, and accordingly switch between *participant* and *non-participant observations* in moving across sites.

Site acceptance varies by research topics. Topics such as safety and health have a lower hurdle for ethnographers to be accepted by the community. Highly professionalised environments, such as the engineers or architects' teams, are more difficult for ethnographers to be legitimated as one of the community; in such context, the *non-participant observation* approach may be a better choice.

### *Ethnographic interview*

*Ethnographic interviews* are loosely structured and dynamically formulated discussions along with the emergent events during the ethnographer's participant observation in the field (Spardley, 1979). Along with the observations, the ethnographer's major activity in the field is chatting around and talking to different people to understand the ecology of the field. In a minimalist style, it was also used as a standalone method, which means formal interview in an ethnographic style (e.g., Dainty et al., 2000). Furthermore, Marcus (1995) mentioned a single-site approach to multi-site ethnography, which is done through ethnographic interviews on personal history to elicit the multiple trajectories and conversations that a participant is originated from and embedded in (e.g., Holub, 1991).

### *Ethnovention, or ethnographic action research*

*Ethnovention* (Van Marrewijk et al., 2010) refers to the effect that the ethnographer acts as a change agent while participating in the field ecology. This is made by either systematically planned intervention or an enlightenment of the focal topic associated with the ethnographer's presence or queries in the field. As a standalone methodology, this is named *ethnographic action research*, used in many social development projects (Tacchi et al., 2003; Hammersley & Atkinson, 2007). For example, our heat stress research in Hong Kong aimed to produce an updated guideline for the construction industry. Before the guidelines were officially issued at the end of the research project, our fieldwork had already raised a wide awareness at all levels of the industry through our discussions with the workers, inquiries to the managers and periodical report to the responsible committee under the Construction Industry Council (CIC) composed of representatives from key industry stakeholders and policymakers. Workers became more conscious about self-management to cope with heat stress; organisations such as the Mass Transit Railway Corporate (MTRC) voluntarily implemented the recommended protocol long before the drafting of the guidelines. In this case, the intervention is not a planned act but a natural 'by-product' of the fieldwork. Cautions should be taken on any unintended intervention effect, however, to make sure not to interfere with the organisation's internal politics which could cause unexpected damage to the related people.

### *'Going native' and de-programming between sites*

An unfavourable end of ethnography is 'going native' (O'Reilly, 2013), a danger that ethnographers are advised to guard against. The origin of this word refers to a phenomenon in history that a colonist fully adopted the life style of the 'uncivilised' local people, whom he was supposed to enlighten, to become one of them, thus lost his superior status and identity. When in the positivist paradigm ethnography assumes an asymmetric power relationship between the ethnographer and the community in the field, 'going native' has a similar meaning to its origin. Here, we have clarified at the beginning of this chapter that ethnography is necessarily built on an equal relationship between the two ends; in this context, 'going native' refers to a status of over-immersion, to the extent that the ethnographer no longer feels anything in the field worth noting down or reporting. As discussed earlier, ethnography is a discourse between two cultures. In the case of 'going native', the two ends of the discourse merge to become one, thus the discourse has no reason to exist. The danger of 'going native' is not about losing the superior status of the observer to become one of the 'savages' in the field, but loss of the discourse, that is, the ethnography.

Thus, the ethnographer needs to plan *deprogramming* sessions to withdraw from their immersion of the site culture and situate oneself back into the readers' world. Deprogramming is also needed in-between ethnographers' movement from one site to another, because the experience in the former site can potentially formulate assumptions of links between symbols and meanings, which will shadow the ethnographer's apprehension of and response to signals in the next site.

## Positioning: The essential practice in multi-site fieldwork

*Positioning*, being 'appropriate' to a specific context, is the essential activity of multi-site fieldwork (Harré & Van Langenhove, 1991; Marcus, 1995: 113). Empathy is the overarching principle for ethnographer's positioning of self across multi-sites and is to be enacted through *Discourse* (with a capital D).

Discourse with a small 'd' refers to the social function of language. Any language in writing and speaking is a *discourse* dispatched from the author to the audience, which carries a social function of constructing and scaffolding six areas of 'reality': (1) meanings and values, (2) activities, (3) identities and relationships, (4) politics, (5) connections and (6) symbol systems (Gee, 1999). A discourse brings with it a social order that sets people and things in a structure of positions. The actual ordering effect is however not determined by the author's intention but how the audience make sense of the message. At the audience's end, language is not the only source of information to make sense of a message. Instead, people make sense of a message with multi-sensory, including perceptions of body language, environment, weather, noise and other stuff (Pink, 2009). Thus, at the author's end, the conveying of a message is not words alone, but a coordination of languages and many other situational factors such as ways of acting, interacting, timing, spatial arrangement, aided by tools and technologies. This is *Discourse* with a capital 'D'.

In the field, the ethnographer engages with 'Discourse work' to enact an identity that is recognised by the community as meaningful in their logic (Gee, 1999).In multi-site fieldwork, each site is unique in terms of time, space, structure, culture, process, power and politic. The ethnographer needs to be sensible to the subtle differences between sites and situations. You need to be aware of who you are speaking to in what situation, and what identity you are presenting yourself to the audience. Because the information you get in return is their discourse to an identity of what they perceive you are. For example, when you are shadowing a safety officer in your fieldwork, what you hear from the workers is what they would tell a safety officer. If at the beginning of an interview you present yourself as a friend of the informant's boss, you will not hear anything that the informant doesn't want his boss to know. In all the visible and invisible sites, the ethnographer must not assume different sites have the same structure or culture. The ethnographer needs to be aware of the assumptions in one's own mind and put down them in order to discover new elements. Empathy is key. As the ethnographer moves across sites, their identity is in constant renegotiation and recalibration.

Below we introduce some essential concepts and principles of positioning to sensitise ethnographers for multi-site fieldwork.

## *Key dimensions of positioning*

### *The natural dimension*

The ethnographer's natural identity is the basis of any positioning work. This is constituted of the ethnographer's personal attributes such as gender, age, race, ethnicity, social class, education, temperaments and experience (Jaju, 2023). In Oswald et al. (2022), the ethnographer was accepted by the site in his natural identity, 'a young researcher in his twenties, and being attached to a university, the researcher assumed an "apprentice-like role"' (p. 379). So was Bresnen (1988) admitted by the field as a student when he was a student.

The ethnographer's natural identity does not have to be literally the same with that of the field. Oswald et al. (2019) mentioned the ethnographer is 'male of white-British origin, and therefore blended in demographically with the majority on the project' (p. 2585). However, in my megaproject study, the Asian female ethnographer in an all-white-Australian-male site had also been successful; and in our second fieldwork, two Asians going around an all-white Australian site were also welcomed (cf., Chan and Jia, 2019). Thus, sameness is not a necessary condition.

More deeply, the ethnographer's thinking logic and personal values, manifested in patterns of react and interact, constitute the foundation of one's identity perceived by the field (Gee, 1999). These attributes are acquired over a personal history. Comparing to the timescale of an ethnographic fieldwork, these personal attributes should be taken as part of the ethnographer's natural identity rather than external symbols that can be planned or configured at the research design stage.

### *The situational dimension*

Bearing in mind that Discourse work is the coordination of one's language with many other situational factors to enact an identity of self and a social order of people and things around; in multi-site fieldwork, the ethnographer needs to be sensitive to the characteristics of place, space and other existing physical and social structures of each site. Research protocols need to be generic and adaptable to multi-site situations. In some situations, the ethnographer can proactively re-organise the physical environment to set an initial structure with the participants as part of the Discourse.

For example, in our fieldwork of heat stress research in 26 construction sites over a summer, we were on a protocol of spending two working days on each site, meeting the participants in a morning session and an evening session in a site office, and going for on-site observation during the day. Take, for example, our first morning session on a new site, the three ethnographers needed to work out with 12 workers a series of activities, including (1) creating a friendly and comfortable atmosphere (2) explaining the research to participants; (3) workers signing the consent form; (4) a brief survey on worker's trade, (5) personal health history, (6) medication and (7) sleep quality; (8) measurement of body height, (9) weight and (10) body temperature, (11) blood pressure and (12) resting heart rate; (13) wearing for each worker two sensors for continuous recording of heart rate and body temperature during the day's work; (14) providing breakfast to the workers who ate it before going to daily work.

All these activities need to be worked out in a randomly assigned site office within a 30-minute session. The offices offered by each site differed from one another: one was a meeting room in an office tower crowded with tables and chairs, another was a cargo container with two small tables at the corner, another a warehouse without chairs. In all these emergent situations, the ethnographers needed to make a swift evaluation on available elements that can be used to re-arrange the space for guiding and structuring the people and work flow and take actions to put the tables and chairs in place before the workers' arrival. A wise spatial arrangement can do a lot of work for you and make a difference to the quality of the data.

Apart from spatial arrangement, *time* is an important factor in situating activities and a sensitivity needed for the ethnographer. As discussed earlier, the same physical space at different 'moments' can be taken as distinctive sites to be matched with different positioning practices. The ethnographer should also be sensible to the *timing* of research activities on site, and watch over what can be done at when. For example, Bresnen (1988) described that he 'caught people for interview as and when they seemed to become available'; and in-between meetings and interviews, he 'plough through documents' in the sub-agent's office. I once met a research manager who complained that his official requests of project documents were repeatedly ignored by the contractor (the field being researched) and concluded that the research could not be done because the contractor was uncooperative. This, in my view, is a mis-defined problem. The problem in this case is not an uncooperative contractor, but a wrong research strategy. By making an official request for every specific project documents, the research team posed a threatening message to the contractor, who would worry about unforeseeable consequences of releasing commercial information to an unfamiliar third party. In this case, the research team only needed to ask the contractor to give access to an ethnographer to their construction site for a few weeks. By establishing personal rapports and mingling with the project team, the ethnographer can then watch over the right timing to sort out the needed data in a non-threatening manner. We'll discuss the timing dimension further in the following sections.

### *The performing dimension*

On top of the natural identity and the situational elements, the ethnographer has an agency to perform out who she or he is through language, behaviour, attitude and sensible responses to various situations. The general objective of performing is to '*be appropriate*' in diverse and emergent situations. The ethnographer needs to practise culturally sensible behaviours that comply with the cultural code of the specific site.

In emergent situations, the ethnographer should be sensible to pick up leads from observations or conversations to identify and follow up chains of events that worth attention. Oswald et al. (2019) described how the ethnographer adapted his behaviours at different sites: when observing on construction site, notes were taken on his mobile phone; when observing the project meetings, notes were handwritten on the meeting minutes. Both behaviours were 'appropriate' in their respective context, but not so if exchanged context. In another scenario, the ethnographer evaluated the situation and decided not to take photos 'as this could raise suspicions amongst the workforce' (p. 2586). In my observation of the megaproject in Australia, when I saw something interesting on site, I would give my mobile phone to the accompanying safety advisor and ask him to take a photo for me. Readers should not apply these decisions literally to their own fieldwork, but should first observe the norms of the surrounding environment and evaluate the situation holistically before deciding what behavioural code is to be practised. The key principle of acting in these situations is to maintain the *trust* between the ethnographer and the field people.

In most of the ethnographic studies, handwritten notes were preferred over audio-recording on concern that the latter might be perceived as intrusive. However, in situations that the site has established sufficient trust of the ethnographer, audio recording is of advantage to enable details of events being captured and micro-level dynamics being analysed in depth, and a feast of details being reported in publications (cf. Van Marrewijk & van den Ende, 2022).

## *Strategies for positioning in a multi-site field*

### *Work with legitimacy*

To remain 'appropriate' in multiple sites and situations, the ethnographer should be conscious about what she or he is legitimated to ask in what situation. In moving across sites and discourses, the ethnographer needs to be sensible and readily shift between participant and non-participant observations, based on the ethnographer's level of legitimacy in different situations.

There is a timing dimension in what questions can be asked or documents requested, conditioned by the ethnographer's legitimacy in the eyes of the community of the field. Some questions are not wrong, but they need to be asked in the right timing. If you ask a question before the participant legitimates you to be in that territory, the question will be perceived as an intrusion or aggression; and the participant will close their heart to you. You will get no further information from the field except what they can present to an outsider. So, a principle for eliciting information from the field is: *Question follows legitimacy*.

Legitimacy is inherent in the ethnographer's perceived identity by the field and shaped by the level of trust from the field. There's not much you can do to change it instantly, but by being there and causing no harm and be helpful when needed, you can build trust with the community, and your legitimacy will increase over time. Finally, there is no unlimited legitimacy. There are territories that simply cannot be opened to you. Watch out this boundary and respect it.

### *Manage trust*

Trust is indispensable in any fieldwork technique that involves interaction between the researcher and the participant. Trust cannot be assumed; it has to be earned. There are three key aspects that help ethnographer to build up trust with the people in the field: (1) ensure *confidentiality* of information, particularly don't transfer information across levels; (2) keep *neutrality* and independence from the field's internal politics to ensure no harm is made to any individual or organisational interest; (3) on top of neutrality and independence, the ethnographer should try to make all 'sites' in the field to *benefit* from the research. For example, the benefit can be making the workers look good to their managers, or best practices of the researched organisation being recognised, or a copy of research report being fed back to the studied organisation. Particularly, the researcher needs to look after the interest of the gatekeeper and make sure the gatekeeper is not damaged or destroyed for leading you into the field. Meanwhile, bearing in mind that the gatekeeper might be standing in a different 'site' from the field people you are talking to, the ethnographer needs to keep an independent position from the gatekeeper and don't transfer other people's information to the gatekeeper. Sensible balance between confidentiality, neutrality and benefit will help building up a long-term research partnership with the industry.

Ethnographers don't often realise that they do damage people in the field through even slightly interruption of the field ecology due to their unawareness of the field politics and the political implications of their activities on the field people. For example, when researchers come to a site with some health check equipment, workers perceive them as external healthcare consultant and therefore feel comfortable to complain about their health problem or job stress at length. What they have talked to the researchers is a normal part of organisational life, in which they are performing their daily jobs. However, if a manager happens to be in the scene, or if any of such information is passed to their managers (either on the manager's request or by slipped words or out of an immature researcher's enthusiasm to improve the workplace condition), the managers often quickly jump to a conclusion that the worker is having a negative attitude at work or being physically or mentally unfit for the work, and in extreme cases it could cost someone's job. Such asymmetric perception and unintended consequence is because the ethnographer tends to treat the field as a community when it is actually a business organisation, invisibly structured by financial incentives, accountabilities, performance monitoring and evaluation, managerialism and workplace politics. Some information that can be freely shared in a community environment may be politicised and taken as performance or fit-for-job indicator by the employer.

### *Create events for data production*

In exceptional situations, the ethnographer may act out of the boundary of 'appropriateness' to create some event to test out people's response pattern. One of the reasons for the traditional ethnography's essential feature of immersion in a site for a prolonged period is to wait for the observable events to emerge over time. In studies of conditions of certain behavioural patterns, the ethnographers sometimes create events simulating the conditions that trigger people's responses to reveal their pattern of behaviour or test a hypothesis of cause-effect relationship (Bjerre-Nielsen & Glavind, 2022).

For example, when observing in the megaproject site in Australia, I once encountered an incident investigation process in which two workers entered into a confined space without signing the registration form, although they had filled in their names and all other required information in the form. The investigator initially decided to literally implement the strict site rules, that is, to fire the involved workers. Trying to understand the situation of the incident, I asked a few 'why-not-this-option' questions to clarify what happened and the decision logic of the organisation. In answering my questions, the safety officer was engaged in a reflective conversation, through which he realised that the true nature of the incident was more about paperwork than a real safety issue. At the end of our conversation, as I finally understood what the incident was about, the safety advisor arrived at a conclusion to reverse his original decision and reduce the worker's penalty to a re-training session. In this case, if the ethnographer had not created a conversation with the field people to clarify the nature of the observed event, what was noted down in the fieldnote would have been an incident that 'two workers engaged with unsafe behaviour and were subsequently sacked from their job'. The intervening questions helped uncover the true nature of the event and the true behavioural patterns of both the workers and the employer. As a by-product, the field people had an opportunity to engage with reflexivity in their decision-making, which led to a 'change' in the field. In my view the 'changed' scenario is more truthful than the would-be scenario, because the would-be scenario was based on an immediate, ungrounded and false impression of the investigator, rather than what was really happening on site. Therefore, the ethnographer's presence, questioning and actions at this level may not be understood as influencing the natural setting of the site, rather, it is a method of ethnographic data production at exceptional conditions.

### *Boundary conditions of positioning*

#### *Boundary of performing*

To what extent should we consciously manage people's impression, or, to what extent should the ethnographer 'perform' in the field? The boundary should be not to the extent of decoupling your performance from your true being such that the self you presented to the field is a fake one. In the experience shared by Jaju (2023), who did her fieldwork in low-income communities in India, the ethnographer deliberately 'dressed down' in order to eliminate the participants' perception on their class difference with her. This performance was however betrayed by the other subtle differences being perceived in her everyday practice on site. Consequently, she was labelled as hypocritic by the field community. On another occasion, she felt obliged to 'invent a partner' and 'lied that he was half-Indian' which 'proved to be immensely beneficial to [her] social navigation – improving [her] relations with [her] male informants and giving [her] more common ground with [her] female informants' (p. 8). This in my view had gone too far. I would suggest don't go so far as to present yourself into a person that you are not. Keep authenticity and sincerity to the field people. If you fake some attributes that is not your true identity, people will sense it and the information you elicit by this fake identity will be more or less distorted.

But what if the researcher does have a gulf of difference from the researched? Three approaches can be chosen from: (1) take the passive observation approach and be less visible, For example, Mahalingam (2022) described his approach to observation in the project meetings as 'a fly on the wall'; or (2) present yourself as an 'empathetic context', which means more than a listener but to bring with you a whole atmosphere of understanding; or (3) present yourself as a student of the field people, as discussed earlier in Bresnen (1988) and Oswald et al (2022). Often, working people are more willing to talk about their thoughts and motives to an 'empathetic context' or a 'loyal disciple' than they would want to talk to their peers.

Finally, there is a limit for every researcher on what one can do and be. Some topics may simply take a different ethnographer to study. You can save the time to do something else that best align with your set of talents – empathy is key. Being aware of an ethnographer's limit, it is important to keep an open mind to the possibilities that multiple realities might exist in the same set of observed events.

### *Behaviours that alienate the ethnographer from the site*

As discussed earlier, the ethnographer's body and mind is the key instrument for data production in the field (Goffman, 1989). Therefore, while in the field, the ethnographer should be consciously managing what messages you are conveying to the people through language and behaviours in coordination with situational factors, and what identity you are enacting yourself as. This cannot be fully controlled at the ethnographer's side, as identity is socially constructed. How and what other people perceive you are also depends on the cultural frame in the participants' mind and how the gatekeeper presented you to the participants in the field.

In the field, the ethnographer must *forget* about the research mission of 'collecting my data'. Just immerse yourself in the field culture, give full attention to the people, treat the people with sincerity, open your senses to perceive and learn the environment around. Never treat the people as 'subjects' of your research. Some researchers have a patronising attitude to the informants on site, which is not in the researchers' awareness but is manifested in all their words and deeds. This will quickly close off the conversation with, and information from, the informant.

If there are more than one ethnographer on the same site, the ethnographers should refrain from talking among themselves. Ethnographers discussing among themselves in academic language in front of the participants will create their own 'site' within the field, which will alienate themselves from the community around. The ethnographers should either remain in the role of a participant or stay passive. A simple structure is to separate the times for observation and reflection, keep the academic reflection to off-site time (Bourdieu, 1977).

In some occasions, the community of the field may already have an established negative stereotype of you before your arrival, created by people who share some identify features with you and come before you. In such cases, you will unfortunately find yourself in a position to bear the consequence of someone else's behaviour that might have damaged the trust. You will be either denied access or, if the field does let you in, the stereotype in people's mind could be easily activated by some signs in your behaviour, and you find yourself surrounded by unreasonable suspicions. An example of such a case can be found in Zikargie's (2022) study of sugar factories in Africa, where the ethnographer had difficulty to access the sites despite having been granted access officially; one of the major reasons was 'the factories have been criticised by researchers and international organisations for damaging the environment, violating the rights of local people, and of generally poor performance' (p. 4).

In such situations, be aware that the field's rejection is not a denial of you but their reaction towards a past experience. You need not blame yourself for an inability to perform in a difficult situation made by someone else, nor the field for treating you by a stereotype that is not you. If you are let in, be patient to practice their cultural code and take time to build the trust. After a period of peaceful presence in the field, you may earn their trust, and may even be able to modify the stereotype in the field culture, to make it easier for ethnographers coming after you.

## Conclusions

This chapter elaborates on planning and conduct of multi-site ethnography, with an empirical focus on the study of firms and project organisations in the construction industry. The strategies and principles in general are transferrable to other fields outside of the construction industry. *Empathy* is the soul of ethnography and the key to draw out ethnographic data from the field. Through the lens of empathy, ethnography the establishment of a discourse between two equal meaning systems, one of the field, another of the audience; in-between is the agency of the ethnographer. Multi-site ethnography is one that recognises the field end of the discourse is an arena of multi-realities. The key challenge to doing multi-site ethnography is constantly and dynamically Discoursing an empathetic self to the diverse communities in the field of modern hierarchical business organisations. As such, the ethnographer's essential activity of fieldwork is *positioning*. The ethnographer's positioning of self in the field is to be built on the natural identity of the ethnographer in coordination with the environmental elements, on top of which the ethnographer has an agency of performing.

In moving across sites, ethnographers need to be sensible about the *timing* dimension in terms of what they are legitimated to ask and how far they are trusted to explore in what situation. The principle is *Question follows legitimacy*. The ethnographer should consciously build up trust with the community of the field, but at the same time be aware that the presence of the trusted researcher could blur workers' cognition of organisational context. The ethnographer needs to carefully manage the situation and the trust to make sure that the information provided by the participants will not become a source of destruction to anyone through the internal politics of the organisation. In exceptional situations, the ethnographer can act out of the boundary of 'appropriateness' to create intervening events to test out the true patterns of the field community, which produces more valid and in-depth data. The performing act should stop at a boundary, that is, not to present yourself into someone that you are not. If the ethnographer's natural identity is too far from what the site culture can perceive as one of them, the ethnographer can present him or herself as an 'empathetic context' or a 'loyal disciple' to the participants, or switch to passive observation.

When the research involves multi-ethnographers, protocols, training and research activities need to be planned for a level of standardisation to ensure consistency of interpretation across contexts, researchers and over time. The structuring and management of ethnographic data need to consider compatibility with future datasets, and adaptability to new analytical techniques or perspectives of interpretation emerging over time.

Limitations are discussed at different levels of the methodology. Firstly, there is a boundary on the ethnographer's agency of 'performing'. Secondly, there is a limit in one's scope of apprehension due to the personal involvement of the ethnographer in the field, e.g., personal interests and emotions attached to one's natural identity or, in the case of autoethnography, their role in the organisation. Thirdly, multi-site ethnography involves a trade-off between time of immersion in each site and the number of sites for study, which implies some potential risk of misinterpretation due to insufficient understanding of the context. Finally, ethnography itself has its limitation in that the ethnographer cannot be everywhere at the same time or fundamentally change his natural identity to experience what another person would see. Ethnographers need to acknowledge these limits. After all, the answer to why people do what they do is plural and exists in multi-realities. In conclusion, this chapter suggests the planning of multi-site ethnography should start from a future vision. Empathy is the foundation of any techniques and strategies, with which the ethnographer must never assume any superiority, but sincerely and honestly position themselves as equal to the people and communities in the field.

## References


Abdirad, H. (2022). Managing digital integration routines in engineering firms: Cases of disruptive BIM cloud collaboration protocols. *Journal of Management in Engineering*, 38(1), 05021012, 1–14.

Akemu, O., & Abdelnour, S. (2020). Confronting the digital: Doing ethnography in modern organizational settings. *Organizational Research Methods*, 23(2), 296–321.

Alvehus, J., & Crevani, L. (2022). Micro-ethnography: Towards an approach for attending to the multimodality of leadership. *Journal of Change Management*, 22(3), 231–251.

Atkinson, P., Doffey, A., Delamont, S., Lofland, J., & Lofland, L. (2014). Editorial introduction: Mapping ethnographic diversity. In P. Atkinson et al. (Eds.), *Handbook of Ethnography* (pp. 1–14). Thousand Oaks: Sage.

Berthod, O., Grothe-Hammer, M., & Sydow, J. (2017). Network ethnography: A mixed-method approach for the study of practices in interorganizational settings. *Organizational Research Methods*, 20(2), 299–323.

Bjerre-Nielsen, A., & Glavind, K. L. (2022). Ethnographic data in the age of big data: How to compare and combine. *Big Data & Society*. https://doi.org/10.1177/20539517211069893

Bornakke, T., & Due, B. L. (2018). Big-thick blending: A method for mixing analytical insights from big and thick data sources. *Big Data & Society*, 5(1), 1–16.

Bourdieu, P. (1977). *Outline of a Theory of Practice*. Cambridge: Cambridge University Press.

Bresnen, M. (1988). Insights on site: Research into construction project organizations. In A. Bryman (Ed.), *Doing Research in Organizations* (pp. 34–52). Abingdon: Routledge.

Brooker, P. (2022). Computational ethnography: A view from sociology. *Big Data & Society*, 1–6. https://doi.org/10.1177/20539517211069892

Chan, P. W., & Jia, A. Y. (2019). Finding the micro in the mega: a multi-ethnographers study into the micro-dynamics of a megaproject in Australia. In *2019 ARCOM Annual Conference Working Paper Series, 5–7 September, 2019,* Leeds, UK. Association of Researchers in Construction Management.

Cury, M., Whitworth, E., Barfort, S., Bochereau, S., Browder, J., Jonker, T. R., Kim, K. S., Krenchel, M., Krenchel, M., Ramsey-Elliot, M., Schuur, F., Zax, D., & Zhang, J. (2019). Hybrid methodology: combining ethnography, cognitive science, and machine learning to inform the development of context-aware personal computing and assistive technology. *Ethnographic Praxis in Industry Conference Proceedings EPIC 2019*, pp. 254–281.

Dainty, A. R. J., Bagilhole, B. M., & Neale, R. H. (2000). A grounded theory of women's career underachievement in large UK construction companies. *Construction Management and Economics*, 18(2), 239–250.

Dohan, D., & Sánchez-Jankowski, M. (1998). Using computers to analyze ethnographic field data: Theoretical and practical considerations. *Annual Review of Sociology*, 24, 477–498.

Falzon, M.-A. (2009). *Multi-Sited Ethnography: Theory, Praxis and Locality in Contemporary Research*. London: Taylor & Francis.

Gee, J. P. (1999). *An Introduction to Discourse Analysis*. Abingdon: Routledge.

Gille, Z., & Ó Riain, S. (2002). Global ethnography. *Annual Review of Sociology*, 28, 271–295.

Glaser, B. G. (1978). *Theoretical Sensitivity*. London: Sociology Press.

Glaser, B. G., & Strauss, A. L. (1967). *The Discovery of Grounded Theory: Strategies for Qualitative Research*. London: Weidenfeld and Nicolson.

Goffman, E. (1989). On fieldwork. *Journal of Contemporary Ethnography*, 18(2), 123–132.

Goldman, A. I. (1993). Ethics and cognitive science. *Ethics*, 103(2), 337–360.

Grabher, G. (2002). Cool projects, boring institutions: Temporary collaboration in social context. *Regional Studies*, 36(3), 205–214.

Grabher, G., & Ibert, O. (2011). Project ecologies: A contextual view on temporary organizations. In P. W. G. Morris, J. K. Pinto, & J. Söderlund (Eds.), *The Oxford Handbook of Project Management* (pp. 175–200). London: Oxford University Press.

Hamid, W., & Tutt, D. (2019). "Thrown away like a banana leaf": Precarity of labour and precarity of place for Tamil migrant construction workers in Singapore. *Construction Management and Economics*, 37(9), 513–536.

Hammersley, M., & Atkinson, P. (2007). *Ethnography: Principles in Practice* (3rd ed.). London: Routledge.

Hannerz, U. (2003). Being there … and there … and there! Reflections on multi-site ethnography. *Ethnography*, 4, 201–216.

Harré, R., & Van Langenhove, L. (1991). Varieties of positioning. *Journal for the Theory of Social Behaviour*, 21(4), 393–407.

Hartmann, T., Fischer, M., & Haymaker, J. (2009). Implementing information systems with project teams using ethnographic-action research. *Advanced Engineering Informatics*, 23, 57–67.

Hofstede, G. (1980). *Culture's Consequences*. Thousand Oaks: Sage.

Holub, R. C. (1991). *Jürgen Habermas: Critic in the Public Sphere*. London: Routledge.

Honderich, T. (2005). *The Oxford Companion to Philosophy* (2nd. Ed.). Oxford: Oxford University Press.

Hopkins, A. (2016). *Quiet Outrage: The Way of a Sociologist*. Sydney: Wolters Kluwer.

Jaju, G. (2023). Extra-terrestrial landings: An ethnographic account of doing ethnography. *Ethnography*, 1–18. https://doi.org/10.1177/14661381221147021

Jia, A. Y., Rowlinson, S., & Ciccarelli, M. (2016). Climatic and psychosocial risks of heat illness incidents on construction site. *Applied Ergonomics*, 53, 25–35.

Jia, A. Y., Rowlinson, S., Loosemore, M., Gilbert, D., & Ciccarelli, M. (2019). Institutional logics of processing safety in production: The case of heat stress management in a megaproject in Australia. *Safety Science*, 120, 388–401.

Lalonde, P.-L., Bourgault, M., & Findeli, A. (2012). An empirical investigation of the project situation. *International Journal of Project Management*, 30, 418–431.

Loosemore, M., Powell, A., Blaxland, M., Galea, N., Dainty, A., & Chappell, L. (2015). Rapid ethnography in construction gender research. In A. Raidén & E. Aboagye-Nimo (Eds.), *Procs 31st Annual ARCOM Conference* (pp. 1271–1280). Association of Researchers in Construction Management.

Mahalingam, A. (2022). How institutional intermediaries handle institutional complexity in vanguard megaproject settings. *International Journal of Project Management*, 40, 320–331.

Marcus, G. E. (1995). Ethnography in/of the world system: The emergence of multi-sited ethnography. *Annual Review of Anthropology*, 24, 95–117.

Millen, D. R. (2000). *Rapid Ethnography: Time Deepening Strategies for HCI Field Research*. AT&T Labs-Research Atlanta.

O'Reilly, K. (2012). *Ethnographic Methods* (2nd ed.). Abingdon: Routledge.

O'Reilly, K. (2013). Going 'native'. In *Key Concepts in Ethnography*. Thousand Oaks: Sage.

Oliveira-Silva, P., & Gonçalves, ÓF. (2011). Responding empathetically: A question of heart, not a question of skin. *Applied Psychophysiology and Biofeedback*, 36, 201–207.

Oswald, D., Lingard, H., & Zhang, R. P. (2022). How transactional and transformational safety leadership behaviours are demonstrated within the construction industry. *Construction Management and Economics*, 40(5), 374–390.

Oswald, D., Sherratt, F., & Smith, S. (2019). Managing production pressures through dangerous informality. *Engineering, Construction and Architecture Management*, 26(1), 2581–2596.

Pink, S. (2009). *Doing Sensory Ethnography*. Thousand Oaks: Sage.

Pink, S., & Morgan, J. (2013). Short-term ethnography: Intense routes to knowing. *Symbolic Interaction*, 36(3), 351–361.

Pink, S., Tutt, D., Dainty, A., & Gibb, A. (2010). Ethnographic methodologies for construction research. *Building Research & Information*, 38(6), 647–659.

Pitsis, T., Clegg, S., Marosszeky, M., & Rura-Polley, T. (2003). Constructing the Olympic dream: A future perfect strategy of project management. *Organization Science*, 14(5), 574–590.

Prasad, A., & Shadnam, M. (2023). Balancing breadth and depth in qualitative research: Conceptualizing performativity through multi-sited ethnography. *Organization Studies*, 44(5), 831–854.

Rämö, H. (2002). Doing things right and doing the right things time and timing in projects. *International Journal of Project Management*, 20(7), 569–574.

Simon, H. A. (1957). *Administrative Behaviour* (2nd ed.). New York: The Free Press.

Spardley, J. P. (1979). *The Ethnographic Interview*. New York: Holt, Rinehart & Winston.

Sydow, J. (2022). Studying the management of project networks: From structures to practices? *Project Management Journal*, 53(1), 3–7.
Tacchi, J., Slater, D., & Hearn, G. (2003). *Ethnographic Action Research*. Paris: UNESCO.
Tutt, D., Pink, S., Dainty, A. R. J., & Gibb, A. (2013). Building networks to work: An ethnographic study of informal routes into the UK construction industry and pathways for migrant up-skilling. *Construction Management and Economics*, 31(10), 1025–1037.
Van Marrewijk, A. (2018). Digging for change: Change and resistance in interorganizational projects in the utilities sector. *Project Management Journal*, 49(3), 34–45.
Van Marrewijk, A., & van den Ende, L. (2022). Shaping interorganizational strategic projects through power relations and strategic practices. *International Journal of Project Management*, 40, 426–438.
Van Marrewijk, A., Veenswijk, M., & Clegg, S. (2010). Organizing reflexivity in designed change: The ethnoventionist approach. *Journal of Organizational Change Management*, 23(3), 212–229.
Vanni, I., & Crosby, A. (2023). Place-based methodologies for design research: An ethnographic approach. *Design Studies*, 85(101168), 1–27.
Vaughan, D. (1996). *The Challenger Launch Decision: Risky Technology, Culture, and Deviance at NASA*. Chicago: University of Chicago Press.
Ventrella, J., Zhang, S., & MacCarty, N. (2020). An international, multi-site, longitudinal case study of the design of a sensor-based system for monitoring impacts of clean energy technologies. *Design Studies*, 66, 82–113.
Zerjav, V. (2015). Design boundary dynamics in infrastructure projects: Issues of resource allocation, path dependency and problem-solving. *International Journal of Project Management*, 33, 1768–1779.
Zikargie, Y. A. (2022). Jala role: Normative practice of collaborative ethnography in a hostile research frontier, South Omo, Ethiopia. *Ethnography*, 1–18. https://doi.org/10.1177/14661381221110051

# 13

# DOING CASE-BASED INTER-ORGANISATIONAL ETHNOGRAPHY

*Evelijn Martinius and Leentje Volker*

*Department of Organisation Studies, Vrije Universiteit, Amsterdam, The Netherlands*

*University of Twente, Enschede, The Netherlands*

**Key takeaways**

- The experience of an inter-organisational ethnography of underground infrastructure projects in the Netherlands provides an excellent basis to develop methodological insights on doing ethnographic studies in construction based on an anthropological view on multi-sited ethnography.
- This chapter offers practical questions to aid ethnographers in the field by encouraging them to reflect on three specific questions related to the people, places and project context of the ethnographic fieldwork situation at hand.
- Three questions aid ethnographers of inter-organisational construction projects while they are in the field:
    - Who, where, what?
    - Say that again?
    - What is going on here (and here, and here)?

## Introduction

On our way to the construction site, Mike and I are standing at the bus stop near the project hub. We are waiting for the shuttle bus to arrive and drive us to the site. Instead of my dark blue corporate uniform that makes me fit right in at the offices, I am wearing gigantic safety shoes that are way too big, a helmet, safety glasses, gloves, and fluorescent jacket. I do not exactly feel like a 'fly on the wall' type of ethnographer. Mike jokes about my 'costume' – as he calls it. 'I'm just hoping to keep warm' I grumble to Mike. The wind is ice cold and heavy grey clouds are gathering. He continues his laughs with me, 'What a researcher has to put up with, to see some action…well it could be good for you, you know, getting outside for a bit …'

DOI: 10.1201/9781003379584-16

The vignette above shows a small excerpt from the field notes of a two-year ethnography of inter-organisational project in the context of a PhD research project into collaboration in inter-organisational project arrangements. The empirical data for this research builds on extensive fieldwork, interviews and document analysis of underground utilities projects in the Netherlands. These were major civil asset projects that included renovation on buried utilities. The first author immersed herself in the daily work inside both of these projects, attending to each project for a year (2018–2019 and 2019–2020). The vignettes in this chapter – that serve as illustration for ethnographic experiences – come from the fieldwork in an aviation project. Incorporating this case within this study provided a unique setting to study how the renovation of underground utilities is of crucial relevance in a setting as 'above ground' oriented as an international airport. This project was a typical mega-project and included representatives from several organisations that temporarily were joining forces to building a new terminal and pier. As such, this case presented a unique opportunity to study an internationally oriented inter-organisational project, with managers facing high stakes. In addition to fieldwork in projects, the first author engaged in broader sector events, took research trips to individual contractor companies, municipalities and public administrations, to become familiar with the broader setting in which the projects operate. The second author has been performing qualitative research in the construction industry for over 20 years. In her work, she is used to combining interviews, documents and observations to create deeper insights into collaborative mechanisms within project organisations and the home organisations of the partners involved in these projects.

In this chapter, our aim is to explore how the inter-organisational setting of construction projects informs the fieldwork. Hence, the question central to this chapter is: *how can ethnographers remain attentive to the characteristics of inter-organisational projects as they are in the field?* We engage with the broader literature on multi-sited ethnography to propose guidelines for future ethnographers who aim to study inter-organisational projects in the construction sector and build on our expertise and experience with ethnographic data collection.

The chapter starts to outline the promise and pitfalls of inter-organisational project ethnography. Then we take these characteristics and draw on the case example of the inter-organisational underground utility project. We offer practical pointers to aid ethnographers by encouraging them to ask themselves to reflect on three specific questions related to the people, places and project context of the ethnographic fieldwork situation at hand.

### *The promise and pitfalls of inter-organisational project ethnographies*

Ethnography has a lot of potential to build knowledge that can help to solve practical problems in the field as well as propel academic discussions (Pink, Tutt & Dainty, 2010; Berthod et al., 2018; Volker, 2019; Oswald & Dainty, 2020). Ethnographers can specifically tease out the dynamics on the micro-level within the project, which are overlooked by quantitative 'hard science' studies (Oswald & Dainty, 2020). Instead of testing hypothesis, ethnographers generate hypotheses. As researchers immerse themselves in a field, they remain open to the lived experiences of professionals. This makes it an excellent method to develop practical value for both academics and practitioners and give back to the sector through research presentations, action-based interventions or reflection reports. Zooming in on the beliefs that guide people's action makes ethnography an especially suitable method to lay bare different organisational cultures that come together within a project. Elements characteristic to the project can mirror cultural dynamics of the construction sector, which is for example known for being result-driven, action-oriented and of great public value (Volker, 2019).

Inter-organisational projects are defined as temporary endeavours (Lundin & Söderholm, 1995) where multiple actors seek to optimise outcomes by combining resources (Jones & Lichtenstein, 2008). Given their organisational format, these projects require collaboration across organisational boundaries between public, private and third sector organisations (Ainamo et al., 2010). Hence, doing ethnographic studies in this context requires an inter-organisational ethnography approach.

For ethnographers of inter-organisational ethnography, the challenge is 'follow the action' when the people, places and project scope are constantly changing (Marshall & Bresnen, 2013). With others, we therefore argue there are three moving parts: the people, the places and the project itself (Berthod et al., 2018; Ibidem). Concretely, this comes down to, for example, the construction site, the project office and the dynamics at the home offices of the collaborative partners in order to capture the dynamics of the project. The challenge can be to zoom in when everything happening at different places, in different constellations and with different understandings of what is going on – all at once (Hannerz, 2003).

To deal with this challenge, we propose to take inspiration from multi-sited ethnography (Marcus, 1995). Ethnography is a research method traditionally cornerstone to anthropology, the study of people and their culture. But in the early 1990s, the traditional anthropologist was challenged by the increasing globalisation of what were previously more secluded, local cultures. The anthropologist Marcus therefore developed a multi-sited approach to ethnography, advocating that anthropologists should study culture in more than one place. Marcus' multi-sited ethnographic method is almost as metaphorical as it is methodologically poignant to studying today's fragmented contemporary construction sector. Davis (2006) pointed out how construction projects have historically included more than one organisation, like the client and contractor organisation. Today, civil asset projects almost always include multiple organisations, who continue their work while temporarily sending out representatives to the temporal project (Sydow et al., 2004). The temporal project organisation then hires a contractor, to construct the work. The geological dispersion of different locations also gives rise to different local cultures: the construction site and the management office each have their own culture.

To attend to these different sites and different cultures is a challenge that inter-organisational ethnographies will have to address. The daily operations inside inter-organisational projects are, for example, in the hands of managers and professionals who have more than one home; their home-organisation and the temporal project (Sydow et al., 2004). Ideally, researchers have an idea of what is going on at the home-organisation that is relevant to the project. How does this organisation define success, for example, or what is the expectation of the project's goal?

Already classical anthropologists like Geertz noted how "the illusion that ethnography is a matter of sorting strange and irregular facts into familiar and orderly categories (…) has been exploded" (Geertz, 1988:1). When answering the question about what expectations about a project might live in the various organisations who collaborate, we are likely to come across different and even contradictory answers. Upon noticing these, what Geertz calls the 'strange facts and irregularities', we might have already found interesting starting point for our analysis. The longer we stay in the field, the more paradoxes and contradictions we run into. What Geertz hints at therefore is that there is hardly *one* way of understanding what is going on in our field. Organising the messy reality of work inside organisations (see also Law, 2004, for a discussion on 'messy' social science) into overly neatly organised categories bears therefore a pitfall, which is that this would hide the messy, contradictory reality that is exactly what makes inter-organisational projects so mesmerising and bewildering. Marcus therefore explicitly critiques earlier expectations among anthropologists that local narratives should add up to an overarching generalisation about 'the' local culture. And so, telling stories from the field therefore begs the questions: "which ones?" (Marshall & Bresnen, 2013:114).

This highlights the importance of collecting field notes and memos. Writing field notes is a strategy to collect data from the field, but arguably also helps the researcher to develop a practice of reflection. First and foremost, writing field notes builds a log file that makes transparent which different sites were attended to. This can help to identify potential blind spots. Second, for ethnographers of inter-organisational projects it is arguably essential that the researcher empathises with the various sides of a story – while developing a voice that sides with none and delays judgement. While getting a sense of the multiple-homes to which our research participants bear loyalty, the ethnographer is arguably 'homeless' – resisting the temptation to set up camp with one particular group. Writing and then reading our field notes can help us to catch ourselves before we become overly invested in one specific perspective. Furthermore, selecting where to guide one's attention in the field is helped by reading and writing (Swartz-Shea & Yanow, 2009). Finally, sharpening one's eye in the field to specific academic themes or discussions, it helps to write memo's (see also Corbin & Strauss, 1998). In memo's scholars try out how academic debates would explain the dynamics in the field, and what additional fieldwork should tap into in order to make a contribution to that specific field.

Because there are so many moving parts to keep track of during an inter-organisational ethnographic study, reflection on what moving parts build the field is very important. The following sections take a closer look at the context of the construction project field.

## *In the field*

> I am quite pleased to finally be able to visit the construction site. Among the project managers I had met nobody that was particularly keen to make time in their schedule to go there. Then I heard about this man who was hired to make daily inspections at the site, and I made it a mission to convince him to take me along. It then took a safety training, a lot of emailing about where and when to pick up personal safety gear, and keeping a flexible schedule so to jump at any opportunity Mike would find to take me with him. Even though he seemed enthusiastic to share his work, and he was obviously proud of it, he works on an entirely different clock compared to office hours. Going with Mike means I have to be up and ready before 7 in the morning, and anticipate to be back at the office at 11am. This meant I would miss the morning 'start up' of the engineering team. But better to miss one of those regular meetings, than this opportunity …

This vignette illustrates how gaining access to a construction project relies greatly on gatekeepers and key informants (O'Reilly, 2005). These so-called 'gatekeepers' have central roles in ethnography, because these people alchemise formal permission into actual access. Gatekeepers give you the badge to enter the office building or put you in an outlook invite. Key informants are the people who hold pivotal information that functions like a key; their insight unlocks new information. Think of documents, insights on company history or jargon. They speak the local language so fluently and yet understand your outsider non-native bewilderment enough to translate what is going on in their perspective.

To negotiate access to a project, it might be helpful to discuss what you will be able to give back to your research participants. Perhaps they have questions themselves that they ponder over or have topics that continue to trouble their organisations. Think for instance of research presentations, focus groups, action research-based interventions or a report. Offering something to give back can be motivate your research participants to take you on board. At the same time, you want to remain open and clear about your position as an independent researcher from the beginning onwards. Research is different from for example consultancy. In each research project, negotiating access can look a lot like relationship building in practice. This is the moment where we develop those initial impressions of the field and rely on the motivation of strangers to let us into their world. While researchers get to know their field, research participants get to know them and the research project.

Keeping in mind that there are multiple organisations involved with the construction project, access will need to be negotiated multiple times, with different representatives. Gaining access is therefore dependent on the willingness of gatekeepers and key informants. To say 'yes' to a researcher is one thing, but to say yes to an ethnographer is another: sitting in on meetings, having access to company presentation, network events, lunches, coffee machine talk; the ethnographer asks much more investment from organisations. This is also why we refer to the people in the field sites as research participants. Gatekeepers and key informants are both research participants, but they have a different function. Gatekeepers provide access – quite literally because they for example entitle an access pass. You can identify these people because of their formal role in the organisation. Key informants provide information: they tell you about the project history, help you connect to other crucial stakeholders in the project or inform you about dominant dynamics at their work place. While gatekeepers can be identified through their formal role alone, key informants emerge as you build your relationship to your research participants. Organically, some of these research participants will become key figures as you find out more about the project.

Fortunately, we generally experienced a lot of hospitality in the construction industry, especially the public client organisations. When meeting with international scholars in field like organisation sciences and engineering, one usually envies our access opportunities and large network among practitioners (Volker, 2019). This usually comes with the opportunity to provide reflection in action for project participants, either directly to the project teams or in their home organisations. This is also one of the reasons why our research projects often include organising additional workshops, symposia and other knowledge-related activities that aim to transfer the scientific insights of our studies. Not only a valuable experience for the participants but also for the researcher in validating the value of the concepts as developed (Volker, 2019).

Once in the field, ethnographers first need to make themselves acquainted with the different local logics. Once aware of these basics can scholars start to use the discrepancies between the construction site, project office and client organisations as significant in their analysis of cultural dynamics within the larger project. From the beginning onwards, we urge a reflexive scholarship (Serra Undurraga, 2020) that acknowledges the connections researchers establish between organisations are selective and subjective, and therefore require transparent reflection on the position of the researcher. Reflecting on the field and the research process, the researcher can then start iteratively refining their research design with new insights from the field.

Based on our experiences in construction projects and inspiration from multi-sited ethnography (Marcus, 1995), we identified three major questions that we think scholars should play with while doing fieldwork in inter-organisational projects:

1 Who, where, what?
2 Say that again?
3 What is going on here (and here, and here)?

### *Who, where, what?*

> 'This is where our all efforts supposedly come together', Mike says, as he swings the gates open that are guarding the construction site. It is early morning, the sun is hardly set. Mike greets the man guarding the construction site, waving our passes. There is a lot going on this morning, Mike explains. After the contractor hit multiple utility cables during their renovation work, the project organisation hired Mike to supervise the construction work. He has to report back to the project management team who is supervising the contractor. 'I am their eyes and their ears, you can say' he explains. 'They don't join you, or come to see for themselves?' I ask him. He does not answer but looks straight ahead. I guess that was a strange thing to assume, project managers visiting a construction site. We walk in resolute steps towards a team of man leaning on their shovels and hanging from the cabin of the excavator. 'Morning folk!' Mike exclaims loudly. 'Let's see how far you are, can you show me your plan of action?' 'You took your sweet time this morning!' one of the men respond. Is he pissed at us, I wonder? But as he jumps off the excavator to greet us, he has big smile on his face. He and Mike shake hands, and pat on each other's back. 'Show me what is going on here, brother' Mike says to him.

This vignette gives an impression of fieldwork at a construction site, where the local culture contrasted with the white board project office. We learn about this through Mike, who is one of the linking pins between these two sites in the project. In ethnographic studies everything depends on the people that you study. This means that it is advisable to identify early on who seem to be the 'key' people in your study, where they work and how they go about their day. What are their core activities? Why do these activities show up in their day? Is there a pattern to how they structure their days, weeks, months?

In our case example, it proved relevant to remain aware not to be overly informed by one perspective on the project. Even when people would share the same home organisation, different hierarchies are at play. In this case, the fieldworker spent most time at the project office and followed one of the key participants in their everyday work. Especially at the start of the research, this helped to become acquainted with the project. However, to the other partners in the project, entering meetings together was, for example, seen as a sign that the fieldworker was 'one of his'.

To bring about a balance, it helps to build relationships with representatives from each partner organisation. This takes some patience. Some people are preferably private about their work. Others are an open book. You will learn about individual's careers, their hobbies, sometimes their families, and eventually the more delicate challenges they face from their home organisation. Building trustworthy relationships takes time, and in the context of inter-organisational projects it may take some additional self-awareness to build an image of a 'neutral' reliable party. This includes not taking stock in any side (or site, for that matter) in particular, not spilling the beans when they are not yours to share, not being overly invested in one party or another and just being nice, flexible and open towards people. Sometime this also requires getting up early in the morning or being dressed differently ….

In addition to the different cultures of construction work and 'desk jobs', colleagues in a project tend to have another 'home' that they are loyal to; typically, people are sent by their organisations to contribute to the realisation of the construction work anticipated in the project. Hence, in inter-organisational ethnography in construction you need to be aware of a triangle relationship, between the construction site, home organisations and the project office. From the relations between these sites emerges an image of what is going on 'within' the project. The following vignette explains this example.

> When Mike has given his approval to the plan of action of the construction workers, we are ready to go back to the project office. Mike and I walk back to the shuttle that will take us back to the project office, a ten-minute drive away from the airport's center. 'What were those other guys doing there, standing around?' I ask Mike later when we enter the bus. 'Oh, you mean those four, standing near? They were just waiting for us to finish probably. They did not want to be involved, apparently.' He laughs. I don't get the joke. Probably Mike reads the puzzling look on my face. Why wouldn't they? 'It's not that they don't want to. Probably they don't speak Dutch' he says. 'They are here for this particular job, and then they move on. On to the next job.' He doesn't share much more, and I'm a bit embarrassed now I realised my own naivety. Of course, I knew about migrant workers in the construction sector, but I didn't realise. 'What about you, what is your next job?' I ask him, covering up my embarrassment. Mike works for a consultancy firm specialised in underground utilities. He tells me he doesn't know yet. Probably the airport will keep him around for a bit longer, until the work is really done. There are not a lot of people with his kind of expertise in underground utility projects, and so far his regular attendance through these so called 'hold points', the project office hopes to control the risks of damages to utilities. As we drive up to the project office, the bus stops. A guard checks the work ID's of the passengers, and then the bus continues. 'Well, here we are again', Mike says, 'back in the stable.'

In this vignette we see how Mike is the linking pin for the project, between the project offices and the construction site. His daily inspections took place at the construction site, and after that he would file his reports back at the project office. Even though Mike comments on coming back to the stable, he says it tongue in check, for he also comments on not knowing how long the project will hire him for. This shows how different timelines intersect in the project: Mike's work is connected to the timeline of one specific work package, and then the composition of the workforce changes again. Moreover, on a daily basis it was quite a challenge for Mike to find synchronicity between the rhythms of the day at the construction site and the project office. Whereas sometimes the construction site is connected to the project offices, in this case the construction site was too small to house all of the management staff. Consequently, Mike had to navigate between not only two locations, but also two different rhythms: whereas the construction site was usually buzzing, at the project office it was typically much quieter. I often saw Mike write his reports alone, behind a computer. Desks around him were typically empty, besides a coat hanging over a chair, or a lunch box on a desk. Project managers hurried themselves from meeting to meeting, sometimes at the project hub, but also often at their partner organisations. If Mike wanted to know what was going on within the project, he needed to keep track of the agendas of project meetings, and especially be there at lunch time. Otherwise, he would be surrounded by 'empty' desks that told him little about what was going on in the project.

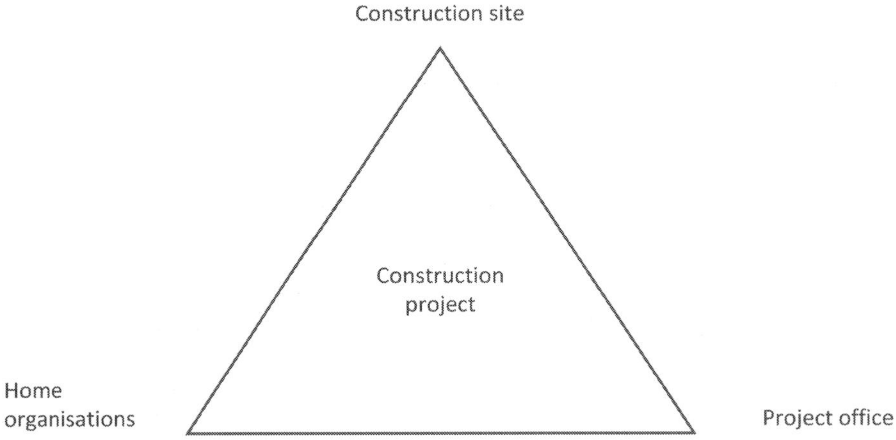


**FIGURE 13.1** Visualisation of the core research sites of an inter-organisational construction project

This example shows how in inter-organisational ethnography in construction you need to be aware of a triangle relationship, between the construction site, home organisations and the project office, which we visualised in Figure 13.1.

Depending on the stage that the project is in, not only the sites of your research might change but also the research participants that you work with. When you are interested in the construction procedure, you will visit those sites relevant to the construction work and schedule in observations with the affiliated professionals. Naturally, the composition of the construction workers changes depending on the stage of the project. To keep track of the variation and differences between your sites, during especially longer times spent in the field, it can help to collect visual documents. Even though walls do not talk – make sure to include your impressions of the key locations in your study in your field notes. Depending on your medium of choice, making pictures of the places, drawing them, making sound recordings or noticing the different smells, colours, materials and energies will automatically lead you to compare or contrast places (Pink, 2015).

When it comes to taking note of the changing compositions of the work force, it might help to list the people you met, each visit. Such logs will allow you to discern who are the 'regulars' and the occasional visitors. In Mike's case, his challenge was to find the right people to keep him updated about the office dynamics as well as the construction dynamics – while the composition of people affiliated with elements of the construction work changed. But most probably, not every project has the same rhythm in turn overs. Note how the people you encounter differ or remain the same. Remember to keep a keen eye on how these dynamics differ between your research sites. And then: who do you keep running into, wherever you go? It might be that few people in the project go to the construction site, for example. Who has the privilege of going where? In this case example, Mike was one of the few who went both to the construction site and to the project office. Writing a memo on questions like these will develop your understanding of place beyond merely 'décor'. Identifying who is prominently present, or absent, will tell you something about the kind of organisational culture that dominates.

Examine also what tools, techniques or technologies people use at these different sites. Tracing the trajectory of a record or software is one approach to getting closer to the organisational dynamics (see Bruni, 2005). Tools, techniques and technologies rarely make sense when examined in isolation. For example, construction work is known to put a heavy load on people's bodies. Despite advanced tools and techniques, accidents continue to happen. What can the protocols for different tools tell you about the safety culture at the construction site? Assessing the risks, physically but also mentally, that people are exposed to is a well-known theme in many construction ethnographies. Their relation to safety might contrast starkly with that of safety managers, who perhaps supervise the enforcement of safety protocols.

Who does what, where and when can be wildly different each day – and still somehow these working days are connected. Perhaps both construction workers and project managers share specific routines related to safety. Shared routines build the rhythms that project members go through. Protocols and routines might even connect the project office and the construction site. It is through identifying the 'mundane' patterns that build a joint sense of the project, that we can start to unravel dynamics within a project culture (see, for example, Sydow et al., 2004).

In brief, continuously taking note of who is present (or absent), in which site, and what goes on in these sites builds a gradual understanding of the everyday patterns that shape daily life inside inter-organisational projects. This can be combined with asking informative questions to the gatekeeper and other people you run into related to 'how' things work, 'who' is involved and 'why' this is the case. Collecting other types of materials and documentation could inform you on the tools, techniques or technologies that are applied on site or which routines are in place at the different sites.

### *Say that again?*

> It is nice and warm at the project office, early morning in January. We are preparing for another visit to the construction site. This time, it feels like it is freezing outside. As we leave the project office and step into the shuttle bus to the construction site, everybody around me is covered in scarfs and beanies. There is a small lining of snow covering the ground. 'Is it not too cold, to go to work today?' I ask Mike. 'There is a protocol for when it is freezing, and for when the ground is frozen. But for now, the work can still continue. Which is good, otherwise we would see delays and additional costs, nobody wants that…'

In our exemplary case of construction work on underground utilities at the airport, there was a noticeable difference between summer and winter. In summer, the construction work continued but the planning was challenged by holidays and leaves. In the winter, the construction work continued – but it was challenged by snow, frozen soil and in some cases even by rain. It was sometimes hard to believe, especially for the supervising management, that welding copper cables took twice as long in the winter than it did in the summer. Surely the contractor was just looking for a way to make more profit? In being there, standing at the pit, seeing the tent that was supposed to shield the bare cables against oxidation, it becomes extremely common sensical. But to project managers who might not even notice the weather, from their warm and dry offices, it is easily forgotten.

Star (1999) noticed how boring and mundane elements are just as informative in ethnography. Star's 'trick of the trade' in articulating those more mundane and possibly even invisible elements of infrastructure work is to identify a master narrative; a single voice that does not problematise diversity (1999). Finding a master narrative is something different from generalising, theoretically or empirically, from your experiences in the field.

With Berthod et al. (2018), we agree that visiting the same people, the same places, and following the same documents or objects around can help to build reliable accounts. For example, following around project documents over time can be an interesting technique to identify patterns of change and stability. Technical drawings of definite designs might seem very 'definite', though in practice these drawings can still change. Following the procedures through which drawings are changed or become 'frozen' can provide a focus point to start unravelling the dynamics between partnering organisations. Following the sequence of design drawings, from a draft to their final stage, also offers you an opportunity to identify if and how there is a change in their shared dynamics. Similarly, interviewing the same people more than once, for instance with a half year interval, means you get to speak to them in different stages in the project. This allows you to refer to assumptions that were prominent in the first conversation and ask if those need to be updated. Moreover, it allows you to see which constants surface amidst the continuously changing dynamics.

Inside the airport project, for example, the role of the underground as a risky place remained surprisingly stable, throughout the different construction stages. The underground was the location where all the planning, designing and negotiating had to come together. As the Dutch soil is typically soft and easy to work with, it was not immediately evident what was so risky about it. 'Listening for the master voice' (Star, 1999) that narrated the story of the underground laid bare examples of 'personification' ('the underground is a risk to project performance'), with a supposedly troubling agenda ('working with the underground is always a hassle'). These unvarnished claims were indicators of the beliefs and assumptions that were considered true within the project – they did not come up just once but came up repeatedly and in rather familiar ways. Identifying local myths or beliefs, like the risky underground, can lead us to better understand a local culture (Geertz, 1988).

Monitoring the changes and constants over a longer period of time can help ethnographers moreover to look deeper and deeper into elements of the work that were not so apparent initially. Through repeatedly interviewing the same people, this process can be built into your research design very deliberately. Asking people not once, but more often will help you to build more refined accounts of your research participants' experiences. At the same time, it should be noted that especially in ethnography, open-endedness and awareness to surprises or detours is embraced as a valuable source for knowledge development. As Star emphasised (1999), incidents are not more (or less) relevant to vivid analyses of the local culture – incidents and mundane patterns simply are informative in different ways.

### *What is going on here (and here, and here)?*

> 'In the end it is about delivering what we have promised.' The project leader of team utilities stands in front of his team. They came together today for their monthly meeting to discuss their progress. 'There is no time nor attention for all the details that each and every one of you is so concerned with' he continues. 'What are details, and what are core concerns in this story?' one of the team members asks him. His manager looks at him and stays silent. 'That is your job to know, is it not?'

In this short vignette of a project team meeting, the dynamics between the project leader and project manager can give the impression that things are a bit tense, even though they were literally and metaphorically 'on the same team'. Initially a bit startled by their way of communicating, after having attended that same meeting a couple of times it became clear to the fieldworker that this kind of interaction did not necessarily imply tension or conflict. It was characteristic to the way people communicated to each other inside the team, with a lot of gusto and not a lot of tact.

Especially when it comes to culture, these impressions of what is considered normal between colleagues are the most obvious when we step into them fresh. At this time the contrast is the biggest. Contrasting between the logics within projects can help to analyse the broader phenomenon like, for example, migrant construction work (see Tutt et al., 2013, for example). The researchers' positionality is therefore a relevant tool of inquiry, so that when entering the field, the researchers initial understanding of the local community will be both enabled and restricted by their own biases and intellectual dispositions (Löwstedt, 2015). Self-reflexive scholarship in sum is another tool to unravel specific dimensions of the internal project culture (Geertz, 1988; O'Reilley, 2005; Pink et al., 2010; Löwstedt, 2015).

Reflecting on impressions with research participants in informal conversations of interviews was a relevant tool to contrast, for example, the researchers' impressions with participants' perspectives on key events in the project. To identify the specifics of 'a' project in comparison to the 'general' construction culture, it helped to inquire the impressions of outsiders, for example, through attending sectoral events and asking there about perspectives on inter-organisational project management with underground utilities.

Sometimes, research participants had strong opinions on what their work was all about, and they would suggest the research took a different focus. Why not look at the tendering procedure, for example? Or at the influence of laws and regulations? These were legitimated questions that helped to refine the relevance and the scope of the research project. Specifically, variations on the ethnographic methods that aim to articulate minority voices or marginalised groups, like diffractive ethnography or abductive research, pay tribute to the role of 'surprises' or 'ruptures' to what is considered the norm (see, for example, Timmermans & Tavory, 2012; Gullion, 2018; Reichertz, 2019). Noting what feels strange, unexpected, surprising, contradictory and weird is a source of information in this line of work. Just like looking for the patterns of action that attribute to the norm of 'how things are done around here', sensing what is considered 'strange' can be indicative of the shared normal.

In sum, while in the field we encourage ethnographers to consider reflecting on the following three fieldwork driven questions:

- *Who, where, what?*
- *Say that again?*
- *What is going on here (and here, and here)?*

Because of the inter-organisational character of a construction project, it is key to clearly identify the people, places and context of the project field at hand. Observation, interviewing and documents tracing are important instruments to analyse possible repetitive patterns and boundaries that could inform you in the things that are going on, not visibly on the different sites, but also underneath the surface. Table 13.1 provides an overview of the actionable suggestions that build on the continuous reflection that the fieldworker is undergoing while performing an inter-organisational ethnography at a construction site.

**TABLE 13.1** Overview of actionable suggestions to build continuous reflections while doing fieldwork in inter-organisational projects

| | *Who, where, what? Identification of the characteristics of the field* | *Say that again? Taking time for repetition* | *What is going on here? Reflection on research findings in context* |
|---|---|---|---|
| People | Identify key research participants from each organisation part of the project, ideally laced with participants from different hierarchical positions | Interview and or observe the same people multiple times – being aware of their double responsibilities | Reflect on your research findings with your research participants |
| Places | Identify the key locations, including the project office, construction site and (representation of) the home office | Observe elements of change and stability at the key locations | Contrast your impressions with 'outsiders' or unusual suspects |
| Project | Identify key activities that bind the project partners to a shared internal logic | Trace how project documents are updated or reviewed | Continuously evaluate the scope of your research: what is 'key' for you, what is out of scope? Does it align with the scope of the project? |

## Conclusion

In this chapter, we have explored the characteristics of doing inter-organisational project ethnography and argued that multi-sited ethnography can be a useful method to inform our inquiries. Inter-organisational ethnographic research has a lot of potential to investigate the everyday work within a project. Multi-sited ethnography starts from acknowledging that the research phenomenon, like 'the project', does not reside in one place or group of people, but instead builds from the direct and indirect relationships they have to each other. Consequently, a crucial tool to shape the scope of this kind of research is a constant reflection on the question which places should be included, which people and which objects. Furthermore, it is important to reflect on the impressions within the context of the research to differentiate between 'the normal' and 'the strange'. It is likely that tracing the organisational culture within inter-organisational projects leads to a kaleidoscopic image. It can be challenging to keep up with the multiplicity and richness within these research sites. To anchor a reflective process alongside our fieldwork can help to build ethnographic accounts that are rich in their complexity, diversity and detail, while maintaining a sense of oversight and structure. Based on three reflective questions we aim to motivate such a reflective ethnographic practice among scholars that study project practices and support their research journey through the intriguing world of construction.

## References


Ainamo, A., Artto, K., Levitt, R. E., Orr, R. J., Scott, W. R., Tainio, R., & Vaara, E. (2010). Global projects: Strategic perspectives. *Scandinavian Journal of Management*, 26(4), 343–351.

Berthod, O., Grothe-Hammer, M., & Sydow, J. (2018). Inter-organizational ethnography: Promises and problems. In *Unconventional Methodology in Organization and Management Research* (Chapter 13, pp. 212–230).

Bruni, A. (2005). Shadowing software and clinical records: On the ethnography of non-humans and heterogeneous contexts. *Organisation*, 12(3), 357–378.

Corbin, J., & Strauss, A. (1998). *Basics of Qualitative Research: Techniques and Procedures for Developing Grounded Theory* (3rd ed.). Sage. https://doi.org/10.4135/9781452230153

Davis, H. (2006). *The Culture of Building*. New York: Oxford University Press.

Geertz, C. (1988). *Works and Lives: The Anthropologist as Author*. Stanford University Press. Stanford, California

Gullion, J. S. (2018). *Diffractive Ethnography: Social Sciences and the Ontological Turn*. Abingdon: Routledge.

Hannerz, U. (2003). Being there … and there … and there! *Ethnography*, 4(2), 201–216.

Jones, C., & Lichtenstein, B. (2008). Temporary inter-organisational projects: How temporal and social embeddedness enhance coordinatioin and manage uncertainity. In S. Cropper, M. Ebers, C. Huxham, & P. Smith Ring (Eds.), *The Oxford Handbook of Inter-Organisational Relations*. Oxford: Oxford University Press.

Law, J. (2004). *After Method: Mess in Social Science Research*. London: Routledge.

Löwstedt, M. (2015). 'Taking off my glasses in order to see': Exploring practice on a building site using self-reflexive ethnography. *Construction Management and Economics*, 33(5), 404–414.

Lundin, R. A., & Söderholm, A. (1995). A theory of the temporary organisation. *Scandinavian Journal of Management*, 11(4), 437–455.

Marcus, G. E. (1995). ethnography in-of the world system: The emergence of multi-sited ethnography. *Annual Review of Anthropology*, 24, 95–118.

Marshall, N., & Bresnen, M. (2013). Tunnel vision? Brunel's Thames Tunnel and project narratives. *International Journal of Project Management, 31*(5), 692–704.

O'Reilly, K. (2005). *Ethnographic Methods*. London: Routledge.

Oswald, D. D. A. & Dainty, A. (2020). Ethnographic research in the construction industry: A critical review. *Journal of Construction Engineering and Management*, 146(10), 1–13.

Pink, S. (2015). *Doing Sensory Ethnography*. London: Sage.

Pink, S., Tutt, D., Dainty, A., & Gibb, A. (2010). Ethnographic methodologies for construction research: knowing, practice and interventions. *Building Research & Information*, *38*(6), 647–659. https://doi.org/10.1080/09613218.2010.512193

Reichertz, J. (2019). Abduction: The logic of discovery of grounded theory - an updated review. In A. Bryant & Charmaz, K. (Eds.), *The SAGE Handbook of Current Developments in Grounded Theory*. Thousand Oaks: Sage.

Serra Undurraga, J. K. A. (2020). Reflexivities as affective ways of relating that produce. *Qualitative Inquiry*, 26(7), 920–930.

Star, S. L. (1999). The ethnography of infrastructure. *American Behavioral Scientist*, *43*(3), 377–391.

Swartz-Shea, P., & Yanow, D. (2009). Reading And writing as method: In search of trustworthy text. In S. Ybema, D. Yanow, H. Wels, & F. Kamsteeg (Eds.), *Organisational Ethnography. Studying the Complexities of Everyday Life*. London: Sage.

Sydow, J., Lindkvist, L., & DeFillippi, R. (2004). Project-based organisations, embeddedness and repositories of knowledge: Editorial. *Organisation Studies*, 25(9), 1475–1148.

Timmermans, S., & Tavory, I. (2012). Theory construction in qualitative research: From grounded theory to abductive analysis. *Sociological Theory*, 30(3), 167–189.

Tutt, D., Pink, S., Dainty, A. R., & Gibb, A. (2013). Building networks to work: An ethnographic study of informal routes into the UK construction industry And pathways for migrant up-skilling. *Construction Management and Economics*, 31(10), 1025–1037.

Volker, L. (2019). Looking out to look in: Inspiration from social sciences for construction management research. *Construction Management and Economics*, 37(1), 13–23.

# 14

# DOING ETHNOGRAPHIC ACTION RESEARCH

*Léon olde Scholtenhuis*

*University of Twente, Enschede, The Netherlands*

**Key takeaways**

- Ethnographic Action (EA) research integrates ethnography with action research steps in a recursive cycle.
- EA enables close-up studies of the dynamics between construction practices and implemented information system technologies.
- EA offers flexibility to develop theories ranging from rich descriptions of wicked problems on fieldsites, and technical descriptions of information system designs addressing the problem, to reflections on implementations of technology in the field.
- EA researchers' roles have cyclic changes throughout the research between intense immersion and minimally intrusive.
- Researchers should be mindful of their role in changing realities during exploration, development, and implementation of new technology.

## Introduction

Ethnographic methods have had limited uptake within construction management research. Particularly within engineering schools at universities, there is less experience with ethnographic research. Traditionally, engineering studies aim to develop technological artefacts and sometimes intervene to improve an empirical process. The discipline of these studies is pragmatic, based on technological design for observable real-world problems. Further, other traditional construction management studies do not involve development. Many of these still are rooted in positivist approaches and may use surveys and statistics to analyse factors' correlations and their causality.

The pragmatic and positivist approaches taken in these types of studies provide useful methods for our domain's methodological toolbox. There is, however, more to benefit from by the construction research community when embracing perspectives that focus on how meaning is socially constructed based on a culture's beliefs and values. This view enables comprehensive research into how people act, talk, and shape meaning within construction organisations. For instance, ethnographic methods provide a deeper understanding of the social aspects of observable real-world engineering problems and the daily practices of workers who deal with issues onsite. This chapter shows how ethnography can be combined with action research to support

DOI: 10.1201/9781003379584-17

scholarly work in construction. It uses the case of information technology (IT) development for the coordination of utility construction projects to show how EA research can be executed.

Especially in the domain of construction information systems, most studies take a technology push approach and present – for example – a data integration method for building and geographic models of infrastructures (Tang et al., 2022), life cycle management data models for buried utilities (Wang, 2021), and methods for automated bridge construction progress monitoring (Puri and Turkan, 2020). While such advanced engineering studies push technological boundaries, their resulting prototypes may align less with the present needs of end users in the field. Such technical research projects typically pass through exploration stages quickly and dive into development stages. As a result, developments may miss out on identifying important nuances that influence the adoption and effectiveness of systems in the field. To embrace such complexities, the literature suggests that researchers develop descriptive theories about technology while it is in actual use by professionals (Orlikowski, 2000).

Engineering and information systems development in construction research could achieve greater impact when technology development choices and evaluation processes are executed within the context of real project settings. Ethnography can support this process through the development of thicker descriptions of the lived experiences of participants, and thus becoming aware of the everyday actions and beliefs of practitioners in ways that are richer than a remote desk study. Further, ethnography allows small cues that otherwise would have been overlooked to be noticed when implemented technology is evaluated. These cues are difficult to observe because the adoption processes of IT on construction sites are informal and emerging (Davies and Harty, 2013). Studies of real-life settings may for example help practitioners and researchers describe how developed prototypes may be limited by scarcity of data, different uses of software systems, and interoperability issues.

A closer examination of the micro-practices that emerge from users that engage with technologies helps unravel the interplay between the many factors influencing technology implementation. The different purposes, theorising modes and researcher roles of ethnography and action research support one another while researchers go through cyclical processes that alternate between invasive and less invasive steps. We elaborate this in the remainder of the chapter, which is structured as follows: the next section explains how ethnography and action research work together within the EA research cycle. The methodologies are first discussed separately, before being re-integrated again in an example of an EA research case. We then continue with a case example of EA work on construction IT before we address three methodological considerations and recommendations for aspiring EA researchers.

## Ethnographic action research in construction

Ethnographic methods can study local work cultures in great depth. This provides a powerful way to obtain a 'deeper understanding of the realities and lived experiences' of the professionals within the industry and through this account for a framing of problems within their specific contexts, and 'the socialities, materialities, and experiences in which they unfold' (Pink et al., 2012). The methods can unveil those details that harder positivist approaches will not be able to include (Oswald and Dainty, 2020).

Researchers gain entry and establish rapport with construction professionals to see them close to the fire, working like they normally do (i.e., without observers being there). Researchers situate themselves, for example, at an engineering office, on a site, or at the locations in between to build trust and acceptance from key informants. Places included in ethnographic work should not only focus on those who adopt new technology at offices and sites but also extend to those developing new construction technologies. This includes the offices of developers of, for example, robotised construction instruments (see Chapter 9).

While being onsite, researchers use observation as a primary data collection method. Observation and listening take place with varying levels of participation: from less intrusive (e.g., shadowing) to fully immersed (e.g., participating in all activities). These modes give the possibility to become embedded within the field. Here, the researcher learns about the patterns that shape a professional culture (Agar, 1996), by developing 'insider views' of the culture and comparing this with the researchers' initial outsider perspective (Jorgensen, 1989, Chapter 1). They then try to grasp phenomena from the viewpoints of the insiders to the local work culture. This 'immersiveness' allows researchers to develop narratives that describe the experiences of the practitioners who 'live' in the local work setting.

One study that applied only ethnography, for example, describes how safety and risk-taking onsite are shaped by individual aspects such as self-regulation and self-confidence as well as workers' community preferences and solidarity (Baarts, 2009). Another example studied industrial ecosystems cases that aimed to increase the reuse of concrete materials onsite. Ethnography helped the researchers identify how actors' roles expanded and emerged, and how communication and collaboration patterns between the actors had to change to facilitate the reuse goal (Harala et al., 2023). The ability to get close to such phenomena is invaluable and hard to achieve with research methods that take a more distant look at practice.

Action research found its way from the 'periphery of mainstream social sciences' during post post-World War II period to an accepted approach in the study of information systems in a human context (Baskerville, 1996). This interventionist approach combines a collaborative analysis of a social situation by researchers and practitioners, with an introduction and evaluation of change (ibid). Just like ethnography, it situates itself amid an empirical setting. Action research, however, also purposefully intervenes in a natural setting. Interventions can be introduced by a construction organisation and observed by the researcher. Alternatively, interventions may be introduced to a setting by researchers themselves.

The approach further goes beyond describing phenomena. Its theories are based on researchers' reflections and explain what was done to improve a situation, how that was done, and why (McNiff, 2016). This participatory approach supports the analysis of human sensemaking processes during a period of introduced change. Often action research is participatory: the researcher then has an active role within the change process itself. Examples of action research are studies of project organisations implementing information systems for data management (Azhar et al., 2010) and improvements of defect management processes by construction supply chains (Taggart et al., 2014).

At its core, EA research is defined as a methodology that combines the identification of local organisational routines through ethnographic methods with a collaborative technological development process between practitioners and researchers (Hartmann et al., 2009). Within the EA cycle, ethnography is initially used to describe complex problems in rich detail. Then, action research develops and implements solutions, after which ethnographic methods again support the evaluation of the implemented solution.

Specifically, EA research follows four iterative steps: (1) ethnographic exploration of the routines in the field; (2) discussion and description of routines with people in the work culture; (3) designing or redesigning an information system prototype to support the routines; and (4) implementation of the prototype by researchers and practitioners. This is followed by another cycle of observations to validate or further improve the designed systems (Hartmann et al., 2009). An example of these steps is presented in the case study below (see Box 14.1). Through methodological integration of methods in EA research, researchers thus thoroughly understand a problem situation within the boundaries of a local work culture, tailor the developed system to the needs of the field, and interpret its impact comprehensively.

**Box 1: Development and Use of a 3D Digital Surveying Technology**

This case from my research team illustrates the steps through which scholars go when conducting EA research on information systems for utility construction. The case involved the development of a new mobile technology that supported land administration work by crews onsite. Researcher Nima worked for six months conducting *ethnographic observations* (step 1). He shadowed two land surveyors as they drove between various fieldsites in the Netherlands to survey as-built pipelines on construction sites. Nima observed how the work setting pressured surveyors to visit construction sites quickly and register recently deployed lines right before trenches had to be backfilled. He observed how a surveyor travelled large distances between sites and often was too late to accurately measure a deployed cable with his traditional GPS tools. Safety protocols pressured workers to close trenches at the end of workdays. This often meant that the crew could not wait long hours until surveyors measured a pipeline in an open trench. Nima *identified a* tedious *working routine* (step 2) comprising different data collection and processing steps. Time pressures and skill shortages often prevented these steps from being executed accurately. Pipelines were surveyed in a closed trench, reducing the accuracy of the measurements. Nima *developed an information system* solution (step 3) that would not require the surveyor's sophisticated positioning tools (e.g., GPS). His mobile solution comprised 3D global positioning and laser scanning techniques. Combined with machine learning, these techniques formed a method that to semi-automatically scanned and developed a 3D model of the pipeline in the trench. To then implement *the information system on the project* (step 4), Nima instructed jobsite managers, who then explored it within a natural setting of use. While providing technological support, Nima *conducted observations* (new EA cycle, step 1) to identify how the new technology caused required changes to data processing skills, software systems, and surveying standards. He furthermore studied how surveyors and jobsite manager reimagined their job identity as they were confronted with the developed solution.

The methods used in EA research are also common in Design Science Research (Johannesson and Perjons, 2014). Design science researchers argue that management sciences tend to develop theories that are too generic or impractical to support the everyday work of managers. They hence claimed that researchers need to connect their theorising work better with the development of solutions to practical problems to reduce the gaps between a theory's development and its pragmatic use in the field (Aken, 2004). EA can hence be seen as a method that tightens connections between theory development and its use in the field.

**TABLE 14.1** Purpose, theorising, and researchers' positions in ethnography, action, and EA research

| | *Ethnography* | *Action research* | *EA research* |
|---|---|---|---|
| *Purpose* | Exploring a work culture | Introducing and evaluating an intervention | Tailoring an intervention to a local context to gain contextual insights about its impact |
| *Theorising process* | Vividly describe lived experiences to culminate in micro theory | Reflect or explain dynamics and mechanisms influenced by change | Describe the developed intervention or explain its impact |
| *Researcher's position* | Immersed in fieldsite as learner/novice | Immersed as a change researcher | Iterating between intense immersive and non-intrusive |

The strength of EA is that it both gathers knowledge and understanding about a natural setting while also trying to change and improve this very local setting through the implementation of new solutions. It tailors technologies to the routines of construction professionals, brings them to actual contexts of use on projects, and evaluates their impact (Hartmann et al., 2009). This is demonstrated in, for example, studies of the adoption of building information models for workers on demolition projects (van den Berg, 2019), and the implementation of information systems to support project management of metro projects (Mahalingam et al., 2015). Because of the immersed position during the ethnographic stages of the EA approach, researchers can evaluate planned change with a greater insider perspective. Table 14.1 summarises these purposes, theorising modes, and positions of the researcher for the distinctive ethnographic and action research methodologies, and their integration in EA research.

While the integration of action research within ethnography provides benefits, it means that researchers may violate a basic premise in ethnography to minimally influence the natural setting being studied (Angrosino, 2007). This is because, in action research, researchers contribute to implementations within the field and so become inseparable from the change process that they study. Transparency towards participants about the goals of each phase hence helps obtain support throughout the different phases in the EA cycle. During the ethnographic stage, researchers want to minimise their influence on the natural setting and use primarily ethnographic interviews, documents, and participatory observations as data collection methods. For this purpose, it may not be necessary to share the objectives of the full EA study extensively to develop an initial diagnosis of the present situation. However, it is required from an ethical viewpoint to be transparent about what is done with participants' data *after* the ethnographic stage. This means that the researcher should be clear from the outset of his EA work about his own roles and the change-making purpose of the study. The collected explorative data, after all, may be used as a starting point for the development of a technological system. Such later action research stages may introduce resistance because participants will be confronted with change.

Below, I provide an extensive example of my work through a first-person narrative to explain how I combined steps in the EA cycle.

## Doing EA: A construction management IT case study

My step into EA research started during my postgraduate period. I studied IT in construction and was mostly interested in how IT intervened with the realities of the envisioned users of the new systems. My literature reviews led me to papers about adoption factors and comprehensive causal models of innovation. Two popular models presented in this literature were the technology acceptance Models I and II. These were developed through positivist approaches that used experiments to correlate and cluster factors that influence the acceptance of technology by users (Venkatesh, 1999, 2000). Even though these models were appealing, they looked mechanistic and gave me little insight into the everyday actions and beliefs that users of technology had. I believed that it would be difficult to use these models to develop a useful and new theory about the use of information systems in my envisioned study of sewer construction projects. Instead of opting for these models – which in my view missed specifics of context that influence the daily of use IT – I wanted to use scientific work that focused on how technology 'really' interacted with daily practices on projects.

A very valuable advice I then got from my supervisors Prof. Timo Hartmann and Prof. André Dorée was to immerse myself more in the fieldsite to develop curiosity for what happens there. They nudged me in this direction when I extensively reviewed the technology adoption literature during the first months of my PhD. One of the supervisors passed my desk making a quick remark that sounded ironic but also contained a message for me and my office mates. He stated: 'Léon, do not read too much, this will make you unhappy!'. The remark initially surprised me. After all, one of the academic activities is to take a deep dive into literature. Or not? My supervisor's remark made me wonder: Did I try too hard to find *the* scientific gap? Were my supervisors afraid that I would get lost while connecting abstract concepts and papers? Or was their message that my research should connect strongly with daily practice, thus, focusing on the empirical issues rather than on expanding high-level theories?

The questions eventually drove me to look at methodologies that connect intimately with practice. I studied the seminal work of scholars such as Spradley (Agar, 1996; Spradley, 2016). As first guides in ethnography, these books helped me understand the importance of developing accounts of the fieldsites, taking note of surprises and being sensitive to mundane situations. Following other examples from the work conducted in my research unit, I planned to combine my ethnographic work with the development and use of virtual design and construction (VDC) software. EA research then soon became my main approach to studying change through developed technology.

The empirical work for my postgraduate research concerned the topic of coordination between utility network owners and contractors of sewer reconstruction works. When my project started, the national road and waterways authority, Rijkswaterstaat, piloted digital technologies for the design of their infrastructures. This led megaprojects to adopt Virtual Design approaches. The smaller and local infrastructure projects, however, seemed less impacted by such technology. I was therefore keen on exploring whether VDC would also be able to support the coordination of those smaller-scale infrastructure works.

My research design followed a flexible iterative approach. As I progressed, its goals and outcomes became more specific. I, for example, did not start my study with a list of construction projects from which I selected a few to observe. I also did not yet have a specific type of VDC technology in mind to use during my project. Only my study of the EA methodology and my curiosity in sewer project coordination and VDC technologies appeared sufficient to enter the field.

Entering the field took place after the chair of my supervisor's industry network (the innovation foundation Stichting Pioneering) invited me to pitch my research ideas to a range of senior managers in regional construction organisations. This pitch brought me to one municipality to conduct my first study. The relationship that I developed with the project team allowed me to reach out to their competitors and colleagues, and to be introduced to a successive project.

While conducting the study, I aimed to be transparent to the respondents about the nature of my work. I always introduced the goals of my study to the new people that I met. I was mindful of the ways in which my studies may impact the work of professionals and tried to be receptive to the critique that professionals could have on the technology that I introduced. The anonymity of participants remained while I interviewed them, took pictures, and recorded conversations. My approach allowed me to spend time in the field without experiencing restrictions.

I conducted ethnographic work on four sewer reconstruction projects in total. My research process went through cycles of observing project management practices, developing rich descriptions of work routines, developing new technological models, and implementing these in the project management practices. The flow between the research steps was gradual – i.e., the transitions were not that distinct and visible. The subheaders below are only inserted to improve legibility.

### *Ethnographic exploration*

Gaining entry at my first construction projects was unproblematic. As a student aged 24, I was perceived as an apprentice by practitioners. My position as a student allowed me to express my curiosity in VDC technology to a senior manager at the local Municipality of Enschede. I explained that I wanted to study how VDC supported coordination between them (the client organisation) and third parties such as utility network owners and contractors. The manager introduced me to his contractor's project manager, who then allowed me to become involved with the project team.

The team permitted me access to their formal progress meetings, shared project designs, minutes, and other planning documentation. I was allowed to take pictures from mundane situations, such as the – in my first impression overwhelming – collection of planning artefacts such as the printed designs, notes, and bar charts lying on a large central table at the entrance of the temporal project office. I also photographed ongoing bilateral planning discussions between municipalities, gas network owners, water companies, and telecoms owners and pictured how the use of flexible editing tools such as pencils, highlighters, and printed documents allowed them to discuss and swiftly adapt their schedules in a non-digital way.

It was acceptable for the project manager that I entered his site office for small chats during coffee breaks. During these meetings, he explained the logistical struggles he faced when deploying construction material and moving equipment around at a small construction site in a busy urban area. He also elaborated on how he prepared schedules and design documents to support the construction site team in this task.

I returned to the construction site multiple days per week to develop my understanding of the schedule and logistical constraints of the project, and how planning artefacts were used to support communication. One morning, I arrived early before a scheduled meeting when I got the opportunity to speak to the senior jobsite supervisor. He explained to me that he usually walked around the construction site before a formal meeting, and he invited me to tag along. The supervisor introduced me to various specifics of his site: he first explained about some old decaying clay pipes that he had to remove from the underground, and the challenges they faced to haul soil and pipelines within the constrained space of the project site. He continued discussing the difficulties of moving the excavators around the small street, and the challenge he faced to align his sewer installation process with re-alignment of the many buried utilities that were owned and constructed by a range of different stakeholders. The conversation helped me understand how spatial challenges and the involvement of many disciplines in the planning efforts complicated construction processes.

While spending time at the offices, attending project meetings, and discussing with professionals, I then took note of how design and planning documentation was stored, displayed, and exchanged. I learnt through the various discussions much of the project's design information had to be collected from different organisations too. Distinctive stakeholders, such as cable owners, pipeline owners, road owners, and their contractors owned separate parts of the project work and they used different systems to manage their project designs and schedules. I jotted in my research diary how the different ways of sharing these design data, schedules, and milestones complicated coordination discussions and led to ambiguity and lack of oversight.

My position at the fieldsite seemed to become accepted and taken for granted after a period where managers and workers had seen me observing them multiple times. Especially later, when I repeatedly met the same professionals on two successive projects, they casually spoke to me before or after meetings. They made jokes about me 'just being there onsite to watch'. They asked questions like: 'When do you actually finish your study', 'Did you graduate already', and 'When will you start a real career in construction?' While these questions often were with good intentions, some of them were meant to be jokes. They gave me a slightly ill-understood feeling, but also gave me the comfort of being accepted as the learner in the situation. The participants seemed to see no harm in me studying them in action.

### *Describing routines in the work culture*

While I was onsite, practitioners were comfortable enough to tell me about their discomfort with existing project planning practices. On all projects, I spoke to managers that experienced challenges with obtaining and integrating the information from all stakeholders, and with updating schedules based on the changing dynamics onsite. One site manager, for example, expressed to me that he often felt that the planning documentation provided by his managers was outdated and incomplete. He elaborated that he sometimes did not use the schedule produced by his manager, because the plans that he kept in his mind could better reflect the quickly changing situations onsite.

Through those interactions, I was able to describe the coordination processes in the sewer construction work through the various perspectives of municipalities, contractors, and utility network owners. I developed descriptions of these processes on the first project and became interested in the use of design and planning artefacts by professionals. I described how the workers used their paper-based designs and spreadsheets with bar charts to support their coordination discussions.

While I knew about a few studies in the literature showing how VDC tools could support this 'traditional paper-based and 2D mode of coordination', little of this was seen in the empirical realities that I observed. Why was this? And what would be needed to evaluate the merit of such technology? These questions led me to the action research steps of my study.

### *Design of an information system model*

Based on my prior reading of building information modelling literature I expected that four-dimensional (4D) visualisation models could be a helpful VDC technology for coordination of sewer projects. I hence requested the client's project manager to allow me time at the end of their next project team meeting to explain my thoughts. He allowed me to give a short demonstration about the features of a 4D model. The video I showed explained how the combination of 3D design models and construction schedules can produce an animation of stakeholders' different construction activities at the project site. I offered the representatives from the client and civil contractor at the meeting to 'study VDC' by developing a 4D-visualisation model for their project and using it during future meetings.

The team positively responded to my proposal and agreed to support (what they called) 'my study project'. I was allowed to join the work planner, project managers, and site supervisors for the duration of the project to develop 'my digital models'. I successfully requested project documentation, and I could enter the site office during business hours to work on my model. I spent a few days together with another intern at this office to master VDC software and explored how the planning artefacts from the project could be digitised and integrated within the virtual model.

As the development continued, I involved more team members that I met at the site. I asked them to review my initial versions of the project's visualisation models. They keenly helped me during model development by, for example, organising a multiple-hour session at their regional office. During the session, we discussed how the existing project schedule was visualised in the model, and what other elements should else be included in the models to support meaningful discussions during progress meetings. We then identified, for example, different types of physical construction objects, objects from the project surroundings, and intangible concepts – such as milestones and delivery moments that had to be visualised to support stakeholders' discussions.

### *Implementing the information system*

After the various professionals helped me develop the first prototype of the model, I got the opportunity to demonstrate it during a formal project progress meeting. The opportunity given to me to demonstrate the VDC technology was valuable for my research, as this was the first time most of the stakeholders saw such a tool being applied within their work practice. During my first demonstration of the technology, I did not intervene much in the natural process of the group meeting. I showed the model after a brief technical description and allowed the team members to freely respond with their thoughts about the technology. To avoid that my data collection hampered stakeholders from freely expressing their views, I decided initially to not use my voice recorder to record the discussions that took place during my first model demonstration.

To still keep track of the discussions emerging from the demonstration of the tool, I was prepared to make handwritten notes on the fly while I went through the demonstration of the projects' model in the VDC software. This resulted in my observation diary being filled with technical descriptions of how the stakeholders thought that various visual elements in the model (e.g., like kerbstone, traffic lights, and pavement layers) had to be shown better using other geometric representations. Further, I wrote about the discussions that took place between the team members about how the 4D model could be used in the future. They discussed how new practices for coordination could be shaped when using the tool for the coordination of sewer projects. They suggested new technology-enabled tasks such as, for example, 4D enhanced integrated scheduling and updating, 4D construction progress tracking, and the exchange of enriched 3D-data models.

### *Continuing the EA cycle on three successive projects*

The relationships that I developed with participants on the first sewer construction project led me to repeat my ethnographic exploration of coordination work, and development of 4D models on three successive sewer construction projects. These were organised by distinctive civil contractors and municipalities but involved similar stakeholder types. My ethnographic work resulted in similar descriptions of how project team members faced troubled alignment and updating of the different parts and versions of planning artefacts going around in the project.

During the three projects, my action research work led me to iteratively develop a method for 4D-model-based coordination of sewer projects. Each time, I introduced myself to the teams as a doctoral student and requested time at the second project meeting to explain my intentions of implementing 4D models. I was engaged with new types of stakeholders to understand how they would relate to the technology. For example, the municipality's digital designer was shadowed to understand how he developed the drawings that served as input for the 4D model. I also helped the project manager collect missing pieces of information that were required to complete the master project schedule for the 4D model. Finally, the organisers or the official planning meetings on all three projects allocated specific slots for me to show 'my 4D virtual version' of their project schedules. I became an accepted student team member and was permitted to record most of the official meetings using voice recorders and handy cams.

### *Outcomes*

During the three projects, my 4D model development took place in the natural setting of project work, together with practitioners. I kept a research diary to track the design changes that I made while developing the different versions of the 4D models. This enabled me to eventually develop knowledge about how 4D models should be tailored to fit coordination needs of construction teams (olde Scholtenhuis et al., 2016a).

Furthermore, I organised additional meetings to show the models to the practitioners. This changed the formats in which project information was exchanged and triggered discussions between practitioners about the need to structure and better exchange fragmented design datasets to embrace 4D models effectively. Eventually, the collected field notes, audio records, and videos served as input for theory development on how collaborative use of the modelling software by project teams helped coordinate in the multi-stakeholder setting of sewer reconstruction (olde Scholtenhuis et al., 2016b).

Altogether, the stakeholders on all four projects warmly welcomed me as a researcher to first learn about their work and then propose one technological model to support their practice. The relationships built helped smoothly connect four successive projects that could be studied in the available time window for my study. This successfully culminated in the development of a method for effective 4D modelling with IT systems and a contribution to theories about how 4D models influenced existing construction coordination processes.

## Methodological considerations

EA research supports a rich and deep investigation of everyday construction management practices and the way in which new technology prototypes (re)shape this. Three key elements come into play when one aspires to follow the approach. These are: theorising, reflexivity, and role awareness.

### *Theorising modes*

EA research combines two ethnographic and action research modes of theory development. Initially, researchers are novices to the fieldsite with fresh perspectives on the studied local work culture. This allows them to be surprised by events and to be attentive to everyday details in the local situation. The photos that they take at these occasions, the notes that they make, and all the dialogues that they record support the researchers' initial sensemaking of the present. These efforts culminate in theoretical models describing, for example, how design and planning artefacts are produced by stakeholders on the sewer reconstruction projects; how these artefacts are being combined and discussed; and how local project dynamics often cause changes, and debates about these very artefacts.

The ability to see these details of work in natural settings may be reduced when more time is spent in the field. Observations will gradually lead researchers to disclose patterns and reach a 'funnel' (Spradley, 2016) stage where the attention is narrowed to specific and direct elements in the investigated setting (Angrosino, 2007). When this directed attention leads to replication of previously observed patterns, researchers become more immersed in the field, and the richness of other everyday events get noticed less.

Furthermore, the action research we undertake will purposefully influence the dynamics at the fieldsite as it aims to theorise about how interventions influence the described natural settings. Based on initial ethnographic observations, the researcher develops assumptions about how the existing practices may be supported by an intervention. During my study of VDC on sewer projects, for example, I assumed that the data integration and visualisation features of 4D software could help stakeholders resolve coordination problems on sewer projects. While I collaborated with stakeholders to build and implement 4D models, I developed theoretical contributions about the effective features of 4D models and the way they support coordination practices. This theorising about the technological model itself and its impact on practice is different from the initial descriptive theory development of natural (unchanged) coordination processes.

### *Reflectivity*

Ethnographic researchers reflect on their own impact of the unfolding events in the field to create transparency about how their interactions with participants have led them to act as they did, and how these eventually shaped findings (Pink, 2020). To describe a local situation, the researchers need to place themselves within the context of work, which often means that they need to reveal the purpose of their research, the duration of their stay, and their specific interests in the work of professionals. This information influences what practitioners decide to show or hide from the researcher, and whether they support the researcher further during action research steps.

In the sewer project case, for example, my contacts at the first project helped me understand coordination problems on sewer projects. I was lucky to be invited to all team meetings and speak directly to project managers of both the clients and contractors. I also got access to many artefacts that were used for stakeholder coordination. I also intentionally decided not to focus on many other interesting phenomena, such as the ways in which uncertainties in underground influenced cost rework overruns; or how strategic behaviour of stakeholders caused delays.

EA researchers need to be continuously aware of their own impact on the unfolding events. Hawthorne effects (Aboagye-Nimo and Emuze, 2017) may occur when participant behaviour is influenced by the presence of the researcher. This reduces the authenticity of the descriptions of the 'normal' practice at a fieldsite. The technology intervention in this chapter also affected practice. Initially, I had to explain the features and expected process impacts of 4D models to stakeholders. These explanations and the subsequent involvement of the practitioners during the development of the 4D models may have influenced the team members' reflections on the merit of the VDC technology. Furthermore, my intervention was purposeful and would not have taken place without my presence. Consequently, reflections on the impact of the technology had to account for my role as an active moderator and designer of the models onsite. It thus remained difficult to make claims about how the coordination processes would have evolved in situations where my role was not that prominent.

In summation, EA requires researchers to be mindful while they collect data. They can do so by avoiding expressing their opinions about how information systems should be used at the fieldsite, by withholding from sharing their a priori knowledge about the technology with participants, and by executing a technology moderator role neutrally (i.e., without trying to engage with project team discussions that the technology triggers). Finally, reflections on researchers' own contributions to the emergence of practices are needed to support transparency.

### *Role taking*

Essentially, ethnographic methods can be minimally intrusive or embrace a participatory approach in which the researcher executes routine tasks as part of the team at the fieldsite to 'blend in'. Action research steps require a more intense and active role approach. Shaping and sustaining clarity about the role of oneself and the informants supports the transitions between these types or roles.

How participation is filled in extends here from being a participant who at most executes routine tasks to learn about the *existing* practice to one of active interventionists (i.e., a producer of a smaller or larger change). Finding the right moment to move between these positions is difficult. Too early swapping may lead to incomplete narratives and challenging descriptions. It may also lead to a lack of support from stakeholders to support interventions (Hartmann et al., 2009). Late swapping, however, poses the risk that the remaining time available on the studied fieldsite no longer allows researchers to introduce and test an intervention. In the presented study, this was resolved by conducting multiple EA cycles on four successive projects.

The first project in this chapter contained mostly ethnographic work: the problem context was explored extensively before I was able to describe the challenges that stakeholders on sewer projects faced to coordinate their work. This description allowed me later to propose, implement, and evaluate the VDC technology. On the successive projects, I could thus move my focus from ethnographic exploration to action research that involved model development and implementation.

Another consideration is to be aware of the visibility of the researcher during the different EA stages. During later stages of the EA cycle, for example, the action research roles support a prominent role on the fieldsite. The researcher then presents their intervention to support the studied practice. However, the phases before and after the intervention with the solution require the researcher to have a less visible presence onsite. At these moments, the researcher is more likely to develop detailed accounts of the dynamics between the professionals at the fieldsite and the technology.

In specific, when researchers follow iterative steps, they may swap between minimally intrusive and intervening roles a few times. For ethnographic stages, these roles range from being a mostly observing participants to more fully integrated participating observers, and complete immersion (Angrosino, 2007). In action research stages, the research intervenes and reflects actively (for example, as a technology moderator) within the context of the fieldsite. In the discussed VDC- case, for example, I took different roles by observing project meetings (observer as participant role), collecting sewer construction planning artefacts (participating observer role), and developing a 4D model of the project later (active interventionist role), to then observing the impact of the model within a project setting (participating observer).

There is not a fixed process that alternates between the different roles. The chosen role may, though, influence the research process and study outcomes. My project, for example, paid specific attention to my role during the implementation of a VDC technology. I carefully listened to participants before making the informed decision to propose the 4D technology. I thus initially tried to be less intrusive in observing all viewpoints about technology (from proponents and critical informants). This would avoid that I later developed a technology that participants only reluctantly accepted as one that was imposed on them by me or the client organisation. Building rapport with informants and remaining clear about the intentions of my study were therefore steps that I needed to undertake with care. My attempt to have a neutral role with respect to technology eventually allowed me to receive different views on how current planning practices conditions (e.g., related to data completeness, lacking software, or digital literacy) would still impede the success of 4D in practice.

## Reflections and recommendations for aspiring EA researchers

Combining ethnography with action research steps beyond descriptive theorising about phenomena in everyday situations, EA uses ethnography in pragmatic ways (see Chapter 10) to support interventions. This enriches the toolbox of methods that construction management researchers benefit from. The exemplary case shows that ethnography supports a grounded exploration of the challenges at fieldsites, leading to the development of bespoke interventions, and consequently, leads to a better understanding of the evolving dynamics between an intervention and its surrounding local context.

I end this chapter with recommendations for aspiring EA researchers about theorising, reflexivity, and role positioning. This is written in the first and second-person perspective to direct attention to the readers who consider EA research.

I first recommend carefully consider upfront which types of fieldsites are being studied, and what elements on that site are specifically of interest. The use of ethnography provided by engineering researchers is more time-consuming than common desk-study approaches. Time is needed to observe the construction site, its physical environment, the artefacts being presented, the language being used, and actions undertaken. To use time wisely, an initial focus on a phenomenon and intervention type is therefore helpful. Only at the start of the EA cycle you will have the opportunity to explore such elements from a detailed unfocused and fresh outsider perspective (Jorgensen, 1989, Chapter 6). This description itself is rich and contains a wealth of new insights that could be standalone publishable work. These foundational descriptive theories resulting from the first ethnographic steps in the EA cycle could later be used to contrast with practices that were subjected to developed interventions.

Secondly, I recommend keeping a detailed record of your experiences onsite, both during explorative ethnographic work, as well as the action research dates. The use of media recordings taken photos, and recordings of meetings) and the use of daily reports and field notebooks will help memorise the fieldsite dynamics before and after the intervention stage. These data sources support reflexive thinking about your impact on data collection (and analysis). It becomes increasingly difficult over time for you as a researcher to reflect on your own impact and your prejudices, especially when you combine the minimally intrusive (ethnographic) and intervention-based (action) research methods. While the fieldsite changes when action research takes place, your ability to reflect on your early fieldsite work thus becomes more difficult. I recommend dedicating specific space in your field notebook to allow for instant reflection on the role of yourself during all fieldwork episodes to avoid overlooking the influence of your role changes on theory development.

During the EA study, the position of the researcher changes: one gradually gains stronger visibility and prominence in the fieldsite. My third recommendation is thus to communicate about your role change transparently with all participants. My coping with this has always been to update the participants by sharing my curiosity about diagnosed challenges onsite, the features of technological interventions I had in mind, and my intentions to study such technology on their project. Often supported by the established rapport during prior stages, participants (the opponents and proponents) then openly reflected with me on the impact that the intervention may have on the fieldsite during and after my stay.

## Conclusion

EA research allows the development of micro theories of change, usually based on short-term intense forms of ethnography, information systems development, and the reflection on the observed implementation process. Where short-term ethnography follows a deliberate and interventional approach (Pink and Morgan, 2013), EA extends this with purposeful development, implementation, and reflection of bespoke information systems.

For many construction management studies that aim at achieving impact through an intervention, ethnographic methods such as participatory observation, shadowing, and ethnographic interviewing can increase impact substantially. By generating more detailed accounts of problem situations, and by allowing detailed co-development of contextual knowledge about change, the method can reduce the gaps between the development of technology, actual implementation onsite and the development of theory about technology in use. At its core, ethnography, therefore, improves context-awareness when planning interventions during action research and therewith provides the opportunity to create better designs *and* to consider those subtle elements that influence its eventual impact on the organisation. Action research furthermore helps increase the relevance and impact of the theories that ethnographic work produces. It conceptualises well-explored elements of the fieldsite as problem space to develop, implement, and reflect on a solution for it. Together, ethnography and action research hence produce rigour and relevance by making technology development more context-aware, while at the same time delivering theory that addresses technology adoption issues at the level of everyday practice.

Based on the presented cases I exemplified how theorising, reflexivity, and role taking and awareness influence EA studies. I recommend spending significant time at fieldsites to build a solid theoretical foundation for successive EA steps, I suggest developing reflexive accounts in field notebooks to allow for assessment of the researchers' own role during the study, and finally, I propose keeping practitioners informed throughout the process about the changing role you have as EA researcher.

## Acknowledgements

This chapter uses a case from the research from Dr. Nima Zarrinpanjeh (University of Twente), and it is based on my previous projects with Prof. Timo Hartmann (TU Berlin) and Prof. André Dorée (University of Twente). I thank them for their collaboration.

## References

Aboagye-Nimo, E., Emuze, F., 2017. Construction safety through housekeeping: The Hawthorne effect. *Journal of construction project management and innovation* 7, 2027–2038.

Agar, M., 1996. *The professional stranger: An informal introduction to ethnography.* Emerald Group Publishing, Leeds.

Aken, J.E., 2004. Management research based on the paradigm of the design sciences: The quest for field-tested and grounded technological rules. *Journal of management studies* 41, 219–246.

Angrosino, M., 2007. *Doing ethnographic and observational research.* Sage, London.

Azhar, S., Ahmad, I., Sein, M.K., 2010. Action research as a proactive research method for construction engineering and management. *Journal of construction engineering and management* 136, 87–98.

Baarts, C., 2009. Collective individualism: The informal and emergent dynamics of practising safety in a high-risk work environment. *Construction management and economics* 27, 949–957.

Baskerville, R.L., 1996. A critical perspective on action research as a method for information systems research. *Journal of information technology* 11, 235.

Davies, R., Harty, C., 2013. Implementing 'Site BIM': A case study of ICT innovation on a large hospital project. *Automation in construction* 30, 15–24.

Harala, L., Alkki, L., Aarikka-Stenroos, L., Al-Najjar, A., Malmqvist, T., 2023. Industrial ecosystem renewal towards circularity to achieve the benefits of reuse-learning from circular construction. *Journal of cleaner production* 389, 135885.

Hartmann, T., Fischer, M., Haymaker, J., 2009. Implementing information systems with project teams using ethnographic–action research. *Advanced engineering informatics* 23, 57–67.

Johannesson, P., Perjons, E., 2014. *An introduction to design science.* Springer, Berlin.

Jorgensen, D., 1989. *Participant observation.* Sage, Thousand Oaks.

Mahalingam, A., Yadav, A.K., Varaprasad, J., 2015. Investigating the role of lean practices in enabling BIM adoption: Evidence from two Indian cases. *Journal of construction engineering and management* 141, 05015006.

McNiff, J., 2016. *You and your action research project.* Routledge, Abingdon.

olde Scholtenhuis, L.L., Hartmann, T., Dorée, A.G., 2016a. 4D CAD based method for supporting coordination of urban subsurface utility projects. *Automation in construction* 62, 66–77.

olde Scholtenhuis, L.L., Hartmann, T., Dorée, A.G., 2016b. Testing the value of 4D visualizations for enhancing mindfulness in utility reconstruction works. *Journal of construction engineering and management* 142, 04016015.

Orlikowski, W.J., 2000. Using technology and constituting structures: A practice Lens for studying technology in organizations. *Organization science* 11, 404–428.

Oswald, D., Dainty, A., 2020. Ethnographic research in the construction industry: A critical review. *Journal of construction engineering and management* 146, 03120003.

Pink, S., 2020. *Doing visual ethnography.* Sage, Thousand Oaks.

Pink, S., Morgan, J., 2013. Short-term ethnography: Intense routes to knowing. *Symbolic interaction* 36, 351–361.

Pink, S., Tutt, D., Dainty, A., 2012. *Ethnographic research in the construction industry*. Routledge, Abingdon.
Puri, N., Turkan, Y., 2020. Bridge construction progress monitoring using lidar and 4D design models. *Automation in construction* 109, 102961.
Spradley, J.P., 2016. *The ethnographic interview*. Waveland Press, Long Grove.
Taggart, M., Koskela, L., Rooke, J., 2014. The role of the supply chain in the elimination and reduction of construction rework and defects: An action research approach. *Construction management and economics* 32, 829–842.
Tang, L., Chen, C., Li, H., Mak, D.Y.Y., 2022. Developing a BIM GIS–integrated method for urban underground piping management in China: A case study. *Journal of construction engineering and management* 148, 05022004.
van den Berg, M.C., 2019. *Managing circular building projects*. University of Twente, Enschede.
Venkatesh, V., 1999. Creation of favorable user perceptions: Exploring the role of intrinsic motivation. *MIS quarterly* 23, 239–260.
Venkatesh, V., 2000. A theoretical extension of the technology acceptance model: Four longitudinal field studies. *Management science* 46, 186.
Wang, M., 2021. Ontology-based modelling of lifecycle underground utility information to support operation and maintenance. *Automation in construction* 132, 103933.

# 15

# DOING ETHNOGRAPHIC VIDEO INTERVIEWING

*Joan Evelyn Ongodia and Timo Hartmann*

*Technische Universität Berlin, Germany, Berlin*

**Key takeaways**

- Ethnographic video interviewing methods can be used to gather and develop knowledge in unconventional working conditions with decentralised workplaces having diverse teams in different geographical locations, before site establishment, or when unforeseeable circumstances impede normal working conditions.
- Ethnographic video interviews allow time, space and location flexibility since experts can show their workplace practices through video in the comfort of their workplaces and video recordings can be archived for future reference.
- The key to successful ethnographic interviewing is interviewer flexibility to support interviewees' freedom of expression and a good understanding of the construction domain to interpret expert sketches and project documents that experts often use to explain their points of view.
- Ethnographic video interviews differ from conventional interviews because they allow interviewees freedom of expression using any format to enable capturing in-depth data that can be used to uncover reusable empirical results for improving future construction projects. For example, this chapter uncovers our approach from tacit expert knowledge to explicit knowledge, represented as a reusable empirical tool for reviewing and updating construction project requirements.

## Introduction

We live in the information age, where knowledge is the most prized currency. Knowledge development, sharing and management is an intricate process that goes through various phases, including socialisation, externalisation, combination and internalisation (Nonaka & Takeuchi, 1995). Knowledge is developed as individuals articulate their personal experiences. Shared experiences are combined during knowledge application, thereby internalising knowledge subconsciously. Doing practical hands-on work increases the effectiveness of knowledge internalisation. Doing practical hands-on work in a social construction environment (Oswald & Dainty, 2020) facilitates knowledge transfer from the expert master to the inexperienced

DOI: 10.1201/9781003379584-18

apprentice (COSENTiAL, 2013) in master-apprentice roles. Beyond knowledge transference, ethnographic methods can support capturing construction knowledge sustainably for future application. For example, this study archived video recordings after ethnographically interviewing ten independent construction experts in the comfort of their different workplaces and before site establishment amidst the COVID-19 pandemic travel and health restrictions.

Construction experts possess invaluable knowledge from their rich expertise and years of experience. A challenge, however, is to capture the rich expertise of experienced experts about workflows and practices to alleviate the problem of under-theorised construction work practices (Pink et al., 2013). Ethnographic interviewing is a method that can capture valuable construction industry insights and knowledge to build and develop theories. In this chapter, we show how we acquired a common understanding among the construction experts we ethnographically interviewed and video-recorded about their workplace practices to ensure project success.

Ethnographic interviewing is a sequential qualitative research method that involves gathering information from an individual as they narrate personal stories that describe events and processes about their work practices (Spradley, 2016). Ethnographic video interviewing differs from conventional ethnography in that it is conducted outside the natural context of a typical construction environment and work setting. Ethnographic video interviewing creates an environment for socialising where an interviewee construction expert articulately reflects and shares their professional experiences through dialogue with the interviewer. The interviewer listens to gather the expert experiences and acquire their articulated knowledge of how they work. By listening to the expert interviewee's personal stories, the interviewer pries into the interviewee's work life, behaviour and how they do things during their work (Spradley, 2016). Since it involves discussing one's life through video, the efficacy of ethnographic video interviewing hinges on taking progressive steps depending on the interviewee's freedom of expression to share stories about their work life.

The following section explains the methodology for ethnographic video interview data collection.

## Methodology for ethnographic video interview data collection

We were inspired to capture expert knowledge, since experts' experiential knowledge from past projects contributes 80–83% of useful knowledge (Charmaz, 2006; Rawlinson & Farrell, 2023) and often ensure successful project processes that secures their competitive advantage (Nonaka, 2013; Styrvold et al., 2019; Mazoor et al., 2021). Experts use their past knowledge and insightful awareness of the construction industry (Oswald & Dainty, 2020) to support the planning of project processes (Raki'c, 2019). Right from the early stages, construction project processes involve reviewing and updating project requirements, or considering how the requirements can be met. Reviewing and updating project requirements accurately depends on awareness of practical construction sequences (Abdelbary et al., 2020). Expert knowledge provides wise insights that can be used to review project requirements quickly from the early stages of construction projects. Expert high-quality reviews are less prone to errors and rework (Abdelbary et al., 2020). However, low-quality reviews by inexperienced individuals are prone to errors, rework, time delays, lost productivity hours and financial losses (Pink et al., 2013), despite error margins and time delays are becoming more stringent in construction projects (olde Scholtenhuis, 2015).

The methodology adopted an open-ended grounded research approach with intensive (Charmaz, 2006) semi-structured ethnographic video interviewing to explicate insightful knowledge from the tacit knowledge of construction experts. Before the interviews, we prepared a detailed interview protocol that included (i) a formal request to interview, (ii) a preamble message at the start of the interview, (iii) the interviewer's request for permission to record, (iv) interviewee's brief introduction, (v) asking open-ended interview questions and (vi) closure. The protocol guided the ethnographic video interviewing process.

### *Selection of interviewees*

The expert interviewees were our project partners with whom we had signed non-disclosure agreements (NDAs) for our bigger collaborative international research project (ASHVIN, 2020). The NDAs restricted our study population and sample space to only our project partners. To capture the experts' knowledge, we selected the method of ethnographic video interviewing because the experts were located in 11 different countries internationally, there was no site establishment yet and due to COVID-19 pandemic travel and health restrictions. To manage these three challenges, we conducted ethnographic video interviews that allowed expert interviewees to show us their workplace practices through video in the comfort of their workplaces. The international setting helped capture diverse knowledge and gather myriad construction experiences across different cultures and national contexts to create an empirical tool for reviewing and updating the requirements of construction projects.

To formally request interviews, we made public announcements and sent email invitations to partners with at least 15 years of industry experience to volunteer as willing participants. We needed 15 years of construction experience since megaprojects last 15 years (Garemo et al., 2015) and our aim was the whole project life cycle experience. In the end, we registered ten willing experts in consulting, contracting and research as interview participants and we conducted the interviews between July 2021 and February 2022 to fit within the availability of our participants. The shortest and longest interview durations were 0.75–2.5 hours, respectively. Each interview commenced with a preamble message introducing the research, requesting permission for video recording, the interviewee briefly introducing their work experience, exploratory semi-structured interview discussions and closing the interview.

### *Conducting the ethnographic video interviews*

We conducted ethnographic video interviews using semi-structured discussions. We used open-ended questions to facilitate the semi-structured interview discussions into deep qualitative research explorations about construction expert experiences and best work practices. The five open-ended interview questions focussed on how, what and which. The interview questions included: (i) Which factors do you consider important to ensure successful construction projects? (ii) What activities do you do under each factor? (iii) How can each important factor and activity you do ensure the success of construction projects? (iv) How do you apply these factors and activities to other construction projects? (v) How do you track, assess, or measure each factor during the construction project process? We strategically arranged downstream questions to follow up on preceding questions intentionally. For example, after they had narrated the important factors, we asked about their activities under each factor. We also adapted downstream questions to earlier preceding questions during the interview to ensure a free-flowing interview discussion.

For example, in the third question, we asked how a specific factor and corresponding activity they had mentioned ensure project success. Free-flowing discussions allow a natural flow of personal stories as interviewees narrate their experiences (Coffey & Atkinson, 1996; Gibbs, 2012).

During interview discussions, we refrained from interrupting the interviewee with questions to encourage storytelling naturally and allow continuous free-flowing narratives. For example, as interviewers, we had set out with a goal and expectation to learn from the individual expert's personal experiences. Their frequent use of 'we' instead of 'I' differed from what we had expected and we resolved to clarify after they had finished discussing a question to re-focus the expert interviewee to tell us about their individual experiences. Meanwhile, as they were discussing, we made handwritten shorthand notes by writing down the word 'we' each time it was used. For example, we made 11 observation notes during the interviewee's narrative corresponding with the underlined 'we' and 'our' in the quote (I) below. The notes helped us as interviewers to make observations of their use of 'we' after Coffey and Atkinson (1996) and Emerald (2023). The notes were reference points for asking follow-up questions to clarify our observations during the ethnographic video interviews.

Quote I: A response about tracking productivity as an important factor they mentioned.

> We closely tracked productivity for our limited-time project by first understanding our client's requirement for safety and preference for the minimal imprint on the skyline to maximize the visibility of surroundings. We realised that the safety guard rails proposed by the client would reduce the visibility of the surroundings and the skyline. Therefore, we had internal brainstorming sessions to discuss alternative ideas for ensuring safety without reducing visibility. From the collective freehand sketches after brainstorming, we created two ideas that we developed into actual design proposals for the client. We were happy that the client adopted our design idea of inbuilt lighting under the railing despite it inflating the project cost and we did not charge for the extra design work. Our job satisfaction was eliminating structural design rework by proactively submitting alternative design proposals early that considered the client's safety requirement and visibility preferences since this approach increases productivity in design by avoiding potential rework during the design ideation phase.

After they had finished discussing the question about tracking productivity as a factor, we asked them three curious follow-up questions adapted from the discussion to re-focus them to explain what they meant by 'we' and why they were not using 'I' in their expert story. For example, we asked them (i) to tell us about their specific experience as an individual in that particular project, (ii) if they could show us a picture of their own freehand sketch from the brainstorming and (iii) to describe specific aspects of the final collective proposal that were their brainchild. We asked one follow-up question at a time with pauses between questions to closely observe their behaviour and to listen without interrupting until the expert interviewee had finished responding. As interviewers, we stopped nudging them after three failed attempts to avoid infringing their freedom of expression and compromising the rest of the ethnographic video interviewing process. We followed the same approach until we discussed all five open-ended questions before closing the interview.

The following section explains ethnographic video interview data analysis to identify emergent insights.

## Methodology for ethnographic video interview data analysis

We grounded the results in the gathered ethnographic video interview data. Without individual I-statements, we had to resolve how the analysis could distinguish the interviewee's personal experience from the collective team experience. Expert articulation using 'we' instead of 'I' made us adapt the unit of analysis to be teams rather than individuals based on the experts' frequent use of 'we'. Briefly, the analysis comprised six steps, including (i) anonymising interviewees, (ii) manually transcribing the video recordings, (iii) reviewing the combined transcript, (iv) thematically encoding the transcript, (v) clustering the codes to form categories and (vi) creating an emergent insight for supporting early stages of construction projects. Each of the steps is elaborated in the following paragraphs.

In step one, we anonymised the expert interviewee using name-pseudonyms to safeguard their privacy and data (Oates et al., 2021). We used name-pseudonyms that maintain the human being to easily follow the stories (Saunders et al., 2015) during the detailed review in the next steps.

In step two, we transcribed the recording two hours after the interview to archive the gathered data in text format. Initially, we used software to transcribe four interview recordings, but the software transcriptions needed to be corrected. Therefore, we resorted to manually transcribing to ensure precise data capture. To do this, we listened to the recording about two hours after concluding each interview. We transcribed sooner than later before forgetting what was said during the interview and before our memory could get contaminated with other thoughts (Swain & King, 2022). We played back the recording several times to clarify and improve our precision to overcome intuition bias and memory recollection (Heritage, 1984; Heritage, 2005).

We reviewed the transcripts in step three to identify any emergent empirical insights. We initially imported the transcripts as ten separate files into qualitative coding software with an inbuilt advanced machine learning model with an auto-coding feature. However, we found no identifiable relationships between the frequent keywords in the word cloud. We merged the transcripts into one file by clustering responses corresponding to the same question to easily identify frequent keywords for thematic encoding (Gibbs, 2012). However, again, we found no identifiable relationships. At this point, we resorted to thematically encoding the combined file manually. Using the combined file made it easier to get a quick overall content overview within the same file than across separate files to distinguish different points of view and identify frequent keywords we identified as thematic codes.

In step four, we used three levels of thematic coding: (i) open line-by-line, (ii) axial and (iii) selective coding to conduct a detailed review of the transcript to find identifiable relationships in the data that could result in emergent empirical insight for construction projects. Using open coding (Gibbs, 2012), we thematically encoded the transcript into 106 different instances of thematic codes based on the keywords that identified them. For example, verbatim Quote I is an excerpt code instance based on 'requirements' as a keyword that we identified as an insightful emergent theme. Requirements were identified as a specific empirical issue based on Phelps' (2008) hierarchy of ambiguity. As interviewers, we also considered requirements interesting since they are important to achieve UN SDG goals 9 and 11 about building safe, resilient and sustainable infrastructure (United Nations, 2022). We used axial coding to exhaustively review and identify additional themes with similar semantic meanings as requirements in the construction industry. To do this, we considered sentences and paragraphs that implied permitting or acceptable ways of doing work in the construction domain. For example, 'standard' in the following Quote II is a relative keyword used to refer to requirements in the construction industry.

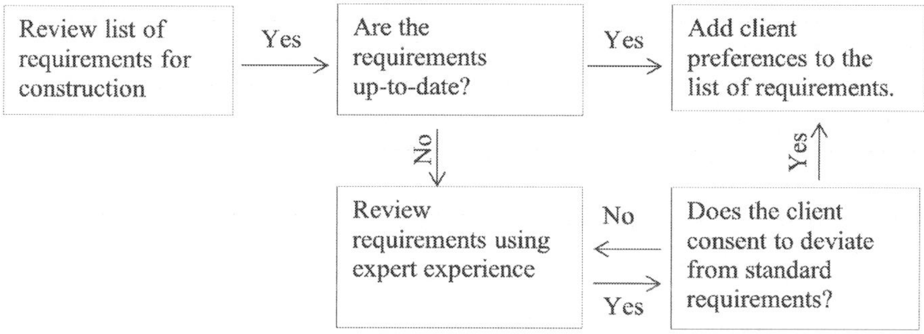


**FIGURE 15.1** Explicit knowledge process flowchart for reviewing and updating project requirements

Similar thematic codes were identified, including guidelines, standards, criteria, codes, specifications, rules and regulations. However, their specific verbatim quotes are not presented in this chapter. In the end, 23 additional instances increased the total sum of requirements thematic code to 31. We reviewed the transcript using selective coding to find phrases extending the requirements argument – for example, Quote II's last two sentences.

In step five, the identified thematic codes were clustered based on related themes to form categories. Requirements were clustered under the 'technical and project aspects' category and other categories included (i) teamwork, collaboration and communication, (ii) business cost aspects and (iii) problems and challenges.

Finally, we used our understanding and domain knowledge conceptualisation to make sense of the gathered interview data to create a simplistic tool that is an explicit form of reusable expert knowledge (see Figure 15.1). To create the simplistic tool, we considered (i) the main category (technical and project aspects), (ii) the major thematic code (requirements) and to extend the empirical issue about requirements, (iii) the counter-argument about client consent and (iv) the challenge of not up-to-date guidelines. We used our domain knowledge to merge the four points (i)–(iv) grounded in the ethnographic video interview data to form the pragmatic tool for reviewing and updating the requirements for construction projects. To use the simplistic empirical tool, an individual starts by listing the project requirements and checking if the requirements are up-to-date or need updating. If the project requirements are not up-to-date, the expert experience can be used to review and update the requirements. However, the client's consent determines whether the experience-based update is admissible. For example, we used the tool to perform a high-level review of project requirements for our project task, including interviewing the client and adding their preferences to the requirements list. We used the agreed list to develop a tailor-made early-stage parametric model. The simplistic technical flowchart for reviewing and updating construction project requirements contributes to the existing body of knowledge by alleviating under-theorisation in the construction industry (Pink et al., 2013) and it provides an initial step towards a unified formal process for reusing expert experience.

The following section highlights the results.

## Results: Emergent insights

From our observations during the ethnographic video interviews, we found the following.

Experts frequently use 'we' instead of 'I' as referring to what they did collectively in their teams. For example, this is evident in the following Quote I and in their response to our curious follow-up questions, they clarified that 'we' refers to their megaproject team since they thrive on a close-knit teamwork culture as their workplace best practice and that it is difficult to isolate individual knowledge from the hybrid project solution that merged myriad personal experiences.

Experts prefer to avoid questioning their experiential stories. For example, we noticed a change in their countenance and refrained from nudging them further after our third curious follow-up question to clarify what they meant by 'we' in their narrative in Quote I.

Client consent is important when reviewing and updating the requirements for construction projects. For example, this is reflected in Quote II's last two sentences, implying that securing client consent can counter adhering to statutory requirements for construction projects, thereby extending the requirements argument.

Quote II: A response about important consideration factors for construction project success.

> Early inclusion of technical investigation details and choices of suitable components eliminates rework and is crucial for slender and lively structures. The current practice undertakes all these integral technical investigations only at the second design phase after the structure has been envisioned and main boundaries are set, thereby making rework necessary to incorporate investigation findings. Project requirements and acceptable limit ranges for the parameters depend on the function and intended user preferences, functional costs and adapting the design to fit within the surrounding environment. Fine-tuning requirements using load monitoring systems to justify individual live load reductions is effective in developing more economically realistic designs. However, the client must give formal consent since there are statutory implications to deviating from the known standard requirements that are known and acceptable.

Experts often drew sketches by hand and show actual project designs, pictures and project files to explain their points of view. For example, when discussing Quote II, the expert interviewee began to draw a freehand schematic sketch of a slender structure with main boundaries using a pencil and paper during the video interview. Such flexibility and freedom of expression, including freehand drawing on the spot, give ethnographic video interviewing an edge over conventional interviews.

Experts prefer to narrate positive work experiences but seldom about negative experiences. For example, only four out of 106 thematically encoded instances discussed negative experiences and project challenges (refer to Quote III). Such depth of expert knowledge about construction problems and challenges was only possible through ethnographic interviewing but cannot be captured from conventional interviewing and questionnaire surveys.

Quote III: Four independent verbatim quotes comprised our 'problems and challenges' category.

> To ensure successful construction projects, we undertake scoping analysis to understand the unique project settings, parties, requirements, client preferences and project-specific constraints such as very built-up areas, a limited-time project, maximum allowable inclination and height, amidst other challenges of outdated information, no tender documents and contractor uncertainty.

To ensure construction success, we are mindful of the type of clients. Public sector clients are much more difficult to work with than contractors and private investor clients because they are rigid, are unwilling to consider win-win compromises, lack clearly defined roles/responsibilities and do not provide details about their projects. These problems generally make contractors more inclined to prefer smaller-scale projects which are simpler and whose mistakes are easier to fix using readily available expertise.

For project success, we mainly consider break-even points, but the challenge of not having all requisite data and information collected early and outdated guidelines is a major project design drawback, technically and economically.

We must use knowledge-based models for informed business sustainability to track productivity and become more efficient during project construction. In the future, we will pay for knowledge instead of per hour. Knowledge-based models will be data-informed using knowledge as data and data-driven approaches will be used. However, this shift is problematic since more business models are based on output per hour, making it difficult to change to another new knowledge-based business model. However, I think we should change our business models into more knowledge-based models if we are going to become more efficient and for informed business sustainability.

The following section discusses the results.

## Discussion

It is difficult to distinguish the personal from collective team experiences since most experts frequently use 'we' instead of 'I' due to their teamwork predisposition, which makes it difficult to isolate individual experiences. Expert's team predisposition underscores the predominant teamwork culture in the construction industry (Kahrens & Früauff, 2018). Expert articulation using 'we' instead of 'I' implies that ethnographic interviewing is not well suited for gathering individual experiences from experts who work in teams. To learn directly from a particular individual expert who is a project team member, an ethnographic interviewer needs to observe the particular individual as they are doing their work collectively in their teams and within their workplaces rather than asking them to narrate their work story. However, there was no site establishment yet during the study duration and experts were located in different countries, which made it impossible to observe particular individuals as they did their work collectively in teams. When a project site is established, we recommend using ethnographic action research (Hartmann et al., 2009), shadowing individual team members and using serious games (Schell, 2010; van Amstel et al., 2016) to gather individual experiences of particular team members and collective teams while they are doing their work.

Expert tendency to draw hand sketches and show project documents while explaining their point of view distinguishes ethnographic video interviewing as a unique method from other types of interviewing. Therefore, an ethnographic interviewer should understand the domain knowledge well to interpret sketches and project documents, be curious and ask questions to learn from the expert's audio and visual presentations. Understanding the domain knowledge is similar to learning the culture and language of people before commencing ethnographic studies about them (Spradley, 2016).

Expert avoidance of negative experiences requires interviewer flexibility and sensitivity to safeguard the interviewee's freedom of expression progressively until the objective of ethnographic interviewing is achieved. Therefore, an ethnographic interviewer should be cautious when prying into the interviewee's life during the interview. On the other hand, the problems and challenges explained during ethnographic video interviews enrich the quality and depth of expert knowledge captured that could otherwise not be possible from conventional interviewing and questionnaire surveys.

During the interviews, experts acknowledged standard requirements that comprised the majority theme and they also emphasised that client-consented preferences can override requirements. For example, Quote III reflects that experts are mindful of the type of client and whether client consent can be obtained to permit deviating from known standard requirements. Recent discussions after the catastrophic Turkey–Syria earthquake (BBC News, 2023) and Grenfell fire (BBC News, 2019) disasters underscore expert arguments questioning the conventional over-reliance on standard requirements and advocacy for client consent to deviate from known standards. As ethnographic interviewers, we uncovered expert advocacy for client consent to bypass the legal implications associated with construction project standard requirements as the established norm and legal authority and we recommend further research in this direction.

## Conclusion

This chapter discussed ethnographic video interviewing as an exploratory qualitative research method to explicate knowledge from construction experts. We conducted semi-structured ethnographic interviews with ten individual construction experts in consulting, contracting and research. Based on our research experience in this chapter, we recommend ethnographic video interviewing to gain deep insights and learn when workplaces are decentralised, sites are yet to be established and in case of travel or health restrictions. The experts shared their experiential knowledge in the comfort of their workplaces, illustrating with hand-drawn sketches and audio-visual project files to explain their points of view. From the manually transcribed data, the main technical and project aspects category comprised mostly thematic codes about requirements. Based on the main category and requirements as the majority theme, we created a novel simplistic tool for supporting reviewing and updating requirements of construction projects. The tool is an explicit form of knowledge representation that reflects the emergent empirical insights translated from the tacit experiences of construction experts. We used the tool to perform a high-level review of project requirements for our project task, including interviewing the client and adding their preferences to the requirements list. We used the agreed list to develop a tailor-made early-stage parametric model. Our technical process flowchart contributes to the existing body of knowledge, an initial step towards a unified formal process for using expert experience to review and update project requirements.

## Acknowledgement

This study was part of the European Union's ASHVIN H2020 research project, funded by the European Union, Grant number 958161. The knowledge contribution of the expert interviewees is greatly appreciated. Any views expressed herein are of the author(s) and neither of the European Union nor Horizon H2020.

## References

Abdelbary, M., Edkins, A., & Dorra, E. M. (2020). Reducing CRR in fast-track projects through BIM. *Journal of Information Technology in Construction (ITcon)*, 25, 140–160.

ASHVIN. (2020). ASHVIN H2020 research project. https://www.ashvin.eu/

BBC News. (2019). Grenfell Tower: What happened? https://www.bbc.com/news/uk-40301289

BBC News. (2023). Turkey–Syria earthquakes 2023. https://www.bbc.com/news/topics/cq0zxdd0y39t

Charmaz, K. (2006). *Constructing Grounded Theory: A Practical Guide Through Qualitative Analysis.* London: Sage.

Coffey, A., & Atkinson, P. (1996). *Making Sense of Qualitative Data Analysis: Complementary Research Strategies*. London: Sage.

COSENTiAL. (2013). Addressing Generational Handover in the AEC Industry. https://cdn2.hubspot.net/hubfs/2287635/Cosential_Master/eBooks/Addressing-Generational-Handover-in-the-AEC-Industry.pdf

Emerald. (2023). *How to Use Ethnographic Methods and Participant Observation*. Emerald Publishing. https://www.emeraldgrouppublishing.com/how-to/observation/use-ethnographic-methods-participant-observation

Garemo, N., Matzinger, S., & Palter, R. (2015). *Megaprojects: The good, the bad and the better.* McKinsey & Company. https://www.mckinsey.com/capabilities/operations/our-insights/megaprojects-the-good-the-bad-and-the-better

Gibbs, G. R. (2012). Thematic coding and categorizing in analysing qualitative data. *SAGE Research Methods*, 38–55. https://doi.org/10.4135/9781849208574

Hartmann, T., Fischer, M., & Haymaker, J. (2009). Implementing information systems with project teams using ethnographic action research. *Advanced Engineering Informatics*, 23(1), 57–67.

Heritage, J. (1984). *Garfinkel and ethnomethodology.* Cambridge, UK: Polity Press.

Heritage, J. (2005). Cognition in discourse. In *Conversation and Cognition* (Hedwig te Molder, Jonathan Potter, eds.) (pp. 184–202). Cambridge: Cambridge University Press.

Kahrens, M., & Früauff, D. H. 2018. Critical evaluation of Nonaka's SECI model. In Syed, J., Murray, P., Hislop, D., & Mouzughi, Y. (Eds), *The Palgrave Handbook of Knowledge Management* (pp. 53–83). Cham: Palgrave Macmillan. https://doi.org/10.1007/978-3-319-71434-9_3

Mazoor, B., Othman, I., & Pomares, J. C. (2021). Digital technologies in the architecture, engineering and construction (AEC) industry: A bibliometric-qualitative literature review of research activities. *International Journal of Environmental Research and Public Health*, 18, 6135.

Nonaka, I. (2013). Ikujiro Nonaka: Winner of the 2013 Thinkers 50 Lifetime Achievement Award. https://www.youtube.com/watch?v=Ynulo288a4M&t=429s

Nonaka, I., & Takeuchi, H. (1995). *The Knowledge Creating Company.* Oxford University Press.

Oates, J., Carpenter, D., Fisher, M., Goodson, S., Hannah, B., Kwiatkowski, R., Prutton, K., Reeves, D., & Wainwright, T. (2021). BPS code of human research ethics. https://doi.org/10.53841/bpsrep.2021.inf180

olde Scholtenhuis, L. (2015). *Enhancing Mindful Coordination in Fragmented Utility Construction Practices: By Developing, Implementing and Evaluating Virtual Design and Construction Models.* Eschende: University of Twente, Gildeprint. https://doi.org/10.3990/1.9789036539845

Oswald, D., & Dainty, A. (2020). Ethnographic research in the construction industry: A critical review. *Journal of Construction Engineering and Management*, 146, 10. https://doi.org/10.1061/(ASCE)CO.1943-7862.00019

Phelps, A. F. (2008). *Lean Information Flow in Complex Healthcare Projects and the Role of the 'integrator' as Information Manager*. University Park: Pennsylvania State University.

Pink, S., Tutt, D., & Dainty, A. (2013). *Ethnographic Research in the Construction Industry* (1st ed.). London: Routledge. https://doi.org/10.4324/9780203834671

Raki'c, M. (2019). Lean BIM-based communication and workflow during design phases. Munich: Master thesis in the Technical University of Munich, Department of Informatics.

Rawlinson, F., & Farrell, P. (2023). Construction site culture: Seeking the optimum methods for an ethnography. PhD dissertation. https://www.irbnet.de/daten/iconda/CIB18873.pdf

Saunders, B., Kitzinger, J., & Kitzinger, C. (2015). Anonymising interview data: Challenges and compromise in practice. *Qualitative Research*, 15(5), 616–632. https://doi.org/10.1177/1468794114550439

Spradley, J. P. (2016). *The Ethnographic Interview*. Long Grove: Waveland Press.

Styrvold, M., Knotten, V., & Lædre, O. (2019). Planning the BIM process in AEC projects. Proceedings of the 27th Research Conference for International Group for Lean Construction (IGLC). 527-528. https://doi.org/10.24928/2019/0208

Swain, J., & King, B. (2022). Using informal conversations in qualitative research. *International Journal of Qualitative Methods*, 21, 1–10. https://doi.org/10.1177/16094069221085056

United Nations Sustainable Development Goals Report. (2022). https://www.un.org/sustainabledevelopment/progress-report/

van Amstel, F. M., Zerjav, V., Hartmann, T., Dewulf, G. P., & van der Voort, M. C. (2016). xpensive or expansive? Learning the value of boundary crossing in design projects. *Engineering Project Organization Journal*, 6(1), 15–29.

# 16

# DOING ONLINE ETHNOGRAPHY

## Reflections on methods and challenges in hybrid environments

*Laura Osburn, Madison Snider and Carrie Sturts Dossick*

*University of Washington, Seattle, WA, USA*

*Siegel Family Endowment, New York, NY, USA*

*Department of Construction Management, University of Washington, Seattle, WA, USA*

**Key takeaways**

- As compared to in-person or hybrid ethnography, online work ethnographies reduce access to informal interaction and evidence of workplace culture while providing researchers with more access to formal meetings and individual interviews.
- Online work ethnographies raise ethical debates surrounding visuals and document collection.
- Online work ethnographies can create challenges for capturing data on hallway conversations and parallel chats and require novel data collection strategies to get access to these interactions.

### Introduction

The COVID-19 lockdown provided an unanticipated opportunity—and necessity—for our research team to develop ethnography techniques when observing people at work in an online environment. In this chapter, we reflect on two ethnographic projects: the first focused on how energy modelers translated energy data to project teams and was primarily in person mjjwith occasional phone meetings; and the second looked at how Facilities and information technology (IT) professionals collaborated on smart building cybersecurity in higher education and was primarily online during COVID lockdown with only the first few months of in-person meetings at the beginning of the study. While the second project was an unusual and unique experience—doing ethnography during a global pandemic and shifting abruptly from in person to online without knowing the duration of this shift—we take this opportunity to share what we have learned about ethnography in online work environments, which now are more common across many industry sectors.

DOI: 10.1201/9781003379584-19

Comparing an in-person ethnography with an online one allows us to highlight how these different types of spaces shape the research. When designing ethnography, we think critically about where and in what contexts we can observe the phenomenon of interest, in our cases we sought to study people at work. Traditional ethnographies include the physical presence of the researcher with observations at the physical work site. As that work moves to online and hybrid settings, ethnographers need to evolve their methods and strategies as in-person and online workplace environments change how you can conduct observations and what you can observe.

Prior to COVID-19, ethnographic research, like many modalities of work and life, had undergone significant changes as a result of digital technology. Ethnography as a method "minimally requires physical and social proximity" (Emerson et al., 1995) to the field site and the subjects. However, achieving this physical and social proximity has been made easier by the ability to use communication technologies to gain access to the field and, simultaneously, has posed challenges to building rapport and understanding of workplace culture both physically and socially. Online ethnography "transfers the ethnographic tradition of the researcher as an embodied research instrument to the social spaces of the Internet" (Hine, 2008). In the past two decades, online ethnography has gained increasing acceptance as an appropriate methodological modality (Pink et al., 2016) and in some cases is the only means for entering the field (Hine, 2017).

The COVID-19 pandemic prompted additional considerations about the state of ethnography. Field sites that would have otherwise been physically accessible became quickly transformed into online spaces. Since early 2020, much has been written about the methodological implications of COVID-19 on ethnographic research across a wide array of disciplines (see Góralska, 2020; Howlett, 2022; Johnson, 2022; Zhao, 2022). This phenomenon has also impacted ethnographic research in the construction industry.

As we reflected on our own ethnographic methods before and after COVID-19, we determined that the research design depends on what types of evidence we need access to address the research questions. Future ethnographers will need to consider the capacities and constraints of online and in-person field sites and strategize their data collection. Due to the digital transformation of workplaces, many workplaces will be hybrid to some extent, and ethnographic methods will require online and in-person methods of data collection. Will online, in-person, or hybrid settings provide access to the data needed to address the research questions of an online ethnographic study? When is physical presence important? What do you miss when meeting online? What do you gain? It is important to be aware of and make intentional choices in terms of what may be lost and gained in each space, and how the environment alters interaction; different questions may lead to collecting data in different types of spaces.

In this chapter, we compare and contrast our own in-person and online ethnography experiences and reflect on what was lost and gained in these different environments. What we found was that one of the challenges with online work was observing informal exchanges. We found they were difficult to access on different chat platforms (e.g., Slack) or unscheduled Zoom meetings as the work culture itself had changed. Most informal communication was either very limited before and after meetings or may have been occurring on other mediums out of researchers' view. In contrast to in-person ethnography, when work moved to online spaces, where did hallway and parallel conversations happen? How did teams form and build rapport with each other? How did people learn from each other? We were often challenged by similar issues as the people and workplaces we studied. Just as the staff who we observed struggled to build rapport with each other, we struggled to build rapport with subjects when the work was online. In contrast, online environments allowed us to connect with interview subjects and collect documents more easily. Furthermore, we found that our presence in online meetings was more neutral (or hidden) than in-person, particularly in non-video meetings where the attendees were often just a list of names and voices. Overall, while there were some benefits to online ethnography, we found it was limiting for answering our particular research questions on informal communication and interdisciplinary collaboration.

While our team had to quickly pivot to online ethnography, what we learned from this project are challenges and opportunities that future ethnographers can consider when entering the field, building rapport, collecting meaningful data, and exiting the field thoughtfully and intentionally. Online ethnography can be the best and sometimes the only way for the researcher to enter the field, as is the case with the study of online communities. However, when online or hybrid ethnography (that is ethnographic research that is done both online and in person) is necessary due to extraneous circumstances (i.e., a pandemic), there are both positive and negative implications on research. We hope that this chapter helps future researchers in crafting ethnographies of work to develop research design methods for all three modalities of online, hybrid, and in-person ethnography and to provide reflections on how online and in-person ethnography changes what you have access to as a researcher and how this access shapes what you can learn from ethnography.

In this chapter, we introduce the two ethnography projects. The first study was a fully in-person ethnographic study of energy modelers. The second, a study of cybersecurity, was a hybrid ethnography in that it started off in person and had to transition to online ethnographic methods as a result of COVID-19 lockdowns. We discuss our ethnography experiences and explore what is lost and gained in three sections: entering the field, data collection, and exiting the field. In this chapter, you can see how we pivoted the research, adapting to what we could accomplish during the extraordinary circumstances of a pandemic and the lockdown procedures many of us experienced. Ethnographic research is an exploration of a time and a place. This is our story.

## The research studies

In this chapter we compare and contrast two ethnography projects: the Energy Modeling Study and the Cybersecurity Study (see Table 16.1).

The Energy Modeling Study was a qualitative research project to study the technical communication and interdisciplinary team dynamics of high-performance building design

**TABLE 16.1** An overview of the two case studies

| | *Energy modeling study* | *Cybersecurity study* |
|---|---|---|
| *Setting* | In-person | In-person and online |
| *Research aim* | Technical communication and collaborative project decision-making on interdisciplinary teams working on high-performance buildings | Formal and informal communication and collaborative work practices of interdisciplinary teams of IT and Facilities workers working to improve smart building cybersecurity |
| *Observation length* | Two years | Two years |
| *Number of participants observed* | 94 | 52 |
| *Number of observers* | 3–4 | 3 |

and construction. We sought to study the translation and synthesis of technical work in the context of multidisciplinary teams. First, we sought to study the translation of technical engineering information to the design team members and the project stakeholders. Second, we sought to study how this technical information was synthesized across broader project decision-making. To observe these communication activities, we decided to focus on watching technical experts (the energy modelers) as they practiced translation and sought to influence design decision-making. We observed these experts in their offices and traveled with them to design meetings with architects, engineers, and at times building owners and operators.

The Energy Modeling Study observations began in April 2014 and continued through March 2016. We observed 6 different energy modelers who worked on 18 different projects. We were in person with the energy modeler at their workplace, but we also joined the energy modeler at meetings in other locations and observed the energy modeler on conference calls. Locations included the energy modelers' offices, architects' offices, building project sites, and in some cases, car rides to and from work meetings.

The Cybersecurity Study was an exploratory research project where the team set out to study the formal and informal interdisciplinary team dynamics that surrounded emerging cybersecurity concerns in the operations and management of large campus contexts. For this ethnography, we sought to study an organization made up of several operational groups that managed a large collection of buildings. In particular, we were looking for established IT and buildings and infrastructure operations and management (Facilities) teams. We were also looking for an organization that (1) was adopting new Internet of Things (IoT) technologies and incorporating these into existing Operations Technology (OT) systems and (2) was addressing cybersecurity associated with these new IoT systems.

Our Cybersecurity Study observations began in the late fall of 2019 with a series of meetings where members of the IT and Facilities groups collaborated on IoT adoption and OT management issues. Three months after starting the ethnography, the organization moved into a COVID-19 lockdown and all meetings moved online. The IT and Facilities teams used Zoom as their meeting platform and shifted their focus to emergency campus management strategies. With this shift, the research team had to shift to an online modality as well. We had the advantage of starting in person with some of the IT and Facilities teams we observed (e.g., the electrical metering and critical infrastructure planning meetings), while we started other team observations fully online (the new capital projects team meetings, which used a mix of Zoom and Microsoft Teams). Because we were observing campus operations management staff, some of these team members continued to work on campus while the majority of personnel were required to work remotely, including our research team. Like our research subjects, the research team pivoted quickly and had to develop new practices and norms in the moment.

### *Entering the field*

We found stark differences in the ethnographic process and final field note output of the two research studies. While both studies started the same, with in-person meetings to begin negotiation into the multiplicity of locations and activities we wanted to observe, the move to online-only interaction during the Cybersecurity Study created a series of challenges (and some opportunities) due to differences in the life cycle of conducting field research: from entering into the field and building rapport with our participants to the types of observational data we could collect, to exiting the field.

Both research studies began by working with an industry expert who was both a member of the research team and a member of the community that we wanted to study. In the Energy Modeling Study, this was an energy modeler who consulted with design teams to support high-performance design. In the Cybersecurity Study, this was a senior director in the owner organization. In both studies, these industry experts were the ones who could set up introductions and negotiate with community members about whether we could observe certain meetings and helped us to determine what specific meetings would be worth observing. Negotiations primarily consisted of meetings between one or more members of the research team and identified community members to discuss the project and the observation opportunities.

In the Energy Modeling Study, our industry expert, who worked with an energy modeler, negotiated to have a researcher shadow him throughout the workday for approximately one year. In the Cybersecurity Study, our industry expert, who worked in cybersecurity on campus, helped introduce two of our researchers to members of campus and vendors who frequently worked with the campus to identify important meetings to observe and planned to introduce us to potential campus workers who would be willing to have us observe them in the workplace. However, unlike with the Energy Modeling Study, the Cybersecurity Study was moved entirely online before we could negotiate shadowing someone at work. This led to several difficulties during the early days of our cybersecurity fieldwork online, primarily, being able to build rapport with our subjects. Furthermore, as an invited guest, the ethnographers still had to go through a gatekeeper (our industry expert member), to gain permission to be invited to other online meetings that concerned topics of interest.

### *Building rapport with subjects*

Building rapport with the different IT and Facilities subjects participating in the Cybersecurity Study was difficult during our time conducting online ethnography. Since we were limited to the time duration of Zoom meetings, we only had very short periods with which to get to know our many subjects. This was in contrast to the rapport building that we did on the Energy Modeling Study, where the researcher would spend all day with subjects: watching the participant work, sharing lunch together, traveling to meetings, and spending time in meetings together.

Building rapport is not only important for gaining trust between the observer and the participant, but also for better understanding who the participant is, their personal drivers, and their perceptions of how others think in the workplace so that they can communicate effectively. For example, with the Energy Modeling Study, rapport began on day one of entering the field. At the end of the partial workday that the ethnographer attended, there was time for the participant to ask the ethnographer about their research and for the participant to share their own personal philosophy:

> I take my leave at 1 p.m. Before I go and after I put away my notebook, Bruce asks about me and my research history and tells me a bit about his own personal philosophy regarding energy modeling of historic renovations. He states that usually people get so hung up on regulations that they can't make the changes they want to make, but he says that these regulations are guidelines with built-in flexibility. His job is to help them think outside of the box.

This bit of conversation allowed the researcher to get to know more about how the participant understood the impact of their work, his personal perspective on the creative challenges building teams faced, and how he situated his work as a means to help building teams overcome these challenges and build more energy efficient buildings. For research questions that focus on how energy modelers translate data, this was a valuable insight into how the modeler interpreted the perceptions of others on teams.

In the Energy Modeling Study, shared moments outside of actual work time improved interpersonal relationships between the researcher and the participant through informal conversations that generated a sense of being "a part of the team." For example, the researcher accompanied the energy modeler and the industry expert on our research team on a three-hour drive to another city for a new project. The ride home was an opportunity to converse about daily life, musical tastes, and share food: the types of moments that build close relationships and a sense of social belonging.

On another occasion, the consultant asked the ethnographer for their opinion on the presentation slides that she and the energy modeler were working on with an architect. The ethnographer initially started to develop an opinion based on past research, but it became clear that the consultant didn't want a research opinion, but the personal opinion of someone who had been a participant at many meetings where the slides and their accompanying presentation had taken place.

> Ashley then turned to me and asked me my opinion. I ended up first mentioning that we had seen in our research that case studies were deemed useful. Ashley then explained she wanted my personal opinion. I noted to the group that I thought some of the slides had far too much text and often text was not explained, leaving one to wonder why it was up there in the first place.

This type of interaction illustrated the depth to which the ethnographer became part of the team and an indication of having built rapport with the subjects of this research.

On the Cybersecurity Study, there were some brief moments of rapport building prior to observations moving online. At one of the first collaborative meetings that we observed, there was time after the meeting to exit as a group, to discuss the meeting with participants, and to engage in informal conversation that included sharing jokes. In the meeting, the ethnographer had discussed how the data would eventually be coded and the use of pseudonyms. One person in the room had stated that he wanted to be called "Dragon." At the end of the meeting, the ethnographer talked to Dragon and had an opportunity to learn more about him and the meaning behind his chosen name:

> I also talked to Dragon, but mostly to tell him he was [going to be called] Dragon. Apparently, it's his nickname that he loves and wants everyone to use. It started out as a joke but now he keeps trying to "push it on people." I told him if he uses it a lot it might mean he'd be more identifiable in the research. He mostly just laughed and said he liked being Dragon.

Once everything moved online, the ability to build the kind of rapport that developed on the Energy Modeling Study became increasingly difficult, leading to the ethnographers feeling more like an invited guest than a part of the team. With Zoom, it was difficult to find time for informal conversations where one could share their personal background and experiences to generate feelings of social belonging. There were some moments prior to meetings where one could enter a Zoom room with participants. During these moments, conversation topics, such as shared issues with hairstyles during COVID-19 and Zoom background choices, were conversational moments where the ethnographer and the participants could build rapport through shared experiences of the pandemic, and share and exchange humor using Zoom backgrounds, as in this field note.

> I logged into Zoom and saw John there looking tired. We talked a bit about how our hair was all getting a bit wild, and John joked that he was now growing a mullet. I told him I was thinking of dying my gray roots bright red. John also liked my Star Trek background (as did Philip). William put on his own background from Blade Runner. He also tried an Alien background, upon which John asked, "is that the Thing?" Philip replied, "wrong alien" and William jokingly pretended to act disappointed in John's lack of sci-fi knowledge.

As the use of Zoom increased, new organizational and technical rules surrounding Zoom settings had an impact on having the time before meetings to build rapport. After an increase in "zoombombing,"[1] the campus instituted rules for default settings that meant everyone would sit in their own private "waiting room" until the meeting's host would officially start the meeting. The loss of rapport wasn't lost on our ethnographers, as seen in this field note of the ethnographer's first "waiting room" experience on the project.

> I'm on Zoom at 8:32, but it appears John hasn't logged in yet as we're waiting for the host to start the meeting. It means he's currently working with the default Zoom settings. I now have a personal Zoom account since COVID started and realizing how important some of those settings are, not just for privacy (avoiding zoombombing discussed in the news), but just to give people a space where they can just enter and start chatting with friends even though you aren't there [physically]. I am hoping that John changes his settings soon as it is always nice to say hello to folks in a room and know you aren't in the wrong place (even in Zoom).

The end of online meetings had even fewer opportunities to build rapport. For most of the field notes during the COVID-19 era observations end in the following way:

> The call comes to a close and everyone says goodbye and logs off.

Rapport loss was even more obvious in capital project team meetings. In these meetings, there were much larger numbers of people either calling in or logging in, and most had their video off, leaving only a name or a profile picture in their place. This would increasingly become the Zoom cultural norm for large meetings, where only the main speaker might have a camera on. In this sense, rapport loss appeared to not only be a problem for our research team in the project team meetings but also a loss of rapport between the meeting participants themselves. This aligns with Viererbl et al.'s (2022) findings that informal online communication tended to decline among colleagues with more distant relationships.

Opportunities for rapport building were no doubt lost in the move to Zoom; however, there were moments that felt more personal, though fewer and further between. In one meeting around the holidays, a few members showed up: Justin, William, and a research team member. Rather than go through the whole agenda as planned, William asked if the researcher wanted to take the opportunity to speak more candidly about the research, which was explained in more depth than previously and set the stage for Justin and William to open up about their concerns about the state of the field more generally. These insights were helpful in contextualizing a lot of what was usually a task-oriented agenda.

In summary, entering the field and building rapport are both more challenging in online ethnography than in-person ethnography in workplace settings. An ethnographer may seek ways to socially connect with the research subjects by arriving early to meetings or finding out how and when they socially interact. During these moments, telling jokes, discussing shared interests, or engaging in informal conversations about the research provide opportunities for rapport building. Researchers will know they have built rapport when their participants are comfortable sharing their opinions, and information about their daily lives, or simply appear to enjoy your sense of humor. However, depending on institutional policies and meeting norms, it may not be possible to gain access to this pre-meeting time or it may not be part of the meeting culture to make space for informal conversation or use cameras that can help foster a sense of closeness and familiarity between researchers and participants.

### *Data collection*

Online ethnography via video conferencing software also has implications for the types of data that we could collect. In the Cybersecurity Study, these implications impacted how we collected data in shared spaces in ways that contrasted significantly with the Energy Modeling Study, where we were located in the same geographical location. For the Cybersecurity Study research questions related to the dynamics of interdisciplinary teams, online ethnography meant that there was a lot of "missing" data: most notably, missing observational data of what people do in the workplace beyond meetings and missing informal communication data that would generally take place before and after meetings.

In the Energy Modeling Study, ethnographers and participants were often in the same room together, creating shared co-presence. Having a shared space also helped to set the scene for any given moment. For example, in one field note, the ethnographer wrote about the energy modeler's shared office space. The field note describes the quiet energy of the place, the emphasis on sustainable design that in many ways reflected the expressed values of the energy modeler and his team, the youthfulness of the workers, and the professional display of specific objects in the room, such as the magazine, "Today's Boiler."

> The office is open plan, and relatively quiet except when a tour group comes through the building. There are posters hanging from the ceiling in front of a few windows that say "power" words (or "inspirational words?") like, "Vision", "Energy", "Design", and "Water". There are only a few people in the office, many of them look very young. Bruce's desk has several magazines laid out on a side table in a way that appears to suggest they are more on display than as references. One is called, "Today's Boiler".

This introduction to the office sets the stage for the types of people the ethnographer would interact with, the values of the firm, and the type of engineering work that the energy modeler conducts. In other field notes, there are detailed scenes of other firms, including a large architecture firm, and hospital building office spaces, which provide some sense of how the organizations expressed themselves through their interior design and decorative choices, and the objects that are most important to the workers that are placed in ways to express themselves and their work. Objects and spaces in offices such as posters, art, physical models, office layouts, and break room designs all convey the workplace culture and give the ethnographer a sense of the organization.

In the Cybersecurity Study, the shift from in-person to online ethnography changed the types of data one could collect on shared space, as well as the general sense of shared space between the ethnographer and participants. For example, before COVID-19, an ethnographer visited an office in a campus utilities building for a meeting on energy metering. The field note describing the space provides information not only about what the office looked like but also the ways the Facilities organization had visually organized campus management and utility zones.

> I arrived at the Plant Operations building around 8:20 am and managed to get to the second floor "bullpen"... The walls of the bullpen were covered with a series of colorful maps of the [university] campus as well as a large aerial photo of the campus. One map depicted the zone areas and their boundaries for landscape maintenance. Another was a facilities services map, also with color-coded zones (e.g., southwest zone, northeast zone, health sciences zone).

Once the ethnography moved online, there was no longer a shared physical room, but a shared online meeting room, which created a different type of co-presence. There was now no way to collect data about workplace objects that may have indicated how participants organized their work. Shared Zoom spaces in smaller collaborative meetings were populated with small rectangular screens, some with video turned off, so that the person was no longer even visually *in* the shared space. Others shared virtual backgrounds in their video that reflected something about them. These backgrounds would usually express a participant's personal, rather than professional, interests (such as a picture of national parks, or a picture of their family canoeing). A few participants used campus images that did provide some connection to the organization where they were employed. Other participants would simply not use a virtual background and you would see them at home in their own personal space of a bedroom, kitchen, porch, or home office.

The type of online shared space we were in also had an impact on the type of data we could collect (Howlett, 2022). Different platforms created different challenges for gathering data on who was speaking. For example, in Zoom, the ethnographer wrote:

> It's sometimes hard to see who is speaking when there is a shared screen view—we were sharing the agenda—and you have to scroll through folks to see everyone if there are more than four people on the call. I eventually put the view for attendees into the one view at a time mode, which means the person speaking will be the only one seen in the far corner of the screen.

In this case, the solution to the problem was to place Zoom's setting to the "one view at a time" mode, which meant only seeing one speaker at a time. In this case, the sense of shared space in a group meeting becomes a shared space with the observer, a single speaker, and an imagined audience.

Furthermore, the choice of video conferencing platform had an impact on the ability to get a sense of shared space and a sense of how many people were actively engaged in the meeting. The norms around participants having their cameras on or off were negotiated both on an individual level (some participants noted being unwilling to turn on their cameras) and a group level, particularly in meetings with a large number of people calling in. It is difficult to parse the cultural and platform differences that play into this dynamic, but in the Cybersecurity Study, the smaller meetings among university staff were mostly held on Zoom, and apart from a select few participants, most people had their cameras on. However, in the larger capital projects meetings with attendance from owner teams, architects, engineering, and construction teams, the norm of having your camera off was much more accepted. This may have been in part cultural and in part due to the use of the Microsoft Teams platform. Microsoft Teams added difficulty to doing an online ethnography in that you could not easily see everyone's faces at once therefore you may have a hard time figuring out who is talking and/or seeing non-verbal reactions like nods or hearing laughter for those who had their audio on mute.

More importantly for our research, online ethnography had an impact on the type of work we could observe. The contrast between our Energy Modeling Study field notes and Cybersecurity Study field notes is quite striking. For the Energy Modeling Study, we were able to sit in the office with our energy modeler from 9 to 5 p.m. throughout the week. During this time, we would observe participants engaging in all types of work. We collected data on how energy modelers worked on their computers, how they used models, and their daily chatter with other employees about a project. We could follow them on site visits, large project team meetings, and attend phone meetings that they had with other firms.

In contrast, in the Cybersecurity Study, we were only able to see how IT and Facilities were collaboratively working on cybersecurity issues in meetings. There were no opportunities to shadow anyone in the field, to see how they conducted their daily work and the everyday conversations about their work with their colleagues. It wasn't that this work on campus had ceased to exist—far from it—but we were unable to observe this part of their work. The most we could capture were detailed descriptions and explanations about the work people had conducted on campus when questions about this work were asked in meetings or we would gather individuals' impressions of this work through interviews.

In one example, a Facilities worker described to a group of IT and Facilities workers in a meeting about having to come to campus to fix a panel that he had difficulties with and the time the task took. He briefly described what he did to fix the panel, but that is all the data the ethnographer was able to gather about the work. There was no way to know who the technician had spoken to or contacted to troubleshoot the panel (if anyone), or how the work was completed.

> James notes a few issues that he has had around the difficulties of troubleshooting some things from home. In one case he had to come onto campus. He tells the group a story about how a panel with controls was not communicating with something due to different subnet addresses …
>
> James then notes it took him probably 5 hours to fix it as he had arrived on campus at 10 am and then left at 3 pm. He said he "tried a couple different types of configurations and tests, but I had to go back a second time to do the testing," which is when "it succeeded."

In this case, missing data on the actual work of troubleshooting a panel on campus was a lost opportunity for meeting our project's goals: to better understand the processes and daily work practices of how IT and Facilities professionals work together during formal meetings and during technical maintenance work. So, while we could answer our research question about how IT and Facilities professionals work collaboratively on Zoom—how collaborative work occurs through storytelling in online meetings—we could not observe the potential collaborative work that may have occurred when Facilities workers physically did maintenance and repair on smart systems.

Perhaps the most important data that was missing in our online ethnography was data on informal communication: the missing "hallway conversations," or "pre- and -post- meeting talk" and "parallel chats" (Viererbl et al., 2022). For our team, those important "hallway conversations" are moments in the workplace before and after meetings, and throughout the day, when colleagues gather to informally discuss meaningful technical and organizational challenges and generate solutions to meet these challenges. Parallel chats, gossip, or banter that occur behind the scenes during a meeting are equally important as they provide insight into a participant's point of view about how a meeting is going. It can highlight challenges at work.

These data were important to collect in the Energy Modeling Study. Hallway conversations provided information about how the energy modelers felt a meeting went and what that might mean for the preparations, and coordinated work required for future meetings. These hallway conversations provided us with insights as to how energy modelers were strategizing their communication and coordinating their work with others to ensure that they more effectively communicated with their data the importance of sustainable designs in a way that would not only resonate with project team members and clients but would also appear logistically feasible for both project team members and clients. Parallel chats provided greater context for the energy modeler's challenges when working on project teams, and their immediate critiques on different aspects of any design ideas raised in meeting conversations.

In the Cybersecurity Study, we missed data on parallel chats. While there are options to have parallel chats during a meeting, such as a direct "chat" with another meeting participant, no one in the meetings we observed used these direct chats with us (although they may have been used in the meetings between other participants where we did not see them). Instead, our senior personnel member, who was an insider of these groups, would occasionally meet with us later to clarify technical jargon and express what he perceived was happening in a meeting. This was more like an informal follow-up conversation that supported the research team's sense-making. For example:

> I wanted to highlight something that surfaced in yesterday's management meeting. You've probably seen this, at least implicitly, in other meetings as well …

With the Cybersecurity Study, we also missed gathering data from hallway conversations. We didn't always hear how our participants felt a meeting went, and we didn't have the ability to collect data on how a meeting might have influenced their work for future meetings or for future deliverables to their management. There were some occasions where informal communication similar to hallway conversations did occur. We were more likely to hear pre- and post-meeting discussions about capital project team meetings during the collaborative meetings of university IT and Facilities professionals. However, we were not able to learn about how our IT and Facilities subjects felt their own collaborative meetings went. This meant that our only insight into how we should understand the meetings, critiques of the meetings, and what they meant for future work was from the perspective of the industry expert on our research team who would reflect on the meetings with us.

The loss of these data left our team with several questions. Are hallway conversations and parallel chats happening on other channels/platforms? If so, how do you gain access to these chats? We found these types of missing data a key limitation of using online ethnography to answer our research questions, and it would require a great deal of rapport building (which was also difficult online) to gain access to these discussions.

### *Artifact collection*

In the Cybersecurity Study, the use of Zoom in conjunction with calendar invitations with attached agendas and other documents made the collection of artifacts straightforward. The attached documents that participants discussed in meetings were easy to download and save in our research database. On Zoom, people would share their screen to show documents, visuals, and other related items. These could easily and quietly be saved by the researcher through various tools where we could take a screenshot and save an image of the meeting to the computer's desktop. However, there were some major drawbacks to this. Unlike seeing visuals at work, it was difficult to see interactions with visuals except for the person screensharing, who may move their mouse or cursor to different parts of the document or image.

In the Energy Modeling Study, our ethnographers were able to see visuals, such as data visualizations on PowerPoint slides, that people were using and how they interacted with them at work. However, it was at times difficult to obtain copies of those visuals and documents. In meetings with architects, contractors, and energy modelers, it would be disruptive to ask permission to take a photo of documents or visuals. There also may not have been enough copies of the shared documents made to provide one to the ethnographer. Often, we had to ask the participant we were shadowing if they could obtain a copy for us: something that did occur but could take multiple reminders or would eventually be forgotten.

Despite the ease of collecting documents online, there were ethical issues surrounding screenshots of artifacts that participants interacted with online, but had not been provided with calendar invites. While our consent form noted that we would collect documents and images, our protocol was generally to ask for permission first. However, asking in the middle of meetings to take a screenshot would be disruptive. While taking screenshots is a less invasive method for collecting artifacts at the moment, a screenshot can also mean that you are collecting images that include the video screens of other meeting participants, often seen in small rectangles in the corner. While most of the images we were collecting were—in conjunction with our field notes—to help illustrate how people interacted with artifacts, it still raises ethical questions around whether and when to ask permission during an online meeting.

### *Ethnographic interviews*

Another form of data collection that was easier in online ethnography was the use of Zoom for ethnographic interviews. As Howlett (2022) noted in her own work, the casualness of operating in an online environment of video-conferencing often allowed participants to express themselves more freely and appear more at ease. We found this true in our own use of Zoom to conduct ethnographic interviews. In the Energy Modeling Study, there was an increased sense of formality with ethnographic interviews as most required finding an assigned room in a private office and the use of a recording device sitting in between the interviewer and interviewee. One ethnographic interview in the Energy Modeling Study attempted to use a less formal environment of an outside table at a local coffee shop chosen by the participant, but this too proved somewhat awkward for both the interviewer and interviewee. The interview (and recording device) was now easily viewed by the public, and in addition, the interview was periodically interrupted by the waiter at the cafe, removing any sense of privacy for the interviewee.

In contrast, the Cybersecurity Study interviews could be done in the peace and quiet of a private Zoom room. Interview participants could speak in the comfort of their own homes, and while participants had to initially agree to the use of the recording device on Zoom at the start of the interview, the blinking red dot that indicated the meeting was recorded could be easily forgotten. This appeared to help participants feel more at ease speaking to the interviewee. In addition, when interviews took place in an interviewee's own home, it added greater confidentiality for the interviewee since their participation in an interview would be unknown to their colleagues. In other words, no one else from work would be around to see the interviewee walking into an office with the researcher, with a recording device in tow.

### *Exiting the field*

Exiting the field is another place where an online ethnography was at odds with typical in-person ethnography practices. With the Energy Modeling project, the ethnographer had time to say goodbye to their subjects on their last day at the office, then moved on to the next field site.

In the Cybersecurity Study, this did not occur and in a way, we never left the field. Once a decision was made to end fieldwork at one Zoom site and focus on another Zoom site, the ethnographers stopped attending the meetings. However, our ethnographers still receive invites to meetings in their calendars, but we remain as ghosts. Our strategy for truly making an exit was to provide our subjects with a presentation and copy of the final report when COVID restrictions had been lifted.

## Conclusions

Engaging in ethnographic research in workplace settings, particularly in construction workplace settings, has many challenges. Workplaces becoming hybrid or fully remote adds more challenges to this work. Drawing on two ethnographic studies about the built environment, an in-person ethnography and an in-person ethnographic project that was transitioned to an online ethnography, we outlined the challenges and opportunities that accompanied these different field sites.

Entering the field in an ethnographic study is a challenge regardless of modalities. The Cybersecurity Study we showcased was particularly challenging as researchers needed to respond to a swift and complete modality shift from in-person to online shortly after entering the field. Compared to the experience of rapport building in the Energy Modeling Study, rapport building in the Cybersecurity Study was much more challenging in an online setting. However, opportunities did arise in which the shared experience of a pandemic and a swift shift in workplace settings became an opportunity for rapport building in and of itself. Future online ethnographers should seek out rapport-building opportunities that are suited to the online spaces that they observe. For videoconferencing meetings, we suggest arriving to online meetings early and engaging in communication that enhances interpersonal relationships, such as jokes, daily life events, and informal discussions about the research.

In addition to rapport building, we found that online ethnography also resulted in a loss of data that was required to answer our research questions. In the Energy Modeling Study, important data on hallway conversations and parallel chats occurred before, after, and between meetings.

While parallel chats may have been taking place during the meetings we observed in the Cybersecurity Study, we were mostly unable to gain access to data on these important informal conversations. While some pre-meeting discussions were captured via online conferencing platforms, this was largely dependent on the settings of the platform and the personal settings of the meeting organizer. Our observations took place during COVID-19, a time when many were transitioning to an online-only workplace culture. As online and hybrid work become more and more common, the locations of online parallel chats and hallway conversations may become easier to identify and become more accessible to researchers. The timing of our online research was such that the practices of cultivating space for parallel chats and hallway conversations were just being developed and therefore difficult to access as researchers. Future researchers should stay alert to where informal communication is taking place online and—through rapport building—request access to these spaces if possible.

In comparing both studies, we can offer some insights for future researchers. As the future of work continues to evolve it will likely become increasingly common that workplace ethnographies will take place in a hybrid setting. Researchers should consider surveying the workplace to understand the communication channels and the platforms that people are using to share information, chat, connect, and otherwise interact. As researchers set out to enter hybrid field sites, it is important to understand where the opportunities and challenges are for building rapport with subjects, gathering information that provides context to their subjects' workplace and work culture, and gathering data in ways that will answer one's research questions. Different video conferencing platforms, settings, meeting types, rapport with subjects, and team dynamics online and in person will impact the types of data you can collect and the types of workplace cultures that you observe. For example, a video conference platform may not record chat sessions before the researcher joins. The researcher then needed to be present early to capture active chats. As we learned in the Cybersecurity Study, doing fieldwork necessitates being able to roll with changes in the field while still doing good science. This necessitates flexibility and a commitment to meeting the participants where they are while strategizing how to best access the data needed to answer one's research questions.

## Acknowledgments

The authors wish to thank our research team colleagues Jessica Beyer and Chuck Benson for their partnership in this project and all of the team members we worked with. This material is based upon work supported by the National Science Foundation (NSF #1932769, "SaTC: CORE: Medium: Knowledge Work and Coordination to Improve O&M and IT Collaboration to Keep Our Buildings Smart AND Secure.")

## Note

1 Zoombombing is when an uninvited user joins a Zoom session and interrupts the session by saying or showing obscene or racist things.

## References

Emerson, R. M., Fretz, R. I., & Shaw, L. L. (1995). *Writing Ethnographic Fieldnotes*. Chicago: University of Chicago Press.

Góralska, M. (2020). Anthropology from home. *Anthropology in Action*, 27(1), 46–52. https://doi.org/10.3167/aia.2020.270105

Hine, C. (2008). Virtual ethnography: Modes, varieties, affordances. In N. Fielding, R. Lee, & G. Blank (Eds.), *The SAGE Handbook of Online Research Methods* (pp. 257–270). Thousand Oaks: Sage. https://doi.org/10.4135/9780857020055

Hine, C. (2017). Ethnographies of online communities and social media: Modes, varieties, affordances. In N. G. Fielding, R. M. Lee, & G. Blank (Eds.), *The SAGE Handbook of Online Research Methods* (2nd ed.). Thousand Oaks: Sage. https://doi.org/10.4135/9781473957992

Howlett, M. (2022). Looking at the 'field' through a Zoom Lens: Methodological reflections on conducting online research during a global pandemic. *Qualitative Research*, 22(3), 387–402. https://doi.org/10.1177/1468794120985691

Johnson, L. (2022). Adapt and adjust: Doing UK-based ethnographic fieldwork during the Covid-19 pandemic. *Medical Anthropology*, 41(8), 763–777. https://doi.org/10.1080/01459740.2022.2106226

Pink, S., Horst, H. A., Postill, J., Hjorth, L., Lewis, T., & Tacchi, J. (Eds.). (2016). *Digital Ethnography: Principles and Practice*. Thousand Oaks: Sage. https://www.researchgate.net/publication/284409800_Ethnography_in_a_Digital_World_In_Digital_Ethnography_Principles_and_Practices

Viererbl, B., Denner, N., & Koch, T. (2022). "You don't meet anybody when walking from the living room to the kitchen": Informal communication during remote work. *Journal of Communication Management*, 26(3), 331–348. https://doi.org/10.1108/JCOM-10-2021-0117

Zhao, Y. (2022). Adapting ethnographic methods in light of the COVID-19 pandemic: Scottish Country dancing. In L. Russell, R. Barley, & J. Tummons (Eds.), *Studies in Educational Ethnography* (pp. 109–127). Dubai: Emerald Publishing. https://doi.org/10.1108/S1529-210X20220000019007

# 17

# DOING AUTOETHNOGRAPHY

## A built environment education case study

*Leon Crascall and Fred Sherratt*

*University of Suffolk, Ipswich, Suffolk, UK*

*University of Colorado Boulder, Boulder, CO, USA*

**Key takeaways**

- Autoethnography needs defining by the individual, '*auto*' just expands the paradigm of what ethnography is.
- Autoethnography is a methodological chameleon, makes for good pedagogy and helps cultivate sound ethical practices; pragmatism can enrich the narrative and elevate praxis.
- Many labels associated with autoethnography can distract, don't waste time pursing criteria to justify the method.
- Don't be too rigid with research design strategies early on, you need to write in order to strategise research.
- Autoethnography makes archival data current, it is important to creatively work with what the data presents us.

### Introduction

Learning and working in the construction industry is rewarding but challenging. Its culture has many habits engrained into its DNA. Long-standing siloes and fragmentation can place the education of students at the mercy of industry; their knowledge and understanding shaped by pre-professional ideologies (Hosein & Rao, 2017; Hosein, 2018) as specific disciplines use specific pedagogies to direct their students, as is the case with architects for example.

Built environment education can be considered a socio-technical system, one with many stakeholders. Employers, regulatory and professional bodies, academic institutions and the learning programmes they design have an impact on students. The performances of graduates are the commonly measured output of how successful the system is overall, yet students have minimal influence on what and how they study.

Within this system, students are largely taught within prescribed curricula; an indoctrinated disciplinary framework is evident within textual artefacts that influence the student experience.

DOI: 10.1201/9781003379584-20

Knowledge and skills can be transmitted through folk pedagogy by experienced academics and practitioners. Folk pedagogy can manifest as a result of teachers re-casting abstractions of their practices, even though these abstractions may not be transferred in ways that align with how students actually learn (Powers, 2017, p. 21). This can be because professional competence in a practical environment does not necessarily translate successfully to educational competences in learning domains. The goals of each do not always align. A dissonance thus exists between academic education and practical training, just as the friction between theory and practice.

For ethnographic and by extension autoethnographic research, built environment education can be a rich space to explore. If as a researcher you are willing to challenge the status quo, the tools to do this in relative safety are available whilst studying, but you may need to seek them out. This chapter will inform the 'seeking out'.

As someone who left school with no qualifications of any merit, who progressed from tradesman to construction management roles, to then re-train in architectural technology, I have taken an unorthodox route through built environment education. As I looked forward to my undergraduate graduation and working in architectural practice, something unexpected happened: I became a senior lecturer. Studying as a mature student brought out an interest in working with other students, the immediacy of my transition from student to educator led to an interest in the relationship between educational theories and practice.

My introduction to and adoption of autoethnography came as a result of initial research training in my professional doctorate, during which I focused more on curriculum design. I explored design science research (DSR), which is a design method that focuses on the study of artefacts. An unexpected turn of events then placed me in the position as 'object' of the study. This orientated me towards using autoethnography.

During this period, I began to review the significant amount of data I had; 30-years' worth. All brought out a scepticism of curriculum itself and antagonised through conversations between different characters, all to challenge a personal belief that built environment education needed to be re-imagined (Crascall, 2021). In pursuing curriculum definitions McKernan (2008, p. 11) offers a selection, this one resonated most:

> A curriculum can become one's life course of action. It can mean paths we have followed and the paths we intend to follow. In this broad sense, curriculum can be viewed as a person's life experience.
>
> *(Connelly & Clandinin, 1988)*

My focus here as we progress is to outline the tools that helped me to explore the unorthodox journey of my built environment education. In adopting this approach, autoethnographical research was able to offer unique insights into this space, enabling analysis of benchmarking, modularisation and fragmentation, raising questions around assessment and evaluations, as well as analysis of experiential models that raised questions of the relationship between theory and practice. Ultimately this research resulted in recommendations for all stakeholders influencing built environment education, informed by the autoethnographically informed critique of existing humanist experiential models and other artefacts to offer a new lifelong learning model and a considered re-imagination of current practices.

This chapter is not a straightforward 'how to guide', it does not provide clear and simple steps to follow to 'deliver' autoethnography. Instead, it presents a case study story of a journey through autoethnography, being 'both a method and a text' (Etherington, 2004, p. 140) this journey is not always linear. Meandering through philosophy, the need for structural support, the language of autoethnography and the terms and labels often found therein, the chapter also highlights the need for strategy to underpin the craft of autoethnography before reaching its conclusions. It is hoped that this chapter provides insights, guidance, support and reassurance for those planning such a journey in future.

## First impressions

> There are many ways one can go on a solo journey to the wilderniss beyond and the wilderniss within. If one understands the complex circles of life and the connectedness of all things, then going outside and going inside is really the same journey.
>
> *(Smith, 2005)*

Autoethnography is a means to connect the 'wilderniss' beyond and within, a tool that guides, but one that challenges you and your lived experiences. All lived experience comes with many complexities and information can become quickly outdated, but autoethnography helps to keep it current in the story of our lives.

I began exploring built environment education with an interest in curriculum design, unaware of autoethnography at the start of my doctoral journey. However, life events during the middle of my doctoral programme brought a change of focus, and scepticism of curriculum itself began to surface. Blessed with 30 years' worth of data relating to my school, technical and higher education, my life became the curriculum and autoethnography the tool to explore it. As 'both a method and a text' (Etherington, 2004, p. 140), it gives an initial sense of freedom to write anything, don't be fooled. Let's explore why.

## Positioning ourselves

Thinking about our own position from the outset is a good place to begin with autoethnography. You have to ask yourself; what is the story to be told? How will you tell it? What data do you have, or are you collecting? And, how does your own position influence your research?

I position myself here as a 'researching professional'. My experience of using autoethnography is within educational research, whilst studying for my professional doctorate.

It could easily be assumed using autoethnography that you are automatically an 'insider' researcher, but it is worth thinking about the questions I posed at the outset. Herr and Anderson (2015) offer a positionality continuum to support such reflection, which I have depicted in Figure 17.1 to summarise their points.

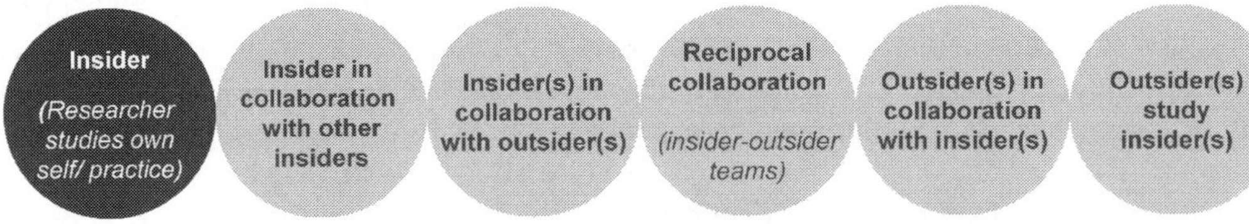


**FIGURE 17.1** Positionality continuum (by author, informed by Herr and Anderson, 2015, p. 40)

## Philosophical orientations

Autoethnography is a shape-shifting tool that moulds itself to the individual and how it is used, tacit knowledge and intuition guide me. I am more 'empeiria' than 'episteme'; 'what I know', 'how I know it' and 'what I've learnt through doing' feel more useful than 'how I came to know' (Hickman, 1992, p. 17).

Typically, autoethnography tends to be orientated through a social constructivist or postmodernist paradigm (Reed-Danahay, 1997; Ellingson & Ellis, 2008), but with a career largely spent growing up in the construction industry I am a pragmatist. To explore the relationship between educational theory and practice, I have done so intuitively through my own experience.

Andrew (2017, p. 48) outlines intuitionism as a philosophical basis for ethical autoethnography, through praxis. Anderson and Glass-Coffin (2016, p. 65) highlight how autoethnographers collect data that requires 'improvisation, experimentation and interpretation'. In other words, autoethnographers first need to appreciate data is everywhere, but that it can just showcase randomness. Rigour in research comes from scaffolding the random to showcase our lived experiences.

Praxiology can be thought of as a 'theory of practical knowing' (Coghlan & Brydon-Miller, 2014), something often overlooked in philosophical discussions. Toyosaki and Pensoneau-Conway (2016, p. 560) suggest this oversight possibly occurs as a result of training to meet 'academic' conventions, continuing to highlight it is in 'doing' autoethnography we gain a better sense of 'self'. Due to its practical nature, the relationship between built environment education and its associated disciplines, praxis was not a philosophical conversation to overlook. It is a highly relevant consideration in any autoethnographical work exploring a practical space.

Yet pragmatism has its critics. Pawson (2013) questions the merits of it methodologically, a difficult beast to contain because it 'bursts evaluation at its seams' (2013, p. 72). I would suggest autoethnography is equipped to address this, in itself or with other pragmatic tools to guide research. Orientation of research is important to highlight here though. If focusing on outcomes of action and not necessarily pursuit of truth, Patton (2015, p. 152) offers useful guidance on emergent design flexibility when strategising writing early on. He continues to suggest pragmatism can aid research evaluation not 'burst' it. My own autoethnography unpacked and analysed my own experiences; however the goal was not to just provide a greater insight into a problem. Instead, the narrative was specifically goal-orientated, with the intention to 're-imagine built environment education'. To support me in my own use of autoethnography – and to ensure things did not burst – I fused it with DSR, 'a problem-solving paradigm fundamentally' (Hevner et al., 2004); as both share an interest in artefacts. DSR is 'pragmatic and solution-focused' (Dresch et al., 2015, p. 56), orientated with a focus on things that 'do not yet exist' (Romme, 2003).

DSR can go beyond scientific realism, 'a need to describe, account for, explain and theorize' (Cherryholmes, 1992), to 'define, suggest, develop, evaluate and conclude, to design and recommend' (Dresch et al., 2015). DSR uses abduction to take intuitive leaps to suggest 'how things should be' and not necessarily 'how things are and how they behave'. Ontologically, DSR and autoethnography help engineer mixed-realities, 'impressions of lived experience that can be naturally occurring or a contrived simulated reality' (Johannesson & Perjons, 2014, p. 52; Adams et al., 2015, p. 84). Crucially, DSR seeks to address practical, 'wicked' problems, where 'the aim is not to find the truth, but to improve some characteristics of the world where people live' (Horst et al., 1973). Autoethnography creatively helped me to do this, whilst still trying to piece things together. Finding appropriate supporting structures can help the autoethnographer greatly in shaping and containing their work.

## Writing approach over 'labels'

Autoethnography is associated with a dizzying array of terms and labels. Chang (2008, p. 48) suggests it reflects a 'diverging evolution of the genre', but as a pragmatist talking about built environment education, I personally find them distracting. 'Lived experience' or 'autobiographical ethnography' (Figure 17.2) works best for me. I found I could get on and write whilst developing my research strategies. Autoethnography is autobiographical by nature, but we need to ensure there is sufficient emphasis placed on the 'ethno' (culture), 'auto' (self) allows us to demonstrate how we are placed within the narrative and how we are guiding it.

From keeping the 'labels' simple you can begin to think about how you want to write. Van Maanen (2011) outlines three 'tales': 'realist', 'impressionist' and 'confessional', or 'expressionist' as Adams et al. (2015, p. 84), phrase it. The latter two focus on first-person narration, 'realist tales' focus more on culture and less on the researcher, 'impressionist tales' tend to place equal focus on both researcher and culture. Something for autoethnographers to consider is how we 'situate the method', consider others in the accounts we provide (Tolich, 2010; Denzin, 2014, p. 7). However, with a significant amount of data and an unorthodox journey through built environment education, I explored a fourth option: a conceptualist approach.

'Conceptualism' offers the means to 're-imagine the role of the artist in culture', a place for 'innovators who create new and often unexpected ideas, perspectives, and experiences' state Adams et al. (2015, p. 88). A conceptualist approach offered me a blank canvas, methodologically I could sit on the fence early on whilst figuring things out (Anderson, 2006). In seeking to re-imagine built environment education through a conceptualist approach, I could be both novel in how I explored it and in how I used autoethnography itself.

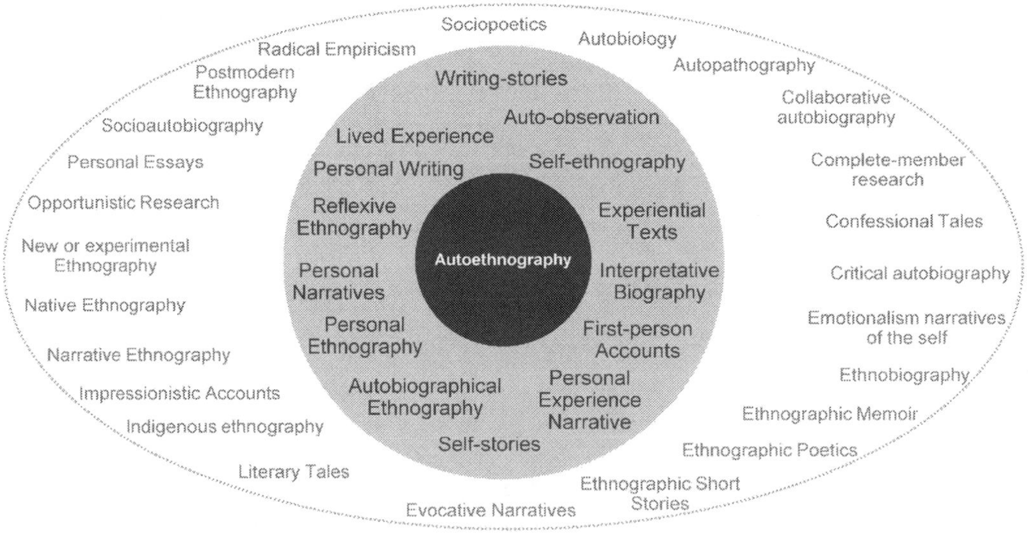


FIGURE 17.2 The many labels associated with autoethnography (informed by Ellis & Bochner, 2000; Chang, 2008, p. 47)

What a conceptualist approach offers in my view is a platform to do many things. It can include the 'confessional, realist, and impressionist tales'; fuse long-running 'evocative' vs 'analytic' conversations, to playfully experiment as your writing develops. It can help when synthesising the complexities of industry and academia relationships with built environment education, there is much to unpack and explore. You can go beyond disciplinary boundaries (Chang, 2016, p. 120), creatively fuse both text and images to re-imagine stories our experiences provide; introduce characters to showcase your voice in different ways.

Van Maanen (2011, p. 105) suggests characters 'must be given names, faces, motives, and things to do if a story is to be told about them'. Mine represented career roles from different time-periods, each offering a different perspective in conversation with one another to aid reflexivity. Goodall (2000, p. 136) says 'the soul of good ethnographic writing' is to pay attention to characters, their personas; how the language opens up and creates intimate relationships between them, that the dialogue between them delivers verisimilitude. They grow as you work creatively with your data, helping both you whilst writing and your audience in how the story resonates with them.

Writing autoethnographically takes time and practice, it can feel like a very unsupportive method initially. It requires content in order to feedback to you if what is written is any good or not, 'writing writes the writer' (Gannon, 2006). It forces you to make decisions, so get comfortable with jettisoning draft work that has taken hours, days, weeks, even months and years to craft.

'Practising' in a formal learning environment, with experienced and supportive supervisors, is a good place to safely write and experiment. Here the relationship between theory and practice can be antagonised, our experiences and industry practices can be challenged, putting us in a vulnerable position. We need to protect others, but we also need protection from ourselves in the accounts we provide (Tolich, 2010; Tullis, 2016, p. 248).

In terms of books that really helped me to write, 'On Writing' by Stephen King (2012) resonated most. The trade references and many 'toolbox' analogies benefitted character voices, focused my writing, and informed my use of autoethnography (Crascall, 2021). To summarise King, short words can be more powerful than dressed up over-intellectualised ones, work by paragraph and not sentence, think about your audience. Go read it.

## Strategising autoethnographic research

Autoethnography is often criticised for 'naval-gazing', requiring 'little fieldwork' and that it is not even research; too 'narcissistic, self-indulgent and solipsistic' (Etherington, 2004, p. 141; Chang, 2008, p. 51); 'lacking theory, non-analytic, has no concepts and is not sufficiently scientific' enough (Denzin, 2014, pp. 69–70), despite it being 'a scientific method of writing' (O'Hara, 2018). These can become valid criticisms if not addressed properly during the research process.

I may have had 30-years' worth of data, but it all needed scaffolding. The lack of perceived fieldwork is replaced with investing more time in crafting accounts we write, embracing vulnerability and second-guessing ourselves ethically on what to include. The ethical focus required in doing autoethnography properly can be overlooked by authors and dismissive critics alike. For me ethical focus becomes a valuable strategic design tool to guide action.

A lack of 'ethno' can be a distinguishing difference between autoethnography and autobiography (Winkler, 2017), which can undermine its value as a research tool. The dizzying labels (Figure 17.2) outline a long-standing issue, autoethnography still 'is not a specific research technique, method or theory' (Hayano, 1979), we have to individually define this ourselves in how we use it.

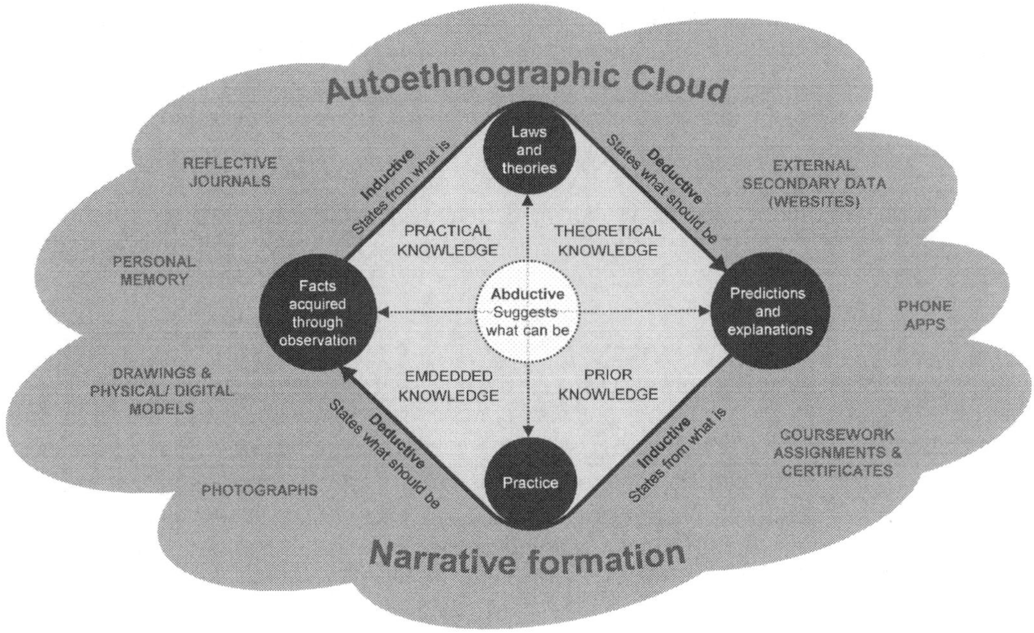


FIGURE 17.3 Fusing scientific thinking with autoethnography to inform the narrative (by author, informed by Chalmers, 2013, p. 50; Dresch et al., 2015, p. 62)

To define our use of autoethnography and make it credible, we have to give attention to the art and science of writing it. This is easier said than done early on, it can feel like many ideas and our data floats around in a 'cloud' waiting to be formed as shown in Figure 17.3.

Autoethnography is both a process and a product (Etherington, 2004, p. 139), it is important to embrace this early. Researchers have to write in order to strategise effectively, writing strategies inform research strategies as you get used to the method; Chang (2008, p. 131) is a useful starting point.

In term of research strategies, I gave a lot of attention to this at different stages. It takes time, the narrative needs space to emerge, being too rigid in the design stage can be unhelpful. I was fortunate that I had a large amount of data in various forms; however, working with it initially meant taking a few wrong turns. I had two stories to address: the subject story which needed to come out of the archival data; and my research story, the one that needed to underpin my use of autoethnography.

Creswell and Poth (2018), Chang (2008) and Patton (2015) all helped me strategically. Patton (2015, p. 49) outlines emergent design flexibility, how it 'requires a high tolerance for ambiguity and uncertainty, as well as trust in what the inductive analysis will yield'. We may write alone, but we share drafts with others. An open-ended approach and what that means, for the method, for writers and their support networks needs to be embraced, especially if using autoethnography for the first time. A sound design strategy and a 'dynamic systems perspective' (2015, p. 46) meant giving attention to process and ongoing change during data collection and fieldwork. This was a compact period with data largely stored together (in my loft), so I could begin to experiment with a range of tools to craft my story.

### *Crafting and experimenting with data in the 'black box'*

Autoethnography is researcher-friendly, it does take you on an emotional rollercoaster, but it is a deeply rewarding and fun way to conduct research in my view. Yes, I was fortunate my first experience of it was whilst studying and I had a careers worth of data, I was also fortunate when it came to writing the bulk of my thesis. Putting a positive spin on the COVID pandemic, I was teaching online at home for over a year and had an office in the garden with everything I needed.

My 'black box' took on a range of different meanings; my office, my brain, cybernetics discussions, a way to describe artefacts relating to built environment education at different levels. It became central to my conceptualist approach.

> A Black Box is something that does something, that one does something to, that does something back – a partner in, as I would say, a dance of agency. Knowledge of its workings on the other hand, is not intrinsic to the conception of a Black Box – it is something that may (or may not) grow out of our performance experience of the box.
>
> *(Pickering, 2011, p. 20)*

Through my 'black box' I began to systematise my thinking, characters began talking to each other. Using autoethnography, I framed theory through conjecture (Chang, 2008), postulating about 'output' and graduate attributes (Figure 17.4). The idea was to use it to achieve my goal at doctoral level, whilst re-visiting my experiences of built environment education. My critique of my own experiences became the platform to abductively suggest what could be, through different voices and perspectives.

Built environment education qualifications are regulated at all level, each has statements that holders should reflect. These statements tend to get lost on their way to the student learning experience, this story became my playground (Figure 17.5).

From the wealth of data that I had and what I could also access in the public domain, I could connect my personal experiences to the culture in which I had been immersed. I had the 'auto' and the 'ethno', the focus became on doing the 'graphy' rigorously. A lot of work went into sufficiently explaining the systematising of my thinking, of scaffolding autoethnography and

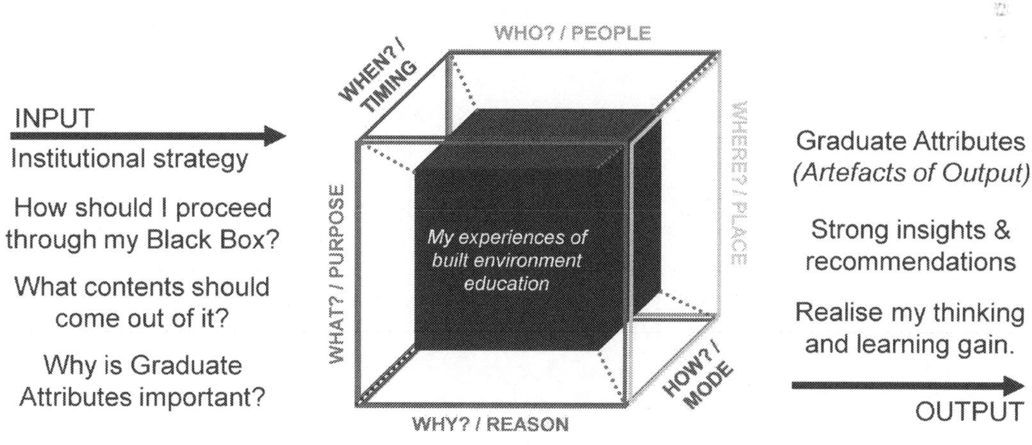


**FIGURE 17.4** My 'ontological Black Box' (image by author)

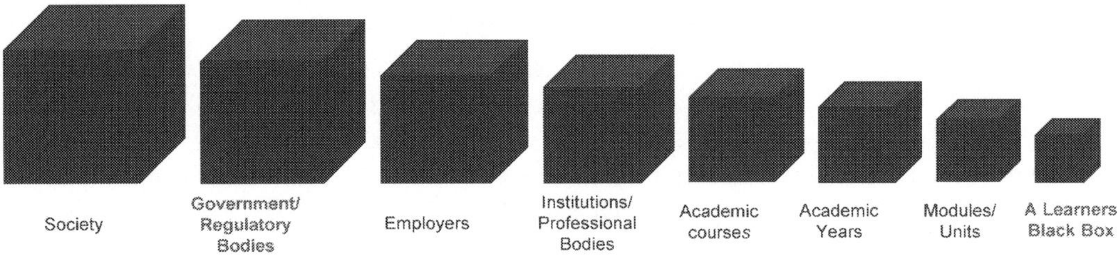


FIGURE 17.5 The 'black-boxes' socio-technical system influencing the learner's experiences of built environment education (by author)

showcasing the science that underpins it. From this the characters were set free to deliver on the 'core ideals' of the method that Adams et al. (2015, p. 102) set out:

- Making contributions to knowledge
- Valuing the personal and experiential
- Demonstrating the power, craft and responsibilities of stories and storytelling
- Taking a relationally responsible approach to research practice and representation

The 'black boxes' mobilised my characters gave them a focus to achieve these 'ideals' through insights drawn from my data.

With characters mobilised, they need to be led by data and what it is says to us (Srivastava & Hopwood, 2009). Inductively this comes through 'experiences, feelings, stories, and happenings' (Ellis et al., 2011), through data in various forms that require care and attention. It is a balancing act bringing past, present and future together in order to go 'beyond the black box' (Clandinin & Connelly, 2000, p. 50). Temporality is something to pay attention to, particularly as the narrative and the characters themselves develop. To help me move backward and forward through time, I had to use a range of techniques to showcase my data (Figure 17.6).

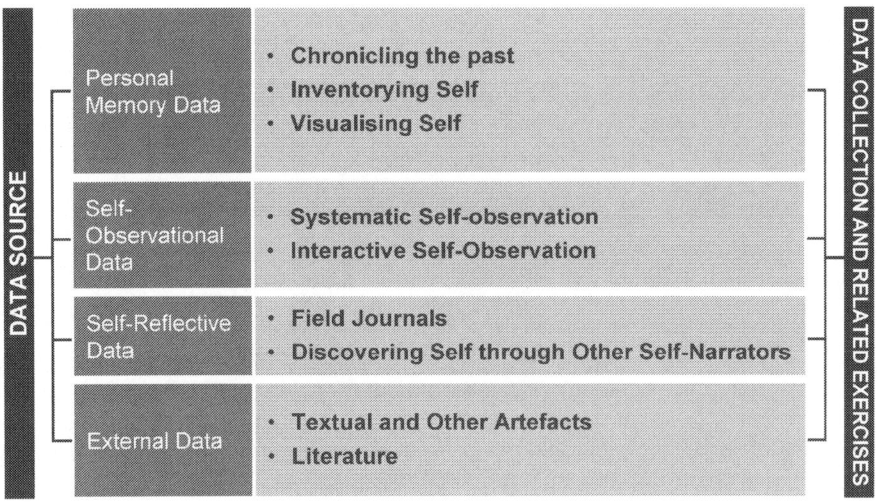


FIGURE 17.6 Data collection and related exercises (image by author, informed by Chang, 2008)

External data from government, regulatory bodies, professional bodies and academic institutions, qualification descriptors from the Quality Assurance Agency (QAA), provided a foundation to support my personal memory data (Figure 17.6), recorded in my personal journals. I also included photographs with other graphics, even school reports all to help the reader appreciate my own evolution. I also used personal data collected through 'lifelogging' within phone apps (Sharples, 2019, p. 55), for location, health, diet, sleep and exercise.

Through both systematic self-observation/reflective practices in my journals, exploration of themes began to develop nuanced with external data outlined above. The most valuable data I had was coursework assignments and feedback from others, I could reflexively use my educator lens to critique the student self over time. This particularly allowed me to explore educational theory against the backdrop of my own lived experience, to evaluate my performance and question my education at different times. But it is important to stress here that personal memory can be over-relied upon when re-living the past; reflexivity through self-reflection/observation can be over-exaggerated (Moors, 2017); external data helps to counteract them, it connects us to 'ethno' to better resonate with the reader and make research valid.

## Ethical underpinnings

Tullis (2022, p. 106) suggests autoethnography makes for good pedagogy. I would agree, where better to experiment than whilst exploring a range of different methods and techniques than in a learning environment? This is a good place for developing the ethical awareness of working with the method.

An incorrect assumption to make with autoethnography and having a focus on 'self' is that ethics is not applicable to the method (Chang, 2008, p. 68). We may be the storytellers, but we are human, our experiences and ownership of any content tend to be shared with others. We have a responsibility to communicate a story, but one that still protects ourselves and others (Clandinin & Connelly, 2000). Despite any suggestions we are 'free of the traditional conventions of writing' (Gergen & Gergen, 2002), I err on the side of caution informed by Tolich (2010) and Tullis (2022) particularly. Ethics can be a valuable intuitive process, if embraced early in our research practice (Andrew, 2017; Poulos, 2021). Empirical observation brings a need to ensure events are presented with 'respect for people, beneficence and justice' that our accounts are fair and honest (Adams et al., 2015, p. 56).

By ethically designing autoethnography from the outset, you can write creatively and safely. I found it elevated my thinking, with my reflective and reflexive practices. Autoethnography can appear to circumnavigate many areas of ethics forms at an institutional level, which is not helpful in periods of vulnerability. Using the method under the banner of 'educational research' you become aware how research can be unethical in its design, planning and the methods used, in the data analysis and in how it is presented (Wellington, 2015, p. 113). What is important to note here is what constitutes 'educational research', because it is more far-reaching than many realise.

The British Educational Research Association (BERA) ethical guidelines are a valuable design tool, they are easily overlooked during ethics training at an institutional level. BERA (2018, p. 1) outlines who their guidance is intended for, it includes those 'studying for a qualification'. Knowing this meant I could write without restraint whilst drafting work, I could experiment whilst being mentored by experienced supervisors during periods of vulnerability. This for me is where autoethnography becomes more than just a method, it's a pedagogical tool to guide ethical action and develop good habits in doing research generally.

| Ethical considerations | Prior to and beginning to conducting study | Data collection prior and during the study conducted | Analysis and reporting of data during this study conducted and when it's published |
|---|---|---|---|
| Responsibilities to participants | • Consider active/ passive involvement; students and colleagues form part of the context of this investigation, but they themselves are not the focus of this research (BERA, 2018, p.6).<br>• Where necessary seek consent from others prior to inclusion or publication (BERA, 2018, p.14).<br>• For transparency, outline how data is being used and treated (BERA, 2018, p.17). | | |
| Responsibilities to sponsors, clients and stakeholders in research | • Outline any sponsorship, funding or who has commissioned research (BERA, 2018, p.26).<br>• Outline conflict of interests, any direct associations, like professional bodies (BERA, 2018, p.18). | | |
| Responsibilities to community of educational researchers | • Despite a focus on our individual experiences you have to consider how you contribute to the community spirit of educational research, that makes a contribution to knowledge and practice but not in a defamatory or unprofessional way (BERA, 2018, p.29). | | |
| Responsibilities for publication and dissemination | • Do not falsify, distort, supress, selectively report or sensationalise research (BERA, 2018, p.33).<br>• Give consideration to areas that may impact self and others in it's authorship (BERA, 2018, p.34).<br>• Consider dissemination to wider research communities and stakeholders; research impact requires consideration at a design stage, can hinder autoethnography during creative synthesis. | | |
| Responsibilities for researchers' wellbeing and development | • We have a moral and ethical responsibility to safeguard ourselves and others, consider proximity of research environment and working environments, particularly fieldwork (BERA, 2018, p.35).<br>• Data management and storage needs to managed in accordance to the UK Data Protection Act (1998) and GDPR, Freedom of Information requests under the Freedom of Information Act (2000) and management of data (BERA, 2018, p.21). | | |
| Ethical Design | Ethical decision-making is an actively deliberative, ongoing, iterative process of assessing and reassessing the situation and as issues arise. (BERA, 2018, p.2) | | |

FIGURE 17.7 Risk-benefit analysis summarising ethical guidelines for educational research (by author, informed by BERA, 2018)

With my ethical position continually being under review, the BERA guidelines were helpful because they explicitly address autoethnography. A simple risk-benefit analysis (Figure 17.8) was a good way to crystallise tacit processes that come whilst writing. Figure 17.8 offered me a simple of way of visualising the risks and benefits both prior, during and after my research had finished. It informed iteration of my research strategies that I could adapt them to the narrative. Figure 17.7 also highlights our responsibilities to others, to the data and how we present and use it.

Fusing autoethnography with DSR I could exploit their shared interest in artefacts, draw on the ethical framework from that field of study. Myers and Venable (2014) outline a proposed set of ethical principles for DSR, the first five principles within Figure 17.8 largely align with what BERA (2018) outline, generally applicable to all forms of research and good ethical practice.

Myers and Venable (2014) outline 'quality of the artefact' (Figure 17.8) that any associated risks in evaluating and testing artefacts should be rigorous, applicable to documentation also.

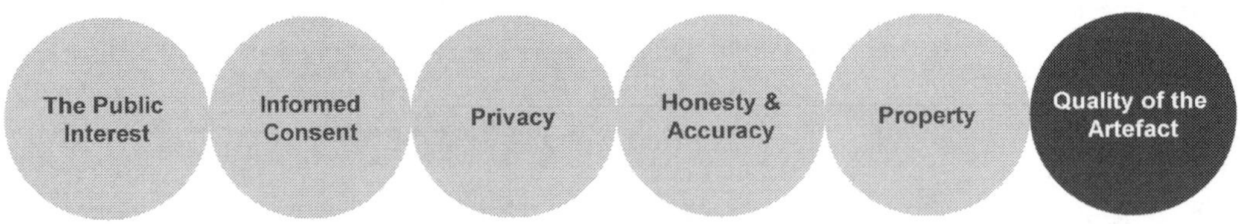


FIGURE 17.8 Myers and Venable's (2014) proposed set of ethical principles for design science research in IS (by author)

I applied this logic to my thesis, an 'artefact' now in the public domain (Crascall, 2021). Figure 17.8 outlines our responsibility for what we put into the public domain, to stakeholders we may impact. Built environment education has many influences at different levels that create many artefacts within one big socio-technical system (Figure 17.5, p. 228). Through critique of the 'system' against my lived experience, I could ethically navigate my way through them.

## Evaluating and legitimising autoethnography

Autoethnography challenges some, it divides opinion and sends barriers up. When presenting it as a method, you can see the deep intake of breath and eyes rolling in some of the audience. We need to appreciate 'auto' just expands the paradigm of what ethnography is (Anderson & Glass-Coffin, 2016, p. 58), which needs some freedom to challenge the status quo (Bochner, 2016, p. 52). We want to be authentic, but we need to attend to reliability, validity and generalisability.

In respect to reliability, this comes in how credible we are as narrators (Ellis et al., 2011). When we use autoethnography, we are only ever creating an impression of a reality re-told differently. There is no value in my view of over-exaggerating the accounts we provide; it just weakens any contribution we go on to make, it undermines the method as a valid research tool. Validity is closely linked to this as we seek verisimilitude. Resonance comes in experiences being believable and lifelike through feelings invoked in our audience (Ellis et al., 2011). It is from this that generalisability comes from the readers response, how it speaks to them about their own experiences (Ellis et al., 2011; Denzin, 2014, p. 70).

Reliability, validity and generalisability have different meanings depending on the methodology in which they are being used, but generally viewed as signals of approval (Wellington, 2015, p. 41). Built environment research is largely rooted in positivism (Zou et al., 2014). Denzin (2014, p. 70) suggests that 'autoethnography cannot be judged by traditional positivist criteria', which I will lightly touch upon.

Built environment education may operate within a largely siloed and disciplinary landscape, but 'education' extends boundaries and realms of possibility. We have now identified its more than likely many students and researchers are doing educational research by 'qualification'. For which we have criteria to explore, which also helps to evaluate and legitimise autoethnography.

Establishing criteria for evaluating autoethnography has run for some time. Bochner (2000, p. 267) suggests it preoccupies us as we seek to demonstrate rigour, but by worrying about how we are judged we can neglect imagination. It can consume you, for me Bochner (2000, p. 269) offers sound advice when he says 'criteria are not found; they are made'. When studying, criteria are already made, disguised within the 'system' (Figure 17.5, p. 228).

Hughes and Pennington (2017) explore the 'practicalities' of problematising, legitimising and synthesising autoethnography. They outline Guba and Lincoln (1989), Starr (2010), Anderson (2006) and Hughes et al. (2012), the latter explores the AERA guidelines but for me it brought into focus the value of what qualification descriptors offer. Figure 17.9 represents the level 8 QAA (2014) qualification descriptor, it took me to descriptors at lower levels with professional body criteria. All artefacts within built environment education, re-communicated in more complex ways by stakeholders en-route to students (Figure 17.5, p. 228).

Without getting too distracted into the mechanics of the 'system' (Figure 17.5, p. 228) that built environment education operates in, Figure 17.9 highlights some areas to unpack briefly.

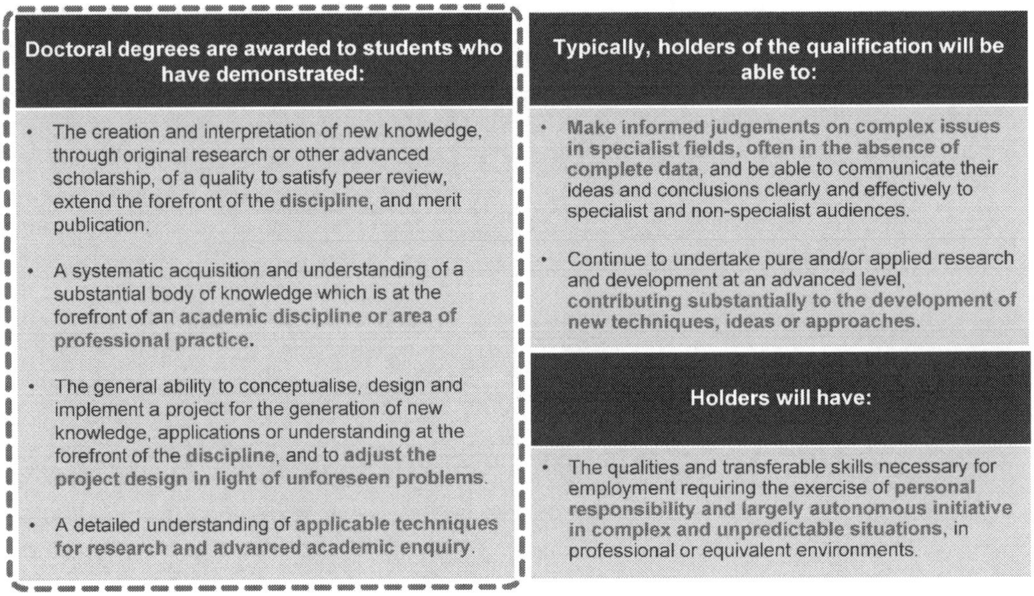


FIGURE 17.9 The Level 8 qualification descriptor for doctoral degrees (by author, represented from QAA, 2014, p. 10)

'Discipline' stokes positivist undertones, particularly their impact on lower-level equivalents, a conversation for another time. What I want to focus on within Figure 17.9 is:

> Adjusting the project design in light of unforeseen problems.
>
> Making informed judgements on complex issues in specialist fields, often in the absence of complete data.
>
> Contributing substantially to the development of new techniques, ideas, or approaches.
>
> Personal responsibility and largely autonomous initiative in complex and unpredictable situations.

Hopefully, this chapter offers enough to demonstrate the above, synergies with the 'core ideals' of autoethnography apparent (p. 228). Taking the initiative through abductive reasoning, helps connect-the-dots with incomplete data and complex problems. We can explicitly demonstrate ethical action, adapt research designs, creatively weave it into the narrative. We can be novel methodologically, with DSR in this instance. Some research can be overly complex as it 'nourishes the paradigm', but dull to read, hopefully a trip into my 'black box' offers food-for-thought.

## Conclusions

Autoethnography comes with an element of stigma attached to it – that it is not really research and delivered poorly it can attract justifiable criticism. As a tool it moulds to the individual, shaped by our own voice(s) and what our experience offers. Aspiring autoethnographic researchers must give attention to the art and science of narrative writing to conduct research rigorously, and to deliver outcomes well. Also, they need to define their use of the method to showcase its merits.

There is a need to be explicit when positioning ourselves to do autoethnographic research, but this presents an opportunity to be creative within the narrative through characterisation. Experimentation and iteration of ourselves can be liberating but does bring vulnerability, embracing ethics and using it to design with early on can be invaluable. From this you can begin to focus on developing strategies to inform research design, be flexible early on whilst developing the characters and their personas. Equal attention must be given to both subject and research stories to demonstrate rigor, otherwise you are going to do autoethnography an injustice and just feed the critics. This can be where pragmatism benefits autoethnography.

The many labels associated with autoethnography can distract, it can stop you from just getting on with writing. How we position ourselves can inform a preferred label, from which we can focus more on the relationship between ‘auto’ and ‘ethno’, pay attention to the science as we creatively craft the narrative.

Verisimilitude comes in confidently communicating ourselves to others, it is where the real thinking needs to be placed in using the method, this takes time and practice to craft. Designing through writing can help to construct thoughts and aid reflexivity but get comfortable with jettisoning drafts in the re-telling of lived experiences. Getting precious over content clogs imagination, it limits the potential of the method to work its magic on us. As a method it offers an opportunity to expand the paradigm of ethnographic research, a toolbox that collects a range of tools and techniques, which help to elevate reflective and reflexive practices and inform praxis.

In trying to justify the method it is important to not be consumed by seeking criteria to validate it, in an educational setting it is likely criteria are already established, seek them out if they are not presented to you.

Autoethnography can be frowned upon within academic circles, but for practitioner research and in my case a DProf, it can be a useful pedagogic tool to aid student learning. Exploring autoethnography alongside other methods presents a good opportunity to test it whilst studying. Do not be afraid to push the boundaries, however, be prepared for push back from those that are dismissive of its merits and what it can offer; status quo should only be a rock band reference.

Autoethnography puts you and the data front and centre, it is important to be true to it in the accounts we provide but creative in presenting ‘truth’. It is a researcher-friendly method, fieldwork may be significantly reduced compared to other methods, but this is balanced up with more time spent crafting what we write.

We can travel through time, utilise an array of tools to visit many locations and different people. These people can all be you, the many locations can be in one place, for it is the imagination that connects writer and reader. Pay attention to the data and you will have more than a mobile phone in the palm of your hand.

## References

Adams, T., Holman-Jones, S., & Ellis, C. (2015). *Autoethnography: Understanding Qualitative Research*. New York: Oxford University Press.

Anderson, A. (2006). *Analytic Autoethnography Journal of Contemporary Ethnography*, 35(4), 373–395. https://doi.org/10.1177/0891241605280449. [online]. Available at: https://journals.sagepub.com/doi/pdf/10.1177/0891241605280449 [Accessed 28 March 2023].

Anderson, L., & Glass-Coffin, B. (2016). Learn by going: Autoethnographic modes of inquiry. In S. Holman-Jones, T. Adams, & C. Ellis (Eds.), *Handbook of Autoethnography* (pp. 57–83). Abingdon: Routledge.

Andrew, S. (2017). *Searching for an Autoethnographic Ethic*. Abingdon: Routledge.

Bochner, A. P. (2000). *Criteria Against Ourselves* [online]. Available at: https://journals.sagepub.com/doi/pdf/10.1177/107780040000600209 [Accessed 20 April 2023].

Bochner, A. P. (2016). Putting meanings into motion: Autoethnography's existential calling. In S. Holman-Jones, T. Adams, & C. Ellis (Eds.), *Handbook of Autoethnography* (pp. 50–56). Abingdon: Routledge.

British Educational Research Association [BERA] (2018). *Ethical Guidelines for Educational Research* (4th ed.). London [online]. Available at: https://www.bera.ac.uk/publication/ethical-guidelines-for-educational-research-2018-online [Accessed 14 April 2023].

Chalmers, A. (2013). *What Is This Thing Called Science?* (4th ed.). New York: Open University Press.

Chang, H. (2008). *Autoethnography as Method*. Abingdon: Routledge.

Chang, H. (2016). Individual and collaborative autoethnography as method. In S. Holman-Jones, T. Adams, & C. Ellis (Eds.), *Handbook of Autoethnography* (pp. 107–122). Abingdon: Routledge.

Cherryholmes, C. H. (1992). Notes on pragmatism and scientific realism. *Educational Researcher*, 21(6), 13–17 [online]. Available at: https://www.jstor.org/stable/1176502

Clandinin, D. J., & Connelly, F. M. (2000). *Narrative Inquiry: Experience and Story in Qualitative Research*. San Francisco: Jossey-Bass.

Coghlan, D., & Brydon-Miller, M. (2014). *Praxeology. The SAGE Encyclopaedia of Action Research* [online]. Available at: https://methods.sagepub.com/reference/encyclopedia-of-action-research/n265.xml?PageNum=651 [Accessed 26 March 2023].

Connelly, F. M., & Clandinin, D.J., (1988). *Teachers as Curriculum Planners: Narratives of Experience*. New York: Teachers College Press.

Crascall, L. (2021). Re-imagining Built Environment Education: Investigating the Pracademic through Autoethnography. Doctoral thesis, Anglia Ruskin University.

Creswell, J. W., & Poth, C. N. (2018). *Qualitative Inquiry and Research Design: Choosing among Five Approaches* (4th ed.). Thousand Oaks: Sage.

Denzin, N. (2014). *Interpretative Autoethnography*. London: Sage.

Dresch, A., Lacerda, D. P., & Antunes, J. A. V. (2015). *Design Science Research: A Method for Science and Technology Advancement*. Cham: Springer.

Ellingson, L. L., & Ellis, C. (2008). Autoethnography as constructionist project. In J. A. Holstein, & J. F. Gubrium (Eds.), *Handbook of Constructionist Research* (pp. 445–466). New York: Guildford Press.

Ellis, C., Adams, T. E., & Bochner, A. P. (2011). Autoethnography: An Overview. http://www.qualitative-research.net/index.php/fqs/article/view/1589/3095 [Accessed 19 March 2023].

Ellis, C., & Bochner, A. P. (2000). Autoethnography, personal narrative, and personal reflexivity. In N. K. Denzin & Y. S. Lincoln (Eds.), *Handbook of Qualitative Research* (2nd ed.) (pp. 733–768). Thousand Oaks: Sage.

Etherington, K. (2004). *Becoming a Reflexive Researcher: Using Our Selves in Research*. London: JKP Publishing.

Gannon, S. (2006). *The (Im)Possibilities of Writing the Self-Writing: French Poststructural Theory and Autoethnography, Cultural Studies, Critical Methodologies (*pp. 474–495), Vol. 6(4). [online]. Available at: https://journals.sagepub.com/doi/10.1177/1532708605285734 [Accessed 20 February 2021].

Gergen, M., & Gergen, K. (2002). Ethnographic representation as relationships. In A. Bochner and C. Ellis (Eds.), *Ethnographically Speaking: Autoethnography, Literature, and Aesthetics [Kindle e-Book]*. Walnut Creek: AltaMira Press.

Goodall, H. L. (2000). *Writing the New Ethnography*. Walnut Creek: AltaMira Press.

Guba, E. G., & Lincoln, Y. S. (1989). *Fourth Generation Evaluation*. London: Sage.

Hayano, D. (1979). Auto-ethnography: Paradigms, problems, and prospects. *Human Organization*, 38(1), 99–104 [online]. Available at: www.jstor.org/stable/44125560 [Accessed 17 March 2023].

Herr, K., & Anderson, G. L. (2015). *The Action Research Dissertation: A Guide for Students and Faculty*. Riverside County: Sage.

Hevner, A. R., March, S. T., Park, J., & Ram, S. (2004). Design science in information systems research. *MIS Quarterly*, 28(1), 75–105 [online] Available at: https://www.jstor.org/stable/25148625 [Accessed 14 April 2023].

Hickman, L. A. (1992). *John Dewey's Pragmatic Technology*. New York: Association of American University Press.

Horst, W. J., Rittel, & Webber, M. (1973). Dilemmas in general theory of planning. *Policy Sciences*, 2(193),55–169 [online]. Available at: www.jstor.org/stable/4531523 [Accessed 26 March 2023].

Hosein, A. (2018). *Employability: Matching professional ideologies and skills* [online]. Available at: https://anesaresearch.wordpress.com/2018/02/05/employability-matching-professional-ideologies-and-skills/ [Accessed 30 March 2023].

Hosein, A., & Rao, N. (2017). *Pre-professional ideologies and career trajectories of the allied professional undergraduate student* [online]. Available at: http://www.tandfonline.com/eprint/RjrVaIfSKX-wNkfy34UtA/full. [Accessed 30 March 2023].

Hughes, S. A., & Pennington, J. L. (2017). *Autoethnography: Process, Product, and Possibility for Critical Social Research*. London: Sage.

Hughes, S. A., Pennington, J. L., & Makris, S. (2012). Translating autoethnography across the AERA standards: Toward understanding autoethnographic scholarships as empirical research, *Educational Researcher*, 41(6), 209–219 [online]. Available at: https://journals.sagepub.com/doi/abs/10.3102/0013189X12442983 [Accessed 20 April 2023].

Johannesson, P., & Perjons, E. (2014). *An Introduction to Design Science*. Cham: Springer.

King, S. (2012). *On Writing: A Memoir of the Craft*. London: Hachette.

McKernan, J. (2008). *Curriculum and Imagination: Process Theory, Pedagogy and Action Research*. Abingdon: Routledge.

Moors, A. (2017). On autoethnography. *Ethnography*, 18(3), 387–389. https://doi.org/10.1177/1466138117723354. PMID:30443197 [online]. Available at: https://journals.sagepub.com/doi/pdf/10.1177/1466138117723354 [Accessed 21 March 2023].

Myers, M., & Venable, J. (2014). A set of ethical principles for design science research in information systems. 53(6), 801–809. https://doi.org/10.1016/j.im.2014.01.002. [online]. Available at: https://www.sciencedirect.com/science/article/pii/S0378720614000081 [Accessed 14 April 2023].

O'Hara, S. (2018). Autoethnography: The science of writing your lived experience. *HERD: Health Environments Research & Design Journal*, 11(4), 14–17 [online]. https://doi.org/10.1177/1937586718801425. Available at: https://journals.sagepub.com/doi/pdf/10.1177/1937586718801425 [Accessed 14 April 2023].

Patton, M. Q. (2015). *Qualitative Research & Evaluation Methods: Integrating Theory and Practice* (4th ed.). Thousand Oaks: Sage.

Pawson, R. (2013). *The Scsience of Evaluation: A Realist Manifesto*. London: Sage.

Pickering, A. (2011). *The Cybernetic Brain: Sketches of Another Future*. Chicago and London: University of Chicago Press

Poulos, C. N. (2021). *Essentials of Autoethnography*. Washington: American Psychological Press

Powers, M. (2017). *Self-Regulated Design Learning*. Abingdon: Routledge.

QAA. (2014). *UK Quality Code for Higher Education – Part A: Setting and Maintaining Academic Standards. The Framework for Higher Education Qualifications of UK Degree-Awarding Bodies* [online]. Available at: https://www.qaa.ac.uk/quality-code/qualifications-and-credit-frameworks [Accessed 1 September 2020].

Reed-Danahay, D. E. (1997). *Auto/Ethnography: Re-Writing the Self and the Social*. Oxford: Berg.

Romme, G. (2003). Making a difference: Organisation as design. *Organisation Science*, 4(5), 558–573 [online]. Available at: https://www.jstor.org/stable/4135149 [Accessed 13 February 2024]

Sharples, M. (2019). *Practical Pedagogy: 40 New Ways to Teach and Learn*. Abingdon: Routledge.

Smith, T. (2005). Going outside to go inside: Frameworks for the solo experience. In C. C. Knapp, & T. Smith (Eds.), *Exploring the Power of Solo, Silence, and Solitude* (pp. 3–18). Boulder: Association for Experiential Education.

Srivastava, P., & Hopwood, N. (2009). A practical framework for qualitative data analysis. *International Journal of Qualitative Methodology*, 8(1), 76–84 [online]. Available at: https://journals.sagepub.com/doi/pdf/10.1177/160940690900800107 [Accessed 17 April 2023].

Starr, L. J. (2010). The use of autoethnography in educational research: Locating who we are in what we do. *Canadian Journal for New Scholars in Education*, 3(1), 1–9 [online]. Available at: https://journalhosting.ucalgary.ca/index.php/cjnse/article/view/30477 [Accessed 20 April 2023].

Tolich, M. (2010). A critique of current practice: Ten foundational guidelines for autoethnographers. *Qualitative Health Research*, 20(12), 1599–1610. https://doi.org/10.1177/1049732310376076 [online]. Available at: https://journals.sagepub.com/doi/abs/10.1177/1049732310376076 [Accessed 19 March 2023].

Toyosaki, S., & Pensoneau-Conway, S. L. (2016). Autoethnography as a praxis of social justice. In S. Holman-Jones, T. Adams, & C. Ellis (Eds.), *Handbook of Autoethnography* (pp. 557–575). Abingdon: Routledge.

Tullis, J. A. (2016). Self and others: Ethics in autoethnographic research. In S. Holman-Jones, T. Adams, & C. Ellis (Eds.), *Handbook of Autoethnography* (pp. 244–261). Abingdon: Routledge.

Tullis, J. A. (2022). Self and others: Ethics in autoethnographic research (2nd ed.). In S. Holman-Jones, T. Adams, & C. Ellis (Eds.), *Handbook of Autoethnography* (pp. 101–113). Abingdon: Routledge.

Van Maanen, J. (2011). *Tales of the Field: On Writing Ethnography* (2nd ed.). Chicago: University of Chicago Press.

Wellington, J. (2015). *Educational Research: Contemporary Issues and Practical Approaches*. London: Bloomsbury.

Winkler, I. (2017). Doing autoethnography: Facing challenges, taking choices, accepting responsibilities. *Qualitative Inquiry*, 24(4), 236–247 [online]. Available at: https://journals.sagepub.com/doi/full/10.1177/1077800417728956 [Accessed 19 March 2023].

Zou, P. X. W., Sunindijo, R. Y., & Dainty, A. R. J. (2014). *A mixed methods research design for bridging the gap between research and practice on construction safety* [online]. Available at: https://www.sciencedirect.com/science/article/pii/S0925753514001647 [Accessed 29 March 2023].

## SECTION IV

# Construction ethnography examples, outcomes, and reflections

# 18

# AN ETHNOGRAPHY OF CO-CREATING COLLABORATION, INNOVATION, AND VALUE

*Ani Raiden and Emmanuel Manu*

*Nottingham Trent University, Nottingham, United Kingdom*

**Key takeaways**

- EAR offers an opportunity to develop 'box-breaking' research as it is built on the cyclical processes of planning, doing, observing, reflecting, and planning again.
- Place is significant in EAR and thus such research should be conducted within the natural settings of the research participants where feasible.
- Strong links with insider collaborators are fundamental for the successful negotiation of access and securing participant engagement for the implementation of EAR.
- Actions should be directly linked to participants' day-to-day activities to prevent them from choosing between engagement in the EAR and the time they spend doing their day-to-day work. Allow time and flexibility to accommodate changes to the planned sequence of activities.
- EAR can empower research participants by giving them a voice, an opportunity to effect change and to contribute to knowledge generation.

## Introduction

In this chapter we discuss an Ethnographic Action Research (EAR) methodology and introduce an innovation-driven procurement (IDP) project that we worked on from November 2019 to September 2023. EAR methodology focuses on the development of a research culture through which knowledge and reflection are made integral to ongoing development. It is place specific and draws on participatory techniques to involve participants into the ongoing processes of planning and acting.

The IDP project sought to tackle the productivity challenge in construction by leveraging collaborative principles, processes, tools, and support mechanisms to drive innovation throughout the procurement process and add value for all the parties involved. Since supply chain collaboration is key to addressing productivity challenges in construction and overcoming wasteful procurement practices (Sarhan et al., 2017, 2018), it was at the heart of the IDP project.

Emphasis was placed on collaborative behaviours, design, and risk management. The focus on behaviours was to equip the supply chain with the soft skills that are necessary for joint problem-solving. These include, for example, clear communication, active listening, taking

DOI: 10.1201/9781003379584-22

responsibility for mistakes, and respect. Further, we focused on how design issues were tackled as there are commonly one of the causes of low productivity in construction (Jarkas et al., 2015; Naoum, 2016). Additionally, focus on collaboration in risk management processes ensured that make-ready (flow) planning (see Ebbs and Pasquire, 2018) is integrated into procurement decisions from the onset. It tackles the reduced performance that arises from 'making do' where the supply chain starts or continues a task without all the standard inputs (Formoso et al., 2011). The collaborative risk management process sought to identify and promptly resolve risks to workflow and productivity before they manifest on site.

The IDP project also had an emphasis on driving value through innovation, backed by a support package for the supply chain to implement innovative solutions that add value for all stakeholders This ensured that the supply chain is equipped to constantly seek out innovative solutions that increase productivity and add value for all stakeholders. The emphasis on value went beyond conventional notions of cost savings for the client and monetary value for the supply chain to encompass the social value contributions as well. Social value in this context refers to the positive social, economic, and environmental impact that any organisation, project, or programme makes on the lives of stakeholders affected by its activities (Raiden et al., 2019). These benefits extend beyond those received directly by the individuals and organisations that generated the value to that created for the wider society (Auerswald, 2009). As part of the IDP project, the supply chain was also incentivised to adopt and embed innovative processes and tools by offering a package that comprised mentorship and consultancy support to realise productivity and added value.

Our focus in this chapter is to discuss how the EAR methodology helped us avoid getting 'boxed-in' and supported the development of a 'box-breaking' programme of research. We show how the 'Plan ≫ Do ≫ Observe & Reflect ≫ Plan' cycle enabled us to integrate research into the project's continuous cycle of planning and doing, and we discuss place as one important and interesting theme.

## Setting out the methodological framing for the IDP project

> Qualitative research does not represent a monolithic, agreed-on approach to research but is a vibrant and contested field with many contradictions and different perspectives.
>
> *(Brinkmann et al., 2020, pp. 23–24)*

This statement by Brinkmann et al. serves as our point of departure. We contextualise our discussion of doing ethnographic research within the broader field of qualitative research and reflect upon what constitutes ethnographic research and how different perceptions of ethnography have evolved. We focus on construction management research and look at the possibilities that ethnographic work has opened for diversifying research in the field as a way of generating rich insights. Our approach to research, and discussion in this chapter, draws inspiration from Alvesson and Sandberg (2014) who discuss the problems with 'boxed-in' research and propose 'box-breaking' research as a strategy forward. Of relevance to our discussion is the 'method box' which means that a specific technique or research approach is privileged, and a researcher leans on their specific methodological approach and applies it without much questioning (ibid, p. 972).

Qualitative research has emerged as a vibrant antidote to the rise and dominance of positivist social science in the mid-20th century (Brewer, 2000, p. 5; Brinkmann et al., 2020). Ethnography and ethnographic fieldwork play an important role in the development of qualitative research. Historically, ethnography has been associated with anthropology, a field within which the principles that still appear important in fieldwork were set out, including the researcher living in the community and among the people who are being studied; learning the specific language of the community; and participating and observing at the same time (participant observation) (Brinkmann et al., 2020: 39). The Chicago school embraced and strengthened the focus on participant observation within the development of urban ethnography (Jaynes et al., 2009).

Anthropological fieldwork methods have now been embraced in many other social science disciplines (Brewer, 2000, p. 13) and in applied fields like construction. There are examples of ethnographic research associated with the more traditional longitudinal fieldwork where the researcher has observed, face-to-face, the everyday interactions of people and groups in their natural settings (see, for example, Paap, 2006; Thiel, 2010, 2012; Koch and Siebken Schultz, 2019; Hirslund, 2021, for such studies in construction). Contemporary approaches to ethnographic work include a wide variety of applications, however. For example, the rapid development and widespread adoption of information and communication technologies (ICT) in the everyday have added to the repertoire of ethnographic methods in the emergence of online/virtual/digital ethnography (see, for example, Hine, 2000).

Ethnography has played a key role in the development of construction management research too (see Pink et al., 2013; Tutt and Pink, 2019). For example, self-reflexive ethnography (Löwstedt, 2015), rapid ethnography (Loosemore et al., 2015; Galea, 2018; Galea et al., 2023), autoethnography (Grosse, 2019), and visual ethnography (Brett et al., 2022) showcase the diversity of practice within construction.

Alvesson and Sandberg (2014) argue that while research specialisation offers many institutional and individual benefits, there are concerns about taking a narrow worldview, developing a rigid mindset, producing fragmented knowledge, unnecessary polarisation, and harmful protectionism. Overly specialised, 'boxed-in' research limits opportunities for imagination, creativity, interestingness, and practical relevance. The continual evolution of ethnographic research prevents it from being 'boxed-in' within the method box. Within the construction management research community then, given that the research tradition has remained loyal to positivist and quantitative study for decades (Pink et al., 2013, p. 2), ethnographic studies have recently begun to open up the field to doing box-breaking research.

For some, box-breaking research means aligning ethnography and other forms of research alongside one another in their strategic frameworks to situate their work. Brinkmann et al. (2020, p. 27) view ethnographic fieldwork as an approach that cuts across most of the paradigmatic differences in qualitative inquiry. Tutt and Pink (2019, p. 476) emphasise that ethnography represents a wide range of established observational and interview-based methods, and it is always inflected by theory. Thus, it is often discipline-specific. This is interesting as where there is variation in the ways different methods are applied, the possibilities of ethnographic research becoming boxed-in within the method box are minimised. Though there lies a danger of ethnographic research becoming boxed-in within a 'domain box' when researchers connect their practice to neatly specialised areas of expertise, for example, construction management. This can become problematic since Tutt and Pink (2019, p. 475) note that interdisciplinary scholarship and practice are key to advancing research that is relevant locally and globally.

Our aim with the IDP project was to be innovative and interdisciplinary and contribute to both generating new knowledge and help improve practice. We wanted to find the 'productive balance between the development, refinement and exploitation of existing knowledge and methods and the exploration of possible new directions' (March, 2005: 8, cited in Alvesson and Sandberg, 2014, p. 977).

Our approach aligns with that of Passos et al. (2012) who employed ethnography and participatory action research, with a view of mobilising the analytical qualities of ethnographic work together, alongside the action orientation and desire to improve organisational work practices from action research. The EAR methodology is well established within ICT. It combines research with project development (Tacchi, n.d). Participatory action research includes three stages of inquiry, action, and reflection in multiple cycles with a distinguishing feature from other forms of action research: namely, involving the participants in the evaluation of changes resulting from the action (Mackenzie et al., 2012). Improvements in the knowledge and understanding gained through the inquiry stage lead to social action whilst reflection on these actions generates new knowledge and further areas of inquiry (Mackenzie et al., 2012).

We employed the 'Plan ≫ Do ≫ Observe & Reflect ≫ Plan' cycle which enabled us to integrate research into the project's continuous cycle of planning and doing (ibid). This way, our view leans on a perspective where different approaches to research, such as participatory action research and critical ethnography, are 'uniquely compatible' (Hemment, 2007).

## Ethnographic Action Research (EAR)

There are four key principles to the EAR methodology (Tacchi, n.d):

- The initiative workers, which in this instance are the IDP project delivery team, develop a research culture through which knowledge and reflection are made integral to ongoing development.
- The research aims, methods and analysis arise from, and then feed back into, a rich understanding of the particular place.
- An EAR researcher draws on participatory techniques to involve participants and workers both as informants and as fellow researchers.
- An EAR researcher listens carefully to what people know from their own experiences and brings this local knowledge into the ongoing processes of planning and acting.

This framework informed the development of our box-breaking research, which was implemented across three stages: (1) Inquiry (plan), (2) action (do), and (3) observation and reflection (observe and reflect), as follows.

### *Stage 1: Inquiry (plan)*

The inquiry stage enables the assessment of the baseline situation before an action is implemented. This is to ensure that changes resulting from the action research process can be evaluated. The IDP project team sought to design workshop content for engaging construction site workers on collaboration, innovation and value, with the aim that these engagements help to address productivity challenges during construction. To achieve this, the IDP project delivery team worked in collaboration with selected supply chain participants to design the workshop content.

Most industry training programmes for the supply chain are designed to meet the requirements of middle and senior management staff in construction without addressing the needs of tradespeople who work on site. This limits the extent to which the site-level supply chains can feel a part of resolving their productivity challenges on site. The IDP project team thus took the approach of providing an opportunity for the supply chain to input into the workshop design to avoid a top-down approach.

Three supply chain representatives from the three firms that will later be involved in the workshops were selected and invited by the IDP project delivery team to support the design of the IDP workshops. These comprised a specialist Drylining Contractor, a specialist Mechanical and Electrical Contractor, and a specialist Ceiling and Interiors Contractor in three separate rounds. During this initial engagement, the IDP project team discussed issues such as supply chain challenges and provided opportunities for their input into the workshop content. This engagement of the supply chain representatives for their input was part of the engagement strategy the IDP delivery team adopted to ensure that the supply chain firms would encourage their workers to participate in the workshops. It was also to ensure that workshop content was reflective of and could help address their challenges. The workshop content was also presented to representatives from a large UK main contractor for further input. All these introduced aspects of co-design into the IDP workshop content with the hope that this becomes well aligned to the realities on site and increases engagement and participation.

The workshops were designed to incorporate elements of gamification. Gamification is the addition of gameplay activities in a non-game context to encourage participants to behave like they are playing a game (Wood and Reiners, 2015). It makes participants take a proactive approach to their learning and are actively driven (Liu et al., 2023). This approach was considered appropriate because the site-level construction workers who participated in the workshops were accustomed to doing things with their hands. They would be more motivated to engage in workshop activities that are hands-on, which gamification enabled. The IDP workshops were hence designed to provide opportunities for the supply chain to interact, learn, and collaborate in suggesting innovations that could improve their own productivity on site and add value for all. This enhanced their engagement, learning and impact.

The four workshops with supply chain participants focused on the IDP principles of collaboration (workshop 1: collaborative behaviours, workshop 2: collaborative design, workshop 3: collaborative risk management), and innovation and value (workshop 4). The aims were to:

1 ensure that useful behaviours are cultivated to support innovative thinking and joint problem solving;
2 challenge the supply chain to engage in design feedback and avoid design problems on site through innovative thinking;
3 identify and feedback on early issues that disrupt workflow on site and seek processes that improve flow as part of fostering innovation; and then
4 explore innovation and how it contributes towards value creation from a wider perspective than just financial value.

Each workshop was designed to incrementally build upon the previous workshop.

### *Stage 2: Action (do)*

The action phase involves a planned and structured intervention where changes in the baseline situation can be closely monitored (Mackenzie et al., 2012). Following each of the three rounds of inquiry with the main contractor and supply chain participants, the four workshops were run with

**TABLE 18.1** Ethnographic Action Research activities with the supply chain

| | *Co-design meetings with main contractor representatives* | *Co-design meetings with specialist contractor representatives* | *Workshop 1: Collaborative behaviours followed by participant evaluation* | *Workshop 2: Collaborative design and the supply chain followed by participant evaluation* | *Workshop 3: Collaborative risk management followed by participant evaluation* | *Workshop 4: Innovation and value in the supply chain followed by participant evaluation* |
|---|---|---|---|---|---|---|
| 17 June 2021 | | Meeting 1 | | | | |
| 14 December 2021 | Meeting 2 | | | | | |
| 31 January 2022 | | | Cohort 1 | | | |
| 8 February 2022 | | | | Cohort 1 | | |
| 15 February 2022 | | | | | Cohort 1 | |
| 22 February 2022 | | | | | | Cohort 1 |
| 4 April 2022 | | Meeting 3 | | | | |
| 12 April 2022 | | Meeting 4 | | | | |
| 17 May 2022 | | | Cohort 2 | | | |
| 24 May 2022 | | | | Cohort 2 | | |
| 31 May 2022 | | | | | Cohort 2 | |
| 7 June 2022 | | | | | | Cohort 2 |
| 21 June 2022 | | | | Cohort 3 | | |
| 28 June 2022 | | | | | Cohort 3 | |
| 5 July 2022 | | | | | | Cohort 3 |
| 20 July 2022 | | | Cohort 3 | | | |

three separate cohorts of supply chain participants for the action phase of the study. The supply chain representatives that fed into the module design also supported the IDP project team to co-deliver the workshops. This collaboration between the IDP project team and the supply chain representatives during the delivery was to manage the risk of the construction workers seeing the delivery team as outsiders during the action phase. It was also meant to increase participation and engagement as one of their own was co-delivering and contributing to running the workshop. A main contractor representative was also present during each of the workshops to relate issues back to the challenges experienced on the construction site. These efforts ensured that supply chain representatives were co-delivering the workshop with the workshop leads. Each round of workshops, illustrated in Table 18.1, led to feedback and experience from the supply chain that was incorporated into the next round to ensure a loop of continuous improvement leading to refinement in impact.

The workshops were held face-to-face in the site office of a live construction project site in Birmingham, UK. This was a £57m project to construct a new five-storey research facility that will provide 12,790 square metres of space for cutting-edge research and improve the research capabilities of the client in the areas of chemical, environmental and bio-molecular sciences. The construction site was chosen because this was considered a familiar work environment for supply chain participants and as such helped to situate workshop content within the exact context where all the issues being addressed manifest. It was also logistically easier for participants from both the main contractor and the specialist subcontractors to attend the workshops on site than, for example, at University or conferencing facilities.

It was important throughout for the project to be meaningful to the many different stakeholders, including the main contractor, the research participants, and the on-site research team. The project benefitted from external source funding, and while this presented some constraints to our ways of working, largely it meant that we had resources (e.g., specific time for the project and the engagement of an interdisciplinary team) that allowed us to think outside of the box. For each of the workshops, the lead presenter introduced the workshop agenda and connected this to the previous workshop where applicable, key issues were introduced and with the help of the co-deliverer from the supply chain, linked to the issues they experienced on site. Then a game activity was used to get the participants to work together in teams, interact and solve problems that related to the workshop topic. These comprised a 'Zim Obelisk' or 'Who Drinks the Water and Who Owns the Zebra' card game for workshop 1 on collaborative behaviours. A Lego design game activity was used for workshop 2 on collaborative design, a flow walks activity (see Ebbs and Pasquire, 2018) game for workshop 3 on collaborative risk management, and a creative connections game for workshop 4 on innovation and value.

The close connection with industry also afforded us the flexibility of focusing on the action research element without undue pressure to produce academic outputs. The last point is important in that much of the knowledge created to date has been practice-based, i.e., we have achieved change in the mindsets of the research participants, and at times this has been radical and transformative (terms that Alvesson and Sandberg, 2014, frequently use in discussing box-breaking research). We observed a major shift in focus away from employing simple and straightforward solutions that rely on technology to important key stakeholders advocating more relational and holistic methodologies to deal with complex problems, such as collaboration. We have also observed a broadened appreciation of value, for example, in the understanding of how important being a good team member is to project success and colleagues' well-being.

Our project embodies the characteristics of 'box transcendence' given the interdisciplinary research team; their differing backgrounds, research interests, and areas of expertise; and the combination of ethnography and action research. A hierarchy of parties involved in the EAR and their relationships is shown in Figure 18.1.

### *Stage 3: Observation and reflection (observe and reflect)*

We have drawn data from many sources in this project and in interpreting the material we have been critically aware of the personal experiences and cultural contexts that play a part in the analysis. One key feature in ethnographic research and our work is situated close-up observation. The researcher from the university who was embedded in all the workshops as a participant observer documented conversations and outcomes of activities that took place, including any observed changes.

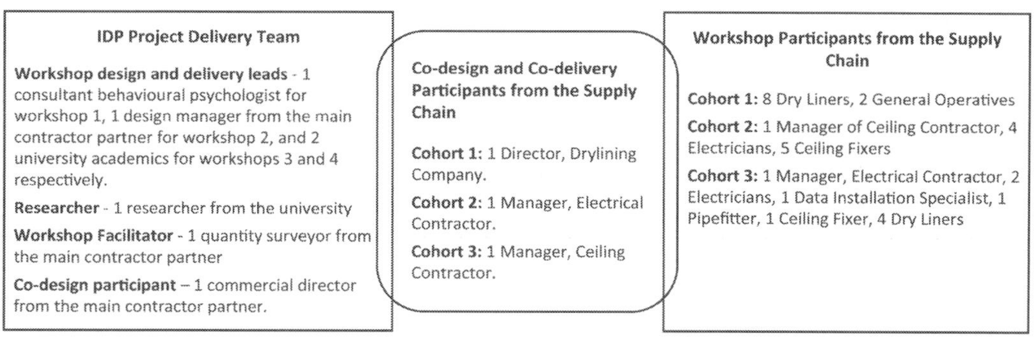


**FIGURE 18.1** IDP Ethnographic Action Research hierarchy of stakeholders and their relationships

All artefacts that were produced by participants during the workshop were also gathered and analysed. Also, at the end of each workshop, participants reflected on what they had learnt and the direct impact that this could have on their work. They provided these reflections by completing a post-workshop evaluation feedback sheet to record what was most useful, what was the least useful, if the workshop will influence how they do work or conduct business and if so, how. After each workshop, the feedback was reflected upon by the IDP project delivery team and used to improve the next workshop.

After round one of the workshops, we further enhanced workshop activities to improve gamification elements. For example, all the risk issues that emerged from the flow walk activity in round 1 of workshop 3 were used to develop two sets of a card game that participants had to correctly sort out into eight different causes (flow constraints). Points were awarded to the teams for both the accuracy and time taken to complete this risk categorisation activity. During rounds 2 and 3 of workshop 3, this gamified flow activity that emerged from round 1 was played before the participants applied it further to identify issues they were experiencing with their work and potential solutions. Another change was that we mixed the trades for round 2 (cohort 2) and round 3 (cohort 3) of the workshops. This was after it became apparent during round 1 with drylining workers that a lot of disruptions they experienced on site were linked to their electrical colleagues due to the interfacing nature of these trades. Thus, the decision to mix these trades for the other 2 rounds of workshops helped to enhance the collaboration between these trades in understanding their unique challenges and finding innovative solutions. Observations during the action and reflections thus fed into planning for further workshop rounds, hence completing the plan, do, observe, reflect cycle across 4 workshops per round and 3 rounds in total (4 × 3 workshop rounds = 12 workshops).

Another notable change that was implemented as an outcome of the reflections was an upgrade to the place that was used for workshops from the construction workers' site office on the ground floor to the office used by senior site management on the first floor. This was to manage issues on power dynamics and the 'us' and 'them' divide between the site workers and site management that was observed during round one as affecting supply chain collaboration. During the round 1 workshop on collaborative behaviours (workshop 1), the tradespeople narrated that meaningful collaboration was difficult to achieve because of the divide that exists between them and the senior site management staff who have the power to effect change. In their view, this divide between the tradespeople and the other site management staff was a general construction industry problem that went beyond the specific project they were building. This was a perception they had internalised that affected their views about collaboration. One example they gave was that even on the site, senior site management had their meetings in the 'posh' offices upstairs of the site cabins whilst they used the ground floor cabins where we were running the workshop. The Quantity Surveyor who facilitated the workshop on behalf of the main contractor explained to them how their own behaviours sometimes contributed to this 'us' and 'them' situation but then promised to negotiate with those in charge for further workshops to be held in the 'posh' offices if that could help break down the barriers. All the other workshops were therefore held in the upstairs offices used by the senior site management team.

## Place

Place emerges as one important and interesting theme within ethnography and participatory action research, and it is integral to the key principles of the EAR methodology. Booth (2015) considers 'what a difference place makes' within qualitative enquiry and challenges us to reflect upon research methodology with a particular view of place, and place gestalt. This includes a duality of research being place-bound and place-making. Research is place-bound in that it is constituted in places; and, research is place-making in that it constitutes places. Place is significant because of the social and political context of the research participants and researchers, and because of geographical, physical, and cultural, dynamics at play, as well as memories, imaginings and emotions attached to a place (ibid, p. 20). In such a way, meaning arises as embodied and place-based.

Research is situated both in that the research methodology and methods are contextualised to a specific place, and the knowledge that is produced contributes towards a better understanding of the place and simultaneously place-making. Relph (2008) reflects on the changes in how we perceive, occupy, and experience a place from 'staying put' to a mobile, transient, and global world. On the one hand, perceiving, occupying, and experiencing a place as part of a mobile, transient, and global world has compressed distance and flattened space. On the other hand, it has broadened and diversified experiences of place. There are elements of deep connection and placelessness, distinctiveness and standardisation, difference and sameness, everywhere. For Relph (ibid) place is a phenomenon of everyday experience, both grounded and open, that precedes all academic concepts and interpretations. It is the connection between self, community, and the world.

Construction is a fascinating context within which to consider this dynamic balance in understanding place. It is a large, complex, and people-intensive industry, which Dainty et al. (2007, p. 3) describe as:

> ...[an] environment characterised by structural fragmentation, a wide diversity of employment practices and an endless succession of short-term projects. Diverse groups of people who are brought together for short periods of time are expected to rapidly establish co-operative working relationships while frequently being engaged on entirely different terms and conditions. Tensions are exacerbated by the need to deliver buildings under evermore-stringent time and constraints.

Further, there are large international and national organisations that operate globally and local micro firms, which differ from one another more than they are alike, not only by size but also by the type of work they undertake, skills and equipment they need, and management style. The general nature of their work – on-site assembly and repair of buildings and infrastructure – and project-based modus operandi provide an element of sameness. Much of the construction work is geographically bound and fixed to the location where the end product is to be located. At the same time, each project is unique, and especially large projects tend to have diverse supply chains.

Paap (2006) discusses the nature of construction work and the inherent politics and insecurities that characterise construction sites as places of work as an integral element of her ethnography. She is explicit about the distinctiveness of the geographical and political place. Her research took place in the Midwest in the USA where she worked as a union carpenter's apprentice. Here differences arise based on the trade, region, and union. Although few construction workers in the USA are members of a union, around 75% of apprenticeships come through union programmes (ibid, p. 18). Elsewhere, apprenticeships are not provided by the unions. In Germany, Austria, Switzerland, and South Korea they are part of a national dual education system. In the UK, an executive agency of the government, the Education and Skills Funding Agency, delivers the apprenticeship programme (King and Raiden, 2022, p. 73). Hence, understanding the specifics of the place is important for appreciating the contextual nuances in the research findings.

Our research for the IDP project took place on the site where the organisations and some of the workers operated or were due to operate as per the sequencing of the construction works. While our research took place on site, our project did not quite align with 'Park's oft-cited mantra to the fieldwork, to go out and get the seat of their pants dirty with real research… or the 'grimy and confusing experiences of doing fieldwork'' (Hillyard, 2010, p. xi). We were in clean site offices, and in fact, access to the ongoing construction works was not possible given health and safety constraints.

Travel to the site took an hour to an hour and a half each way. This was 'everyday' for the participants (construction workers and their managers as well as the QS and the designer) but a notable departure from the normal commuting arrangements for the academics. This was beneficial in that the travel time provided useful time for a shift in mindset, from 'business as usual at the University' to experiencing the practical reality of the participants. Start and finish times were aligned with the site schedule, with arrival on site early 7:30 to 8 am at the latest, and finishing mid-afternoon. This was clearly preferred by the participants. However, the early starts were challenging to accommodate for the academics with caring responsibilities and resulted in some juggling among the team and a couple of stressful mornings. As academics who did not have a 7:30 start, we were subjected to aspects of the regular routine of site workers who arrived on the construction site very early. The anthropological nature of ethnographic research does mean that the researcher is exposed to aspects of the reality of the life of those being studied. Paap (2006) also discusses the sameness of construction sites as places of work in terms of the limited job security; the need for workers to continuously prove their value as effective and efficient workers; and the many ever-changing hierarchies by trade, job roles, and individual qualities that indicate to workers their place, often implicitly, that are prevalent features of construction work in many countries (Dainty et al., 2007, p. 3). Issues with gender, race, and class permeate both, the negotiations of difference and sameness. This leads us to consider the insider/outsider dynamics, which became prominent during the project and on which there is much discussion about in relation to ethnographic research. The next section is dedicated to this.

Researchers applying an EAR methodology should always be ready to adapt their routines so that they can be fully embedded into the natural environment of those being studied. The EAR researcher should also pay specific attention to the influence of place on the issues being studied.

### *Insiders and outsiders*

The ethnographer is traditionally 'detached', with the risk of being seen as taking an outsider's view into an organisation or a setting. Clawson (2005, pp. 243–244) discusses the complexity of the role of an ethnographer, specifically highlighting the impossibility of an ethnographer to always blend in. Gender is one interesting theme that is relevant in construction management research given the high proportion of men working in construction and more equal gender-balance in the academic research community; however, it is not within the scope of this chapter (see Chapter 4 for further discussion on gender).

Insider research refers to research that is undertaken by members of organisational systems and communities in and on their own organisations, and it may also involve collaboration between insiders and outsiders (Brannick and Coghlan, 2007, p. 59). Key benefits of insider research draw on the rich and complex insights, knowledge, and understanding that the insider has of the research context and matter at the heart of the study.

Action research provides a fruitful base for insider research (ibid, pp. 65–66). For the IDP project, having team members with insider access to the organisation, and the site, was instrumental in us gaining primary and secondary access: access to the site and to the participants and relevant documentation. They were also able to work the system and organisational politics to negotiate and secure the change in the place for the workshops that would normally have been only used for senior managers' meetings. This was seen as an important signal of commitment to the project and alleviated the us-them divide as we discuss earlier. To an extent, the insiders' preunderstanding of the relationships the main contractor, the research sponsor, has had with the subcontractors that were the research participants was also helpful in critically reflecting on the observational data.

Passos et al. (2012) report on the challenges that 'participant observation technique also requires the ethnographer to have both, an insider and outsider posture' (p. 13) and Brannick and Coghlan (2007, p. 71) note: 'as action research is a collaborative, negotiated process, the two concurrent roles [insider and outsider] are constantly under review' (role clarification added). Given that there were both insiders and outsiders in our research team, and the team members met all together regularly, a balanced perspective was possible. No one individual had to deal with the challenges of dual-role and role-conflict alone.

There were insiders in the sense that the main contractor had explicitly signalled their sponsorship for the project and two members of the on-site research team were employees of that organisation. However, only one member was a 'complete member' (after Brannick and Coghlan, 2007, p. 66) in that they were employed by the main contractor and also worked on the site where the workshops took place. There were three independent academics, employed by a University, in the team who were outsiders.

Researchers applying an EAR methodology will need to ensure that they have team members with insider access who can collaborate with them as 'outsiders' to blend into the organisational setting, navigate any organisational politics, and secure the level of access and engagement required for the research.

### *Time*

The participatory nature of the EAR methodology implemented during the IDP project required significant time commitments and engagement from the workshop participants. Consequently, they were mostly faced with an 'earn versus participate' decision: to either focus on earning an income by working and thus not participate in the study versus not earning and instead participating in the workshops. This is because most of the participants were self-employed tradespeople on daywork rates and time spent during the workshops impacted on time spent finishing their work on site.

This challenge was addressed by leveraging the high level of collaboration between the main contractor, the supply chain and the IDP project team, and the alignment of the workshop content and activities with the daily challenges they experienced on site. Participants were briefed before their involvement that time committed to the workshops could be beneficial over the longer term by equipping them to avoid common disruptions to workflow and improve their productivity.

The longitudinal nature of the on-site study, which spanned a period of 13 months, can pose a challenge to the continued involvement of research participants in a dynamic construction setting. In one instance, a workshop had to be rescheduled to accommodate the unavailability of some participants. This rescheduling impacted the usual workshop sequence from 1 to 4 during round 3. The sequence for the round 3 workshops was altered to workshops 2, 3, and 4 before workshop 1, which had to be rescheduled to the end on the 20th of July 2022 as shown in Table 18.1.

Researchers applying an EAR methodology will need some flexibility to accommodate changes to the planned sequence of activities for when it becomes necessary in such a dynamic environment. It is also important to consider how this might impact deadlines for funded projects.

In summary, the lessons learnt from our experience developing and employing the EAR methodology for the IDP project can be detailed as follows:

- Strong links with insider collaborators are a must for implementing the EAR methodology. This helps to secure access and negotiate and achieve the required participant engagement, facilitates reflection to planning and doing activity, ensures action can take place and is relevant, and can help with deeper and contextual insights in critical reflection.
- There is a need to carefully consider place, and associated insider-outsider dynamics and time constraints.
- There is a need to be adaptable to accommodate changes to the original schedule of planned activities due to the dynamic nature of the setting within which the action takes place.
- Researchers should aim for 'box-breaking' and avoid 'boxed-in' research.

## Conclusion

In this chapter, we have provided a reflective account of implementing EAR with construction site workers from different supply chain trades as participants in a live project environment. The aim of the study was to embed principles of collaboration, innovation, and value throughout the supply chain during procurement and to ensure that they continuously adopted processes and practices that reflected these principles. This was part of a funded research project called the IDP project.

At the heart of the IDP project was the need to improve productivity by driving change in the way supply chain firms in the construction industry are engaged to deliver projects. It is this need for change and continuous improvement and the opportunity to engage with construction workers on a live project that led to the choice of EAR as a participatory research methodology. Our work pairs ethnography with action research by integrating the 'Plan ≫ Do ≫ Observe & Reflect ≫ Plan' cycle of action research with the analytical qualities and place specificity of ethnographic work. It offered us the opportunity to develop 'box-breaking' research by building on the cyclical processes of planning, doing, reflecting, and planning again. This also meant that as new knowledge emerged during the workshops, it led directly to changes that shaped the day-to-day practice of the construction workers.

We learnt that three elements are key in executing EAR: place, insider/outsider views, and time. Place is significant in EAR and thus researchers that seek to adopt this methodology should ensure that it is conducted within the natural settings of the research participants where feasible. Achieving this will require strong links with insider gatekeepers' who can collaborate in negotiating access and securing participant engagement. Unlike other research methodologies, EAR requires that participants are actively involved in planning, doing, reflecting and planning again, and the inability to achieve this level of engagement throughout the research process can undermine its success. Another consideration for promoting high levels of engagement is to ensure that the design of EAR is directly linked and relevant to participants' day-to-day activities. This prevents the research participants as active stakeholders from having to choose between engagement in the EAR and the time they spend doing their day-to-day work. It also implies that enough time and flexibility will have to be built into the research to accommodate changes to the planned sequence of activities due to the live and dynamic nature of the work environment.

EAR has the potential to offer significant benefits for construction management research in that it empowers research participants by giving them a voice, and an opportunity to effect change and contribute to knowledge generation within their natural setting. It also offers the opportunity to break away from boxed-in research, which can sometimes limit imagination, creativity, interestingness, and practical relevance.

## References


Alvesson, M. and Sandberg, J. (2014) Habitat and habitus: Boxed-in versus box-breaking research, *Organization Studies*, 35(7), 967–987.

Auerswald, P.E. (2009) Creating social value, *Stanford Social Innovation Review*, 7(2), 50–55.

Booth, K.I. (2015) What a difference place makes: Place gestalt and some methodological thoughts, *Qualitative Enquiry*, 21(1), 20–27.

Brannick, T. and Coghlan, D. (2007) In defence of being "native", *Organizational Research Methods*, 10(1), 59–74.

Brett, R., Thomson, D. and Dainty, A. (2022) Exploring craft in construction with short-term ethnography: Reflections on a researcher's prior insight, *Construction Management and Economics*, 40(5), 359–373. 10.1080/01446193.2022.2046827

Brewer, J.D. (2000) *Ethnography*, Buckingham: Open University Press.

Brinkmann, S., Jacobsen, M.H. and Kristiansen, S. (2020) Historical overview of qualitative research in the social sciences. In Leavy, P. (Ed.) *The Oxford Handbook of Qualitative Research*, Oxford University Press, pp. 17–42.

Clawson, L. (2005) 'Everybody knows him': Social networks in the life of a small contractor in Alabama, *Ethnography*, 6(2), 237–264. 10.1177/1466138105057558

Dainty, A., Green, S. and Bagilhole, B. (2007) *People and Culture in Construction*, Abingdon: Taylor & Francis.

Ebbs, P.J. and Pasquire, C.L. (2018) Make ready planning using flow walks: A new approach to collaboratively identifying project constraints. In *26th Annual Conference of the International Group for Lean Construction*, Chennai, India, 18–20 July 2018, pp. 734–743.

Formoso, C.T., Sommer, L., Koskela, L. and Isatto, E.L. (2011) An exploratory study on the measurement and analysis of making-do in construction sites. In *19th Annual Conference of the International Group for Lean Construction*, Lima, Peru, 13–15 July 2011, pp. 1–10.

Galea, N.R. (2018) Built for men: Institutional privilege in the Australian construction industry, Unpublished PhD Thesis, University of New South Wales.

Galea, N.R., Powell, A. and Salignac, F. (2023) The role of homosociality in maintaining men's powerfulness in construction companies, *Construction Management and Economics*, 41(2), 172–182.

Grosse, H. (2019) An insider's point of view: Autoethnography in the construction industry, *Construction Management and Economics*, 37(9), 481–498. 10.1080/01446193.2018.1501156

Hemment, J. (2007) Public anthropology and the paradoxes of participation: Participatory action research and critical ethnography in provincial Russia, *Human Organization*, 66(3), 301–314.

Hillyard, S. (Ed.) (2010) *New Frontiers in Ethnography*, Bingley: Emerald.

Hine, C. (2000) *Virtual Ethnography*, London: Sage.

Hirslund, D.V. (2021) Brokering labour: The politics of markets in the Kathmandu construction industry, *Ethnography*, 22(4), 495–514. 10.1177/1466138119886601

Jarkas, A.M., Al Balushi, R.A. and Raveendranath, P.K. (2015) Determinants of construction labour productivity in Oman, *International Journal of Construction Management*, 15(4), 332–344.

Jaynes, G.D., Apter, D.E., Gans, H.J., Kornblum, W., Horowitz, R., Short, J.F. Jr, Suttles, G.D. and Washington, R.E. (2009) The Chicago school and the roots of urban ethnography, *Ethnography*, 10(4), 375–396.

King, A. and Raiden, A. (2022) Degree apprenticeships. In Raiden, A. and King, A. (Eds.) *Social Value in Practice*, Abingdon: Routledge, pp. 72–106.

Koch, C. and Siebken Schultz, C. (2019) The production of defects in construction – An agency dissonance, *Construction Management and Economics*, 37(9), 499–512. 10.1080/01446193.2018.1519253

Liu, S., Ma, G., Tewogbola, P., Gu, X., Gao, P., Dong, B. and Wu, Y. (2023) Game principle: enhancing learner engagement with gamification to improve learning outcomes, *Journal of Workplace Learning*, 35(5), 450–462. 10.1108/JWL-11-2022-0160

Loosemore, M., Powell, A., Blaxland, M., Galea, N., Dainty, A. and Chappell, L. (2015) Rapid ethnography in construction gender research. In Raiden, A. and Aboagye-Nimo, E. (Eds.) *Proceedings 31st Annual ARCOM Conference*, 7–9 September 2015, Lincoln, UK: Association of Researchers in Construction Management, pp. 1271–1280.

Löwstedt, M. (2015) 'Taking off my glasses in order to see': Exploring practice on a building site using self-reflexive ethnography, *Construction Management and Economics*, 33(5–6), 404–414. 10.1080/01446193.2015.1066021

Mackenzie, J., Tan, P.-L., Hoverman, S. and Baldwin, C. (2012) The value and limitations of participatory action research methodology, *Journal of Hydrology*, 474, 11–21.

Naoum, S.G. (2016) Factors influencing labor productivity on construction sites: A state-of-the-art literature review and a survey, *International Journal of Productivity and Performance Management*, 65(3), 401–421.

Paap, K. (2006) *Working Construction: Why White Working-Class Men Put Themselves –and the Labour Movement – in Harm's Way*, Ithaca: ILR Press.

Passos, C., Cruzes, D.S., Mendoca, M. and Dyba, T. (2012) Challenges of Applying Ethnography to Study Software Practices, *ESEM*12*, 19–20 September, Lund, Sweden.

Pink, S., Tutt, D. and Dainty, A. (2013) *Ethnographic Research in the Construction Industry*, Abingdon: Routledge.

Raiden, A.B., Loosemore, M., King, A. and Gorse, C. (2019) *Social Value in Construction*, Abingdon: Taylor & Francis.

Relph, E.C. (2008) *Place and Placelessness*, London: Pion.

Sarhan, S., Pasquire, C., King, A. and Manu, E. (2018) Institutional waste within the UK construction procurement context: A conceptual framework, *The Engineering Project Organization Journal*, 8, 36–64.

Sarhan, S., Pasquire, C., Manu, E. and King, A. (2017) Contractual governance as a source of institutionalised waste in construction: A review, implications, and road map for future research directions, *International Journal of Managing Projects in Business*, 10(3), 550–577.

Tacchi, J. (n.d) *Ethnographic Action Research Training Handbook*, http://ear.findingavoice.org/intro/index.html [accessed 12 May 2023].

Thiel, D. (2010) Contacts and contracts: Economic embeddedness and ethnic stratification in London's construction market, *Ethnography*, 11(3), 443–471. 10.1177/1466138110370496

Thiel, D. (2012) Builders, bodies and bifurcations: How London construction workers 'learn to labour', *Ethnography*, 14(4), 412–430. 10.1177/1466138112463656

Tutt, D. and Pink, S. (2019) Refiguring global construction challenges through ethnography, *Construction Management and Economics*, 37(9), 475–480. 10.1080/01446193.2019.1643570

Wood, L.C. and Reiners, T. (2015) Gamification. In *Encyclopedia of Information Science and Technology*, Third Edition, Hershey: IGI Global, pp. 3039–3047.

# 19

# EXPLORING HEALTH AND SAFETY COMMUNICATION IN CONSTRUCTION WORKGROUPS

*David Oswald and Helen Lingard*

*RMIT University, Melbourne, Australia*

**Key takeaways**

- We draw upon two studies that have investigated health and safety (H&S) communication within frontline construction workgroups. In total, 60 different workgroups, ranging from 3 to 13 members, across 15 different construction sites, were examined through capturing social networks and over 200 hours of participant observation.
- These findings challenge research on H&S communication that does not adequately consider how it is being defined within the context (i.e. reporting unsafety versus discussing how to do safe work).
- This chapter provides a case example of how ethnographic research can be performed to reveal theoretical insights that could otherwise go unnoticed. It raises questions surrounding the definition of H&S communication, provides insights into the contextual factors influencing how construction workgroups communicate and effective communication methods.

## Introduction

Effective communication is widely understood to be essential for a healthy and safe construction workplace. H&S communication can be multi-directional in terms of the organisational hierarchy (up, down, and horizontal) and is multi-dimensional in terms of the delivery methods (written, verbal, and visual) possible. With respect to direction, Kines et al. (2010) concluded that top-down verbal safety communication from construction foremen to workers had a significantly positive and lasting effect on safety levels. While Olive et al. (2006) suggested bottom-up communication is important for raising H&S issues and supporting structures should be considered that encourage asking questions. Lingard et al.'s (2019a) study stressed the importance of intra-group communication for creating a shared understanding of the priority placed on H&S; and Oswald and Lingard (2019) highlighted the emergence of horizontal supervisor-to-supervisor relationships and communication between different subcontractor workgroups for mature H&S frontline leadership. H&S communication can be delivered: verbally (see Kines et al., 2010);

DOI: 10.1201/9781003379584-23

through documentation, such as safe work method statements (see Borys, 2012); or visually through, for example, signage, which could be especially useful for among multinational migrant workforces (see Bust et al., 2008).

While there has been previous work in H&S communication in the construction industry this area is still evolving. For instance, in 2013, Alsamadani et al. noted that informal safety communication within crews had yet to be studied. Despite now being over a decade old, this still gives an indication that research theory into construction H&S communication is still developing. For instance, even H&S communication in the construction setting is not readily defined.

This study firstly explores the definition of H&S communication in the construction industry context. Further, we explore the assumptions related to H&S communication, and how the context influences communication ties and frequencies on construction sites. Thus, this research aims to explore:

- what H&S communication is; and
- how the context shapes H&S communication.

This study builds on previous work by the research team (see, for example, Oswald et al., 2022), which explored frontline leadership behaviour and effective ways of verbal communication in an informal setting. This study focuses more on the structures, in the form of communication channels and the way workgroups are organised, to explore how this can influence the way H&S is communicated. The industry is a complex setting with multiple contractors, trades, and nationalities working in a dynamic context, where the physical layout of the workplace is always changing. Hence, the organisational structure can be dynamic as new trades start and as others leave when their work finishes. This context is important for understanding H&S communication within, and between, construction workgroups.

## H&S communication

Safety and health are theoretically and practically very different (Sherratt, 2018), yet they are almost always married together as one. Safety has always been dominant in the relationship, in terms of receiving attention in both practice and theory. Safety can be traditionally defined as: 'not in danger or at risk' (Cambridge Dictionary, 2024). Within this traditional definition, H&S communication is more likely to relate to issues of non-compliance, such as holding people accountability (Dekker, 2016), enforcing rules (Weichbrodt, 2015), and reporting lines for unsafe issues (Oswald et al., 2018). However, there is academic work which challenges these assumptions (that are rooted in this traditional definition). Hollnagel (2014) and Dekker (2015) argued that safety should be viewed as the presence of a positive capacity, rather than an absence of negatives (e.g. errors, accidents, and incidents). For example, Gherardi and Nicolini (2002) explained from their ethnographic study that safety was one of the characteristics of an action which denoted it to being competent. Hence, safety was part of 'doing' a successful action, rather avoiding something undesirable ('the condition of not being in danger'). They described safety as: 'a form of organisational expertise, or in other words, a form of knowing that is sustained in the organisation by the interaction among various collective actors, and in the organisational field by the interactions among firms in the industry and its institutions' (Gherardi & Nicolini, 2000:10). Other scholars within the field of resilience engineering have proposed the definition of safety as 'the ability to succeed under varying conditions' (see Hollnagel, 2015). They suggest that instead of the focus being on reducing the number of dangerous or adverse outcome, there should be a focus on enhancing the ability of systems to succeed under varying conditions. Considering there are current debates on the definitions, and therefore assumptions, of what 'safety' is; it is not surprising that the phrase 'H&S communication' is not clearly or readily defined. However, effective H&S communication has been widely accepted as very important for H&S management within workplace. Within the construction industry, the way H&S is communicated within frontline workgroups has been a topic of interest for many researchers (see, for example, Alsamadani et al., 2013; Lingard et al., 2019b). It is regarded as crucial for having healthy and safe construction sites and can come in many forms, including verbal (Lingard et al., 2019a), written (Borys, 2012), visuals (Tezel et al., 2015) and through translators (Oswald et al., 2019).

One unique feature of organisations that has a profound impact on the communication process is its structure (Greenberg & Baron, 2010). The organisational structure refers to the formally prescribed pattern of relationships that exist between various units of an organisation (Ivancevich et al., 2011). Within construction, the structure is complex with multiple contractors and transient work. At the frontline, construction foremen, site superintendents, supervisors, and leading hands have an important role within the organisational structure. Typically, they are at the connection between principal contractor and various subcontractors and are a critical conduit through which senior managers' commitment to H&S is communicated to workers.

Communication and good interpersonal skills have been identified as an important competence for construction supervisors (Hardison et al., 2014). For instance, Cigularov et al. (2010) reported that when workers feel they are able to talk openly and freely to their supervisors about H&S, then there are safer work practices and fewer instances of work induced bodily pain. Despite being a frequent and essential part of construction H&S, Kaskutas et al. (2013) found that the majority of foremen in their study had never been taught how to deliver safety messages. From a review of 176 severe accidents, Winge et al. (2019) found that immediate supervision was a key causal factor connecting management and worker actions. Further, workers are reported to model their supervisors' H&S behaviour (such as wearing protective equipment) and develop similar H&S values to those of their supervisors (Maierhofer et al., 2000). As most workers have little contact with senior managers, they are more likely to be influenced on a day-to-day basis by interactions with members of their immediate workgroup, including their supervisors.

## Research methods

During the study, there were 60 different workgroups, ranging from 3 to 13 members, across 15 different construction sites. The participant observation involved understanding the everyday work of the construction workgroups, with a focus on the supervisor's role in H&S communication within the group. The 200 hours of participant observation were supplemented with a social network analysis of the 60 workgroups, and three one-hour focus groups at the three largest project sites.

Participant observation has been a hallmark for both anthropological and sociological studies (Kawulich, 2005). In participant observation the researcher is exposed to an environment, such as a construction site, and learns through the instruction of other members within the setting (Rooke et al., 2004). The researcher spent time with the construction workers, engaging in short ad-hoc discussions and observing their everyday work and communication practices. Following engagements with participants, initial field notes would be noted, during the course of the action. As memory is frail and selective, there is a need for note-taking in the field where possible (Denscombe, 1998). The initial notes made would be written up into more extensive field notes as soon as possible after leaving the research setting. To complement the participant observation (and focus groups), this research used social network analysis to illustrate a visual understanding of how H&S communication is shared within workgroups, consisting of the frontline leaders and the workers they led. The frontline leaders and workers were asked to indicate the frequency of H&S communication with others within their workgroups. This included: how regularly they provide H&S-related information to each member of the group and how often they received H&S-related information. To understand why communication occurred the way the workgroups perceived, participant observation was undertaken across the sites.

The comparative dimension of the short-term multi-sited approaches can capture valuable insights and understanding and has previously been successfully utilised within the construction industry (see, for example, Tutt et al., 2013; Oswald & Lingard, 2019). At the various project sites, the researcher had a gatekeeper (usually a H&S advisor), who helped ease access, updated the researcher with happenings, raised 'goings-on', and helped put workers at ease with the researchers' presence. Being attached to a University, the researcher was perceived by workers as in a student or apprentice type role, which helped also put the participants at ease, as the researcher had a lack of power or knowledge in their respected trades. Lofland et al. (2022) describe it as 'acceptable incompetence' – a role that can be very productive (Murchison, 2010). The data collection and analysis were a concurrent process, where data was collected and analysed thematically until saturation was reached. The academic work outlined in the prior literature section that advocates safety as a presence, rather than an absence (see Gherardi & Nicolini, 2002; Hollnagel, 2014; Dekker, 2015), has theoretically informed the analysis of understanding what H&S communication is.

The ethnographic excerpts in the following sections are representational and are based insights that emerged from the short-term research engagements, relevant to the vertical H&S communication from effective internal and external workgroup communication.

## What is H&S communication?

> Every operation, every task, everything, you will say information that is about safety and you might not even realise it.

Construction Labourer

> It is 7 am and John, the site superintendent (subcontractor supervisor), gathers the workgroup, including members of the principal contractor and other subcontractor supervisors, for the toolbox talk. This may have been somewhat surprising considering he was an employee of a subcontractor within the wider organisational structure. Yet, while he didn't have the highest rank in terms of the structural hierarchy, it quickly became very clear he was the one with the most responsibility. In the toolbox talk, he identified potential hazards, specifically highlighting the live traffic on a very busy road in the central business district. Hence,

the day had started with formal verbal H&S communication through the pre-start meeting and the site induction; but what followed was dominated by informal communication methods, and John was central to most of it. It was raining, and had been for the past few days, which had blurred the road markings for drilling (that had been painted a few days ago). I watched as John scurried around the construction site investigating potential drilling locations, talking on his mobile phone, and wrestling a tape measure between foot and hand to guestimate distances. He was busy.

Then there was the arrival of the non-destructive drillers (NDDs). The NDDs started placing noise barriers around the fencing on the site in preparation for their works. Once they started working, due to the noise, the workers used a variety of shouts and hand signals to communicate. They made symbols with their hands and arms to illustrate tools that they would then pull out the van. Later I asked one of the drillers about their use of hand signals. He explained: 'I can't hear Steve so the only way to communicate is hand signals. For instance, two fingers like this means put the spanner in, and depending on which way I rotate my hand and arm lets him know which way to go in.' I observed as John also used a combination of hand signal and elevated voice to communicate verbally over the noise. He connected separate sub-groups of workers, and acted as a centralised point of contact for H&S information (see Superintendent in Figure 19.1). In terms of the organisational hierarchy, he was a subcontractor and not at the top of the organisational structure; yet in terms of the management, he had the most responsibility, and as can be illustrated from the sociogram (see Superintendent in Figure 19.1), he was central to H&S communication.

The site changed shape and form in many ways throughout the day, with various workers, plant and machinery coming and going, as well as the site layout physically changing shape. Throughout these changes John was in constant informal verbal communication with the various employees from different employers. It appeared most of the communication

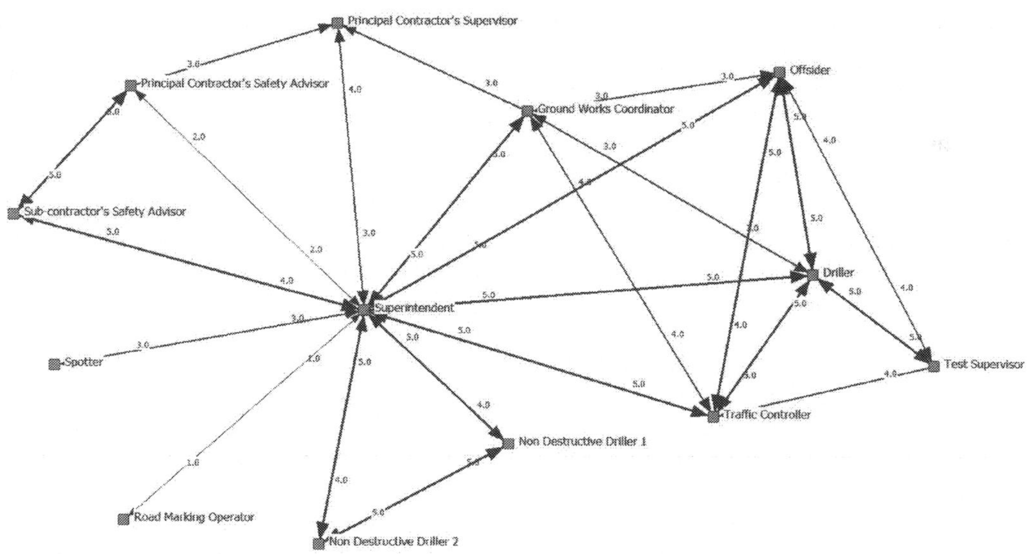


**FIGURE 19.1** A social network illustration of John's (superintendent) H&S communication ties[1]

> was task-related. John was always busy on his phone, helping set up the site, directing, helping traffic in and out the site, checking work permits or identifying drilling locations. I witnessed messages were being communicated in different ways, but the informal verbal communications (e.g. phone calls, and day-to-day site-based dialogue) were the most common, with visual communication (e.g. hand signals) only being used at times of high noise between drillers. There were also formal H&S messages were communicated in documentation (e.g. SWMS and logbooks) and formal verbal communications (e.g. tool-box talks, inductions).
>
> John was central to communication between subcontractors, and most of the communication was task-related, about how to do the job. But is this H&S communication too? Discussions around helping set up and organise the site to get the job finished may not initially seem to be directly like H&S communication, but it was paramount to ensuring a safe workplace. As one labourer put it: 'All day every day you are talking about safety. Every operation, every task, everything, you will say information that is about safety and you might not even realise it.'

This extract is an early field note taken on construction site, which describes and illustrates how the responsibility at the frontline does not necessarily reflect the organisational hierarchal structure. Typically, those higher in the organisational structure have greater responsibility, yet in this case, one of the subcontractor supervisors leads the operations, while principal contractor employees, and other subcontractor supervisors, were not central to the operations. It also raises the question of what is H&S communication? The sociogram demonstrates that workers felt they were discussing H&S frequently and there were many different ties throughout the day. As one supervisor put it:

> All day every day I'm talking to the guys about safety and most of the time you think you are talking about the job, but you are also talking about safety. We are doing a lot of excavation work at the moment, and they guys also throw ideas back at me. Like when we are going over permits, or if they thought there was a service cable in the area.

Another supervisor referred to H&S communication as task-related information:

> Safety communication depends of what task we are doing. For example, at the moment we are removing light poles – so I will be talking about things like risk of falls, where the services are and plant movements. The guys give me feedback as well. Remind me about procedures, like if we need an extra spotter for a task.

However, other workgroups perceived H&S communication in a negative sense, explaining that they only communicated about H&S when there was an issue. For example, one subcontractor worker explained that:

> We only talk about safety if there is an issue. If something is not right, we will generally tell Geoff [supervisor] but other than that, we don't really talk about safety.

This was also acknowledged by those in a leadership role. For instance, one site manager stated:

> ' When it's raining, there's water everywhere, and your work is unclean, you know messy, s**t everywhere, the workers are complaining to the supervisors or communicating, saying, "well this isn't right, I'm not happy with this."

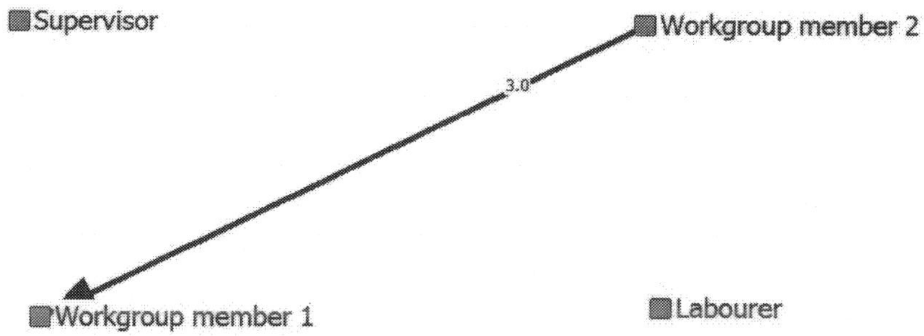


FIGURE 19.2 A social network diagram of a workgroup that only communication H&S messages when safety problems emerged

Figure 19.2 provides an example of a workgroup that only communicated when safety issues arose; hence there were very few communication ties within the workgroup. Workgroup member 2 felt he needed to raise H&S issues with workgroup member 1 once a week; but other than that there was no need to communicate H&S.

Other workgroups also communicated in this way, only sending H&S messages when H&S was not at an acceptable standard. It was not the most common response from participants but showed that H&S communication could be either viewed as raising H&S issues or could include more positive H&S messages and H&S-related discussions on completing job tasks.

## How does the context influences H&S communication?

The frontline is complex – we sometimes even have subbies under subbies.

H&S Professional

> I stood in a group of five subcontractor supervisors and leading hands. On previous site visits, I had noticed how noise would sometimes lead to an increase in hand signals. There was no drilling on this site; but I did observe a leading hand extensively using a radio. Curious, I enquired. Our conversation was constantly being disturbed by his radio, as calls came in and he responded. I enquired how often he uses his radio, and he responded 'I'm never off it. It's essential for keeping everything flowing across the site. I couldn't do my job without it.' Curious about the importance of his radio, I asked what the latest call was about. He explained there was a delivery truck on its way, so he was making sure the guys over there (he points to the site area over the road with live traffic) were aware. He was going to try and get around to help guide the truck in and visually inspect it was in a safe condition to enter the site, as they had sent a few trucks away for safety concerns – one recently was a truck with material that wasn't surrounded with adequate handrails. He set off to make his way across the road.

The ways workgroups communicated internally depended on the context they were working in. In this case, the large-rectangular-shaped construction site along a railway line meant

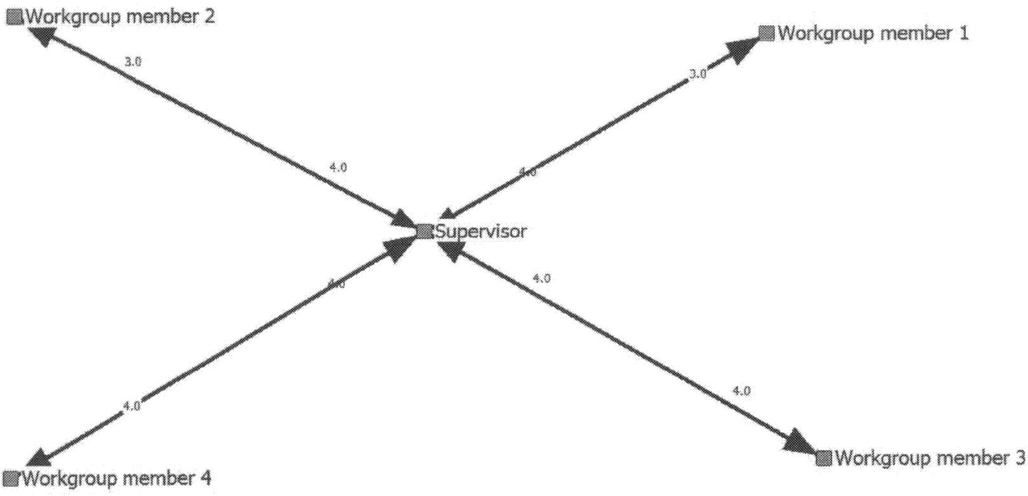


FIGURE 19.3 A construction workgroup that only communicated H&S through their supervisor due to their isolated work locations away from other workgroup members

that leading hands optimised their radios to communicate across long distances (e.g. on how to safely deliver materials). As well as geographically large sites, communication barriers could also occur if workers were in different site areas. These did not necessarily need to be geographically large in terms of distance, as for example, some workers of the same trade would operate on the same floor, but in different rooms, within an apartment building. Figure 19.3 provides an example of a network where workgroup members were in different physical areas on the site, so they communicated only through their supervisor.

As well as the location of the works, the type of works could also influence workgroup communication. For example, a subcontractor leading hand explained that during one-off works, more communication was needed to keep safe:

> It is a bit more complex because we are following an angled road, so there are six levels here. Because of all these levels, it is about keeping on top of everything, so there is a bit of extra care and communication needed. Like that cable shouldn't be going down a level – that shouldn't happen because if there's an emergency and you need to turn the power off, you need to get down a floor.

There were many ways workgroups communicated both internally with each other and externally with other workgroups and employees. H&S communication was very important considering the dynamic construction site, which consists of various hazards and risks, and many workgroups within the same area undertaking very different tasks. The structure of the workgroup also would influence how H&S messages were communicated. Figure 19.4 shows an example of a network in which the one supervisor worked across multiple trades and workgroups and consequently delegated most H&S messages to the leading hand.

In other cases, there would be subcontractors under subcontractors, as one H&S professional explained:

> The frontline is complex – we sometimes even have subbies under subbies. As far as we [Principal Contractor] are concerned they are all working for the subcontractor that we employed.

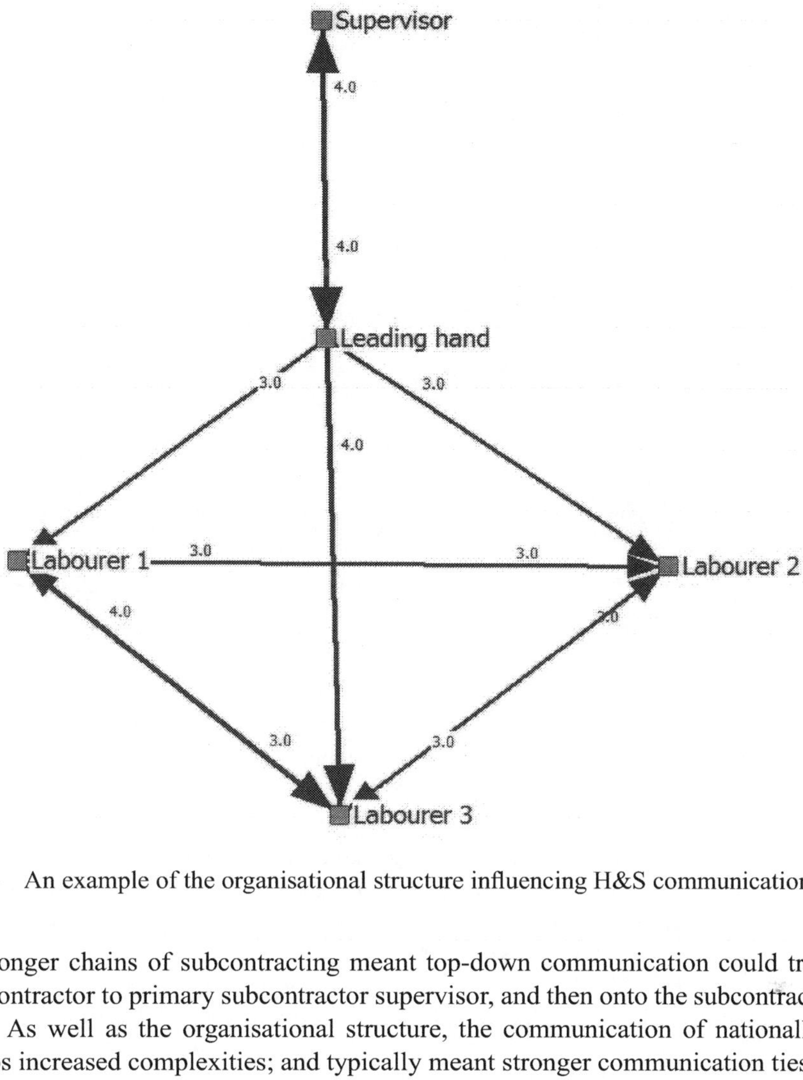


**FIGURE 19.4** An example of the organisational structure influencing H&S communication

These longer chains of subcontracting meant top-down communication could travel from principal contractor to primary subcontractor supervisor, and then onto the subcontractor's subcontractor. As well as the organisational structure, the communication of nationally diverse workgroups increased complexities; and typically meant stronger communication ties were observed between workers of the same nationality. For example, in the workgroup sociogram shown below, workgroup member 1 and workgroup member 2 were foreign workers whose mother language was not English. They understood English but confirmed with the researcher that they spoke in their native language with each other to improve their H&S understanding. Hence, in their communication tie, they provided and received H&S communication to and from each other more than once per day (5), which was the most frequent activity between any two members in this workgroup (Figure 19.5).

The frontline leaders acknowledged that language barriers were sometimes a challenge when trying to communicate H&S messages to workgroups without fluent English skills

***Foreman:*** We have some issues communicating messages to the foreign workers. They always nod and say yes. I kept asking them to do two tasks, but they weren't getting done. So eventually after a couple of weeks, I stopped the job, and then the

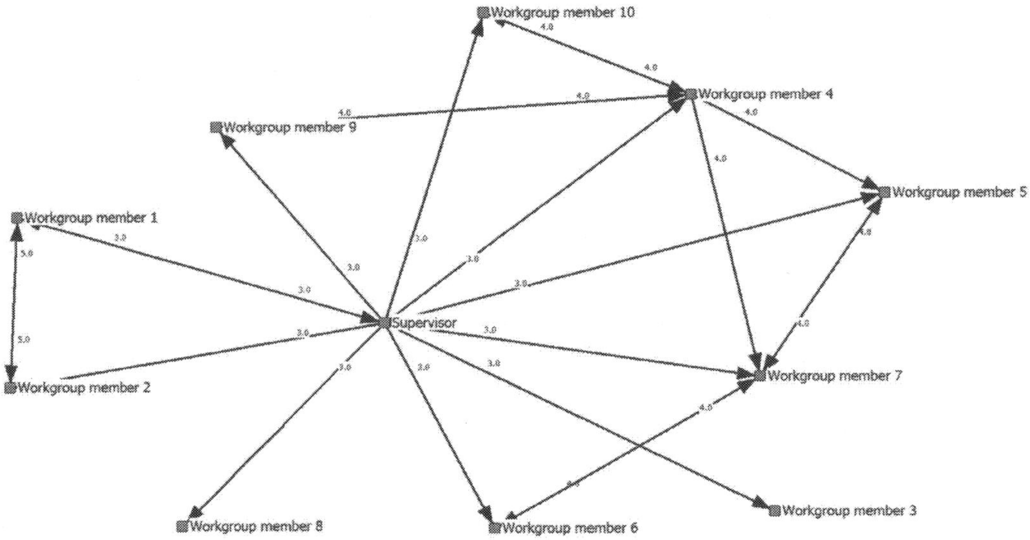


**FIGURE 19.5** A workgroup showing two members (1 and 2) communicating H&S frequently in non-English language to try to stay safe

> translator came to me to ask what was going on. They had some idea of one of the tasks, but no idea about the other. Language is a problem for getting messages through to the foreign guys; laziness is the problem for getting the English-speaking guys to act on them.

Hence, communication barriers could be language-based, as well as geographically based, if workers were located in different site areas. The observation and social network findings revealed several factors that influence how H&S communication was delivered in workgroups, in terms of the number of communication ties and their frequency. These included: the type of work they were undertaking, their trade, their site location, their mother language, the stage of the job, their workgroup's structure, and the type or frequency of H&S communication accepted as the norm.

## Discussion

H&S communication was understood and perceived to have different meaning to construction workgroups. Some perceived it as the messages they received at the pre-start meetings, identifying potential risks and hazards; others believed it was so intertwined within messages of how to do the tasks; and some understood it in a negative sense, where it was viewed as sharing H&S issues or accidents/incidents that had occurred. As such, receiving lots of H&S communication could be perceived negatively, rather than being a characteristic of a competent action (see Gherardi & Nicolini, 2002). Workers would explain they needed to send H&S frequently to particular group members, suggesting they were not working to the required or expected H&S standards. As well as the underlying assumptions and definition of what H&S communication is, the context had an important role in shaping the levels of H&S frequency and ties. At the frontline of construction projects, the organisational structure is complex with multiple contractors and trades working together.

Relevant factors that influenced levels of H&S communication included the type of work they were undertaking, their trade, their site location, their mother language, the stage of the job, their workgroup's structure, and the type or frequency of H&S communication accepted as the norm.

Theoretically, the different perspectives on the meaning of H&S raise questions about what is being measured in H&S communication research studies. Those perceiving H&S communication was about raising issues and incidents typically had lower frequencies (of H&S communication) than those who perceived H&S communication as information sharing for the safe undertaking of tasks. This could be clearly demonstrated by the sociograms in the results with some workgroups communicating very infrequently and only when a H&S issue emerged; and other workgroups communicating H&S information constantly to assist with completing tasks safely. This raises philosophical questions surrounding the definition and assumptions associated with H&S communication. Traditionally safety is defined as 'the condition of not being in danger' (Cambridge Dictionary, 2024). Yet other definitions emerging from the field of resilience engineering prefer the definition of safety as 'the ability to succeed under varying conditions' (see Hollnagel, 2015). The first traditional definition more aligns with the view that H&S communication is about identifying and discussing relevant risks and hazards (e.g. in prestart meetings). The second definition is more linked to those that viewed H&S communication as messages associated with undertaking a task. The raising of issues, incidents, and accidents is arguably the opposite of H&S communication and resembles more closely with the unsafe definitions of being exposed to danger or risk.

The frequency and number of communication ties was also dependent on many contextual factors. For example, if the type of work was new, unexpected or involved work with other trades, then more H&S communication may be necessary than work that was routine. The type of work undertaken by specific trades would also dictate the location of tasks. Some subcontractor trades worked physically next to each other, whereas others were spread across the site and used their supervisors as a focal, and sometimes only, point of H&S communication. In these cases, or when frontline leaders were overseeing multiple different workgroups, the use of the radio was more common. Workers explained that the stage of the job would also dictate the number of H&S messages perceived to being required. Tasks that did not involve high-risk activities such as excavation work, moving plant or work at height potentially did not require as many H&S messages to be sent and received among workgroups.

The organisational structure can exert an influence on the safety culture of an organisation (Hopkins, 2019). Hence understanding how the organisational structure 'plays out' within social context is an important area of H&S knowledge that is not well understood in research terms. The organisational structure would often depend on the levels of subcontracting. Within some cases, work was subcontracted by the subcontractor. This meant both top-down and bottom-up H&S communication had further to travel and through multiple subcontractor supervisors as focal points. Hence, it could be argued that there is a greater risk that H&S messages will not reach their intended recipients, which could link to why subcontracting has been noted for its adverse H&S influence within the construction industry (see Manu et al., 2013). Communication within the organisational structure also depended on the multinational nature of workgroups. Workers who spoke the same (non-English) language would translate and confirm the English H&S messages with each other, to strengthen their understanding. Hence, horizontal levels (worker-to-worker) of H&S communication appear to increase in these situations. Multinational workgroups can have very complex and innovative ways of communicating. For example, Tutt et al 2013c found a multinational construction workgroup had 'created their own language' to stay safe. Considering language has a fundamental role for those learning safety (Gherardi & Nicolini, 2002), avenues to communicate top-down H&S messages within multinational workgroups could also consider the use of visuals (Hare et al., 2013), or formally recognised translators (Oswald et al., 2019) to help improve success rates of H&S message transfer. It is also worth noting that there will be other factors linked to the demographic of the workgroups that may emerge in different social contexts. For instance, the trade crews within this study were all male; yet previous work has found that mixed-gender crews had different safety communication patterns than the more homogenous all-male crews (Allison & Kaminsky, 2017).

## Conclusions

The research in this chapter has demonstrated that H&S communication could be considered as raising awareness of a particular hazard, alerting an individual about an unsafe behaviour that was being undertaken, or simply explaining how a task was to be undertaken appropriately. Yet, the interpretations of what H&S communication is appeared different within construction workgroups. Some believed they communicated about H&S all the time, even when they weren't realising, as discussing how to complete tasks was safety as well as good planning. Other workgroups explained they only needed to communicate about H&S, when there was an H&S issue. Their interpretation of H&S was more aligned the opposite of H&S, when unhealthy or unsafe conditions were present, and individuals needed to be alerted. These perceptions are arguably opposing, i.e. reporting unsafety or planning and discussing safety. Theoretically, this draws the important question of what H&S communication is, especially for the use of perception surveys to measure H&S communication. Future research work should aim to have a clearer understanding on which definition is being perceived.

The work also reveals that H&S communication ties and frequency within workgroups are very context specific. It is dependent upon the type of work they were undertaking, their trade, their site location, their mother language, the stage of the job, their workgroup's structure, and the type or frequency of H&S communication accepted as the norm. Hence, when considering and comparing levels of H&S communication the context of the workgroup and worksite should be included. A more nuanced understanding of the context can help identify ways to improve H&S communication channels and frequencies. This work provides insights into the contextual factors influencing how construction workgroups communicate, which is not well understood within academic literature.

This chapter has provided a case example of doing ethnographic research within the construction context. The benefits include the deep insights on meaning, in this case, on the how H&S communication is interpreted by different construction workgroups. This depth of understanding can be captured by an ethnographic approach, showcasing the importance of this research approach in complementing other lines of inquiry within the field.

## Note

1 This social network data was collected by asking each group member about their H&S communication ties and frequency within the workgroup. Note the number closest to each group member indicates the frequency of out-going H&S communication (5 = more than once per day, 4 = once per day, 3 = once per week, 2 = biweekly, 1 = once a month).

## References


Allison, L., & Kaminsky, J. (2017). Safety communication networks: Females in small work crews. *Journal of Construction, Engineering and Management*, 143(8), 04017050.

Alsamadani, R., Hallowell, M., & Javernick-Will (2013). Measuring and modelling safety communication in small work crews in the US using social network analysis. *Construction Management & Economics*, 31(6), 568–579.

Borys, D. (2012). The role of safe work method statements in the Australian construction industry. *Safety Science*, 50(2), 210–220.

Bust, P., Gibb, A., & Pink, S. (2008). Managing construction health and safety: Migrant workers and communicating safety messages. *Safety Science*, 46(4), 585–602.

Cambridge Business English Dictionary (2024). Cambridge University Press. https://dictionary.cambridge.org/dictionary/english/safety

Cigularov, K. P., Chen, P. Y., & Rosecrance, J. (2010). The effects of error management climate and safety communication on safety: A multi-level study. *Accident Analysis & Prevention*, 42(5), 1498–1506.

Dekker, S. (2015). *Safety Differently: Human Factors for a New Era* (2nd ed.). Boca Raton: CRC Press.

Dekker, S. (2016). *Just Culture: Balancing Safety and Accountability (*2nd ed.). London: CRC Press.

Denscombe, M. (1998). *The Good Research Guide for Small-Scale Social Science Projects*. Buckingham: Open Press University.

Gherardi, S., & Nicolini, D. (2000). The organizational learning of safety in communities of practice. *Journal of Management Inquiry*, 9(1), 7–18.

Gherardi, S., & Nicolini, D. (2002). Learning the trade: A culture of safety in practice. *Organisation*, 9(2), 191–223.

Greenberg, J., & Baron, R. A. (2010). *Behavior in Organizations* (10th ed.). Upper Saddle River: Pearson/Prentice Hall.

Hardison, D., Behm, M., Hallowell, M. R., & Fonooni, H. (2014). Identifying construction supervisor competencies for effective site safety. *Safety Science*, 65, 45–53.

Hare, B., Cameron, I., Real, K. J., & Maloney, W. F. (2013). Exploratory case study of pictorial aids for communicating health and safety for migrant construction workers. *Journal of Construction Engineering and Management*, 139(7), 818–825.

Hollnagel, E. (2014). *Safety-I and Safety-II: the Past and Future of Safety Management*. Surrey: CRC Press.

Hollnagel, E. (2015). RAG – the resilience analysis grid. In *Resilience Engineering in Practice: A Guidebook* (pp. 275–296). https://erikhollnagel.com/onewebmedia/RAG%20Outline%20V2.pdf

Hopkins, A. (2019). *Organising for Safety: How Structure Creates Culture*. Sydney Cch Australia Limited.

Ivancevich, J. M., Konopaske, R., & Matteson, M. T. (2011). *Organizational Behavior and Management*. New York: McGraw-Hill.

Kaskutas, V., Dale, A. M., Lipscomb, H., & Evanoff, B. (2013). Fall prevention and safety communication training for foremen: Report of a pilot project designed to improve residential construction safety. *Journal of Safety Research*, 44, 111–118.

Kawulich, B. B. (2005). Participant observation as a data collection method. In *Forum Qualitative sozialforschung/forum: Qualitative Social Research* (Vol. 6, No. 2). https://www.qualitative-research.net/index.php/fqs/article/view/466

Kines, P., Andersen, L., Spangenberg, S., Mikkelsen, K., Dyreborg, J., & Zohar, D. (2010). Improving construction site safety through leader-based verbal safety communication, *Journal of Safety Research*, 41, 399–406.

Lingard, H., Pirzadeh, P., & Oswald, D. (2019a). "Talking safety": Health and safety communication and safety climate in subcontracted construction workgroups. *Journal of Construction Engineering & Management*, 145(5), 04019029.

Lingard, H., Zhang, R., & Oswald, D. (2019b). Effect of leadership and communication practices on the safety climate and behaviour of construction workgroups. *Engineering, Construction and Architectural Management*, 26(6), 886–906.

Lofland, J., Snow, D., Anderson, L., & Lofland, L. H. (2022). *Analyzing Social Settings: A Guide to Qualitative Observation and Analysis*. Long Grove: Waveland Press.

Maierhofer, N., Griffin, M., & Sheehan, M. (2000). Linking manager values and behavior with employee values and behavior: A study of values and safety in the hairdressing industry. *Journal of Occupational Health Psychology*, 5, 417–427.

Manu, P., Ankrah, N., Proverbs, D., & Suresh, S. (2013). Mitigating the health and safety influence of subcontracting in construction: The approach of main contractors. *International Journal of Project Management*, 31(7), 1017–1026.

Murchison, J. (2010). *Ethnography Essentials: Designing, Conducting, and Presenting Your Research* (Vol. 25). Hoboken: John Wiley & Sons.

Olive, C., O'Connor, T., & Mannan, M. (2006). Relationship of safety culture and process safety. *Journal of Hazardous Materials*, 130(1–2), 133–140.

Oswald, D., & Lingard, H. (2019). Development of a frontline H&S leadership maturity model in the construction industry, *Safety Science*, 118, 674–686.

Oswald, D., Lingard, H., & Zhang, R. P. (2022). How transactional and transformational safety leadership behaviours are demonstrated within the construction industry. *Construction Management and Economics*, 40(5), 374–390.

Oswald, D., Sherratt, F., & Smith, S. (2018). Problems with safety observation reporting: A construction industry case study, *Safety Science*, 107, 35–45.

Oswald, D., Wade, F., Sherratt, F., & Smith, S. (2019). Communicating health and safety on a multinational construction project: Challenges and strategies. *Journal of Construction Engineering & Management*, 145(4), 04019017.

Rooke, J., Seymour, D., & Fellows, R. (2004). Planning for claims: An ethnography of industry culture. *Construction Management and Economics*, 22(6), 655–662.

Sherratt, F. (2018). Shaping the discourse of worker health in the UK construction industry. *Construction Management & Economics*, 36(3), 141–152.

Tezel, A., Koskela, L., Tzortzopoulos, P., Formoso, C., & Alves, T. (2015). Visual management in Brazilian construction companies: Taxonomy and guidelines for implementation. *Journal of Management in Engineering*, 31(6), 05015001.

Tutt, D., Pink, S., Dainty, A. R., & Gibb, A. (2013a). Building networks to work: An ethnographic study of informal routes into the UK construction industry And pathways for migrant up-skilling. *Construction Management and Economics*, 31(10), 1025–1037.

Tutt, D., Pink, S., Dainty, A. R., & Gibb, A. (2013b). 'In the air' and below the horizon: Migrant workers in UK construction and the practice-based nature of learning and communicating OHS. *Construction Management and Economics*, 31(6), 515–527.

Tutt, D., Pink, S., Dainty, A., & Gibb, A. (2013c). "We've got our own language". The communication practices of migrant workers in the UK construction industry. In *Ethnographic research in the construction industry* (pp. 40–57). Routledge.

Weichbrodt, J. (2015). Safety rules as instruments for organizational control, coordination and knowledge: Implications for rules management, *Safety Science*, 80, December 2015, 221–232.

Winge, S., Eirik, A., & Mostue, B., (2019). Causal factors and connections in construction accidents. *Safety Science*, 112, 130–141.

# 20

# AN ETHNOGRAPHIC ACCOUNT OF CIRCULAR DEMOLITION PRACTICES

## Realising reuse potentials

*Marc van den Berg*

*Department of Civil Engineering & Management, Faculty of Engineering Technology, University of Twente, Enschede, The Netherlands*

**Key takeaways**

- This study uses ethnography to explore how demolition contractors can realise reuse potentials during circular demolition projects and, as such, moves the scientific discourse from identifying barriers to revealing solutions.
- Applying a process perspective to ethnographic data is particularly suitable to illuminate how practices unfold over time: demolition contractors engage in identifying, harvesting, and distributing reusable building elements.
- The detailed ethnographically grounded process framework presented here may offer inspiration to promote the adoption of circularity and lead to transformative action in construction.

### Introduction

Rethinking conventional demolition processes is essential to realise reuse potentials. Obsolete or abandoned buildings must often make way for new constructions through demolition. This practice essentially changes buildings from assets into liabilities, with "valuable construction materials commonly hauled away as waste" (Leigh & Patterson, 2006, p. 217). Globally, mountains of waste are being generated, with enormous quantities being landfilled (Wu et al., 2014). The management of construction and demolition waste has, accordingly, become a priority for most environmental programmes around the world, such as in China (Joensuu et al., 2020), Australia (Dalton et al., 2023), and Europe (Gálvez-Martos et al., 2018). Circular economy agendas, such as that of the European Commission (2020), promote maximising material reuse and recycling through piece-by-piece demolition (also called deconstruction) (Lynch, 2022). That alternative practice recognises that buildings consist of many different elements each with their own specific life-cycles and that "all interact dynamically in space and time" (Pomponi & Moncaster, 2017, p. 710). Some of those elements will thus still have reuse potential at a building's end-of-life phase.

DOI: 10.1201/9781003379584-24

This warrants detailed ethnographically grounded understandings of how demolition contractors may realise those reuse potentials.

Reuse is a preferred resource strategy under circular economy agendas. The circular economy has been promoted as an economic system where materials never become waste and nature is regenerated (Ellen MacArthur Foundation, 2013). The alternative economic system positions production and consumption within a cyclical, rather than linear, flow model (Korhonen et al., 2018). This is "most frequently depicted as a combination of reduce, reuse and recycle activities" (Kirchherr et al., 2017, p. 221). Circular economy agendas promote these activities over traditional resource strategies where materials are discarded with or without any attempt to recover energy (i.e., incineration respectively landfilling) (Ghaffar et al., 2020; Yu et al., 2022). Recycling and reuse strategies can reduce both waste and new production. These two strategies nevertheless differ in the extent to which embodied energy is preserved. Recycling involves reprocessing elements into raw materials for new products; reuse concerns recirculating an element in its original form for a similar function (Lansink, 2017). Even though more environmental benefits can be achieved with reuse, recycling is still a predominant strategy in construction industries (Zhang et al., 2022) and studies into recycling greatly outnumber studies into reuse (Ghisellini et al., 2018; Ginga et al., 2020).

The uptake of reuse in construction appears limited due to a wide range of barriers. These have been studied by many. Park and Tucker (2017), for instance, recognised economic, institutionalised, and political reluctance to reuse construction materials. Deconstruction can generally be more expensive than traditional demolition, primarily when the sales of reclaimed elements and avoided waste disposal fees do not offset additional labour costs incurred (Dantata et al., 2005; Coelho & De Brito, 2011). Institutional reluctance can be traced back to the conservative nature of the industry (Kanters, 2020), consumer preferences for new materials (Cherian et al., 2022), and limited information sharing between stakeholders (Wijewickrama et al., 2021). Political barriers differ internationally, but generally concern limited support from government policies, regulatory instruments, and fiscal actions (Knoth et al., 2022; Munaro & Tavares, 2023). As these studies suggest, the barriers associated with reuse appear plentiful, even though their presence may vary across projects. Local variations in demolition practices are, however, typically not accounted for in these studies. This leaves the question how demolition contractors could navigate such reuse barriers.

Ethnographic research methods are particularly suited to answer that question. These methods can capture deep and fine-grained understandings into complex social problems. They deal with the study of social environments through some degree of researcher involvement in those environments. Ethnography has been developed in social anthropology and sociology, which its literal meaning of "writing about the people" hints at (Ingold, 2014). Its potential to provide deeper insights into construction industry practices was recognised later as well (Phelps & Horman, 2009; Pink et al., 2012). Pink et al. (2010) argued that ethnographic research could advance construction site research by departing from dominant positivist traditions that hinder understanding of social action from actors' perspectives and by introducing more theoretically sophisticated approaches to comprehend site work. Ethnographic research sheds light on work practices through intensive participant observation on site. Ethnography variants, such autoethnography (Grosse, 2019) or ethnographic-action research (Olde Scholtenhuis et al., 2016), can illuminate complexities that otherwise remain hidden, undertheorised, or poorly understood (Oswald & Dainty, 2020). Ethnographic research is thus suitable to unpack the social and material complexities of circular demolition practices over time.

This chapter sets out to do so. It applies a new process perspective on two circular demolition projects during which extensive ethnographic participant observations were conducted. The methodological choices, including some best practices for doing ethnography, are laid out first. The next section presents a process framework that illuminates how demolition contractors engage in realising reuse potentials through three essential and sequential activities. The chapter culminates in suggestions for upscaling reuse in construction and reflections on how ethnography helped developing these suggestions.

## Methodology

This chapter aims to explore how demolition contractors can realise reuse potentials during circular demolition projects. To that end, I revisited the ethnographic datasets of two earlier studies in which we generated (Van den Berg et al., 2020) and tested (Hulsbeek & Van den Berg, 2022) a proposition for predicting whether or not a demolition contractor will recover a building element for subsequent reuse. This proposition links together three conditions in a statement of expected relationship, namely that a building element will only be recovered for reuse when the demolition contractor: (1) identifies an economic demand for the element, (2) distinguishes appropriate routines to disassemble it, and (3) can control the performance until integration in a new building. The proposition helped to understand when reuse is realised, but not how that is done. In attempting to shed new light on the latter, I applied a process perspective – which is useful to unveil the dynamic and interconnected nature of organisational phenomena (see Langley et al., 2013) – to the existing ethnographic datasets. At the same time, I respond to a call for "greater researcher reflexivity" in ethnography (Oswald & Dainty, 2020, p. 1) through elaborating on methodological lessons learned.

The ethnographic datasets relate to two demolition projects with circularity ambitions. The first project dealt with a nursing home, the second with a block of duplex houses. The projects were characterised by circularity ambitions and, as such, were originally selected because of their "revelatory" nature (Yin, 2014, p. 52). Both projects were located in the Netherlands, a country that finds itself in an early circularity transition phase (Giorgi et al., 2022). The country has high recycling rates (Zhang et al., 2020) and ambitions to realise a fully circular economy by 2050 (Dijksma & Kamp, 2016), but developments in that direction are still slow (Wiarda et al., 2023). Within this context, the two demolition projects are remarkable because of their circularity ambitions. Both projects strived to maximise reuse. In the nursing home project, reuse of most building elements was specifically enabled by the modular, temporary design of the building. The two-story building had been constructed by a system builder that was selected to deconstruct it after a service-life of approximately five years. That firm subcontracted a long-term partner demolition contractor in turn to realise this. The system builder then planned to reuse most of the modular building elements in constructing a school. Alternatively, in the duplex houses project, reuse targets were specified during the tender process. This way, a social housing organisation had selected a demolition contractor for the "circular demolition" of a block of 96 duplex houses dating from the 1960s. Demolition of the first 48 houses was studied there.

Participant observations during the selected demolition projects were the main source of data. Acclaimed "potentially a revolutionary praxis" that produces new knowledge through being and action, this core method of ethnographic research involves both observing and participating in a focal phenomenon (Shah, 2017, p. 45). Although different degrees of researcher engagement are possible, the two studies adopted both a mode of "active participation" in which the leading researcher attempted to fully learn doing the same work as the demolition practitioners that were being studied (Spradley, 1980, p. 80). I was the leading ethnographic researcher in the theory-generating (nursing home) project and had a supervisory role in the theory-testing (duplex houses) project. Participant observations were conducted for approximately 250 hours (by me) and 300 hours (by Hulsbeek), which corresponds with viable timescales for doing ethnography in project-based industries (Pink et al., 2010). We were granted access to the demolition sites after successfully completing a health and safety training. During our work on site, we sought to understand what demolition practitioners were doing through participating in their daily activities (Musante & DeWalt, 2010). This entailed participation in demolition work (such as removing ceiling tiles or cutting electrical wiring) and social activities (like joining lunch breaks and sharing jokes), all balanced with recording our observations through taking notes and pictures. We gradually gained trust of the workers involved, which reduced the risk that the informants would change their behaviour due to the presence of an outsider (Oswald et al., 2014). Furthermore, semi-structured interviews with site personnel to obtain their views on the practices performed complemented our detailed experiences of being engaged in circular demolition.

A process perspective was applied to develop a rich understanding of how reuse potentials are realised over time. I used ethnographic guidelines of Phelps and Horman (2009) to iteratively order, refine, and link the collected data into a process framework. After immersing myself in the datasets again through re-examining the fieldnotes and pictures, I started with ordering the data chronologically to discern steps in realising reuse potentials. Variations across types of building elements were labelled differently. For instance, when I noticed that the demolition contractors used different sales channels, I ensured that the data analysis labels reflected these differences. I further refined emergent concepts through comparing and categorising these labels. Finally, through writing a thick description, I linked these concepts together into a process framework (see Van den Berg, 2024) that reveals how reuse potentials are realised in circular demolition.

## Findings

Demolition contractors organise three sequential activities to realise reuse potentials: identifying, harvesting, and distributing reusable building elements. These activities can be executed in various ways, using different mechanisms, depending on the elements at hand. This section presents ethnographic insights of such mechanisms for each of the three activities.

### *Identifying reusable elements*

Realising reuse potentials starts with identifying which building elements are actually reusable. Demolition contractors therefore first need to know what elements a building actually consists of. They will want get an understanding about the existing conditions of the elements. Yet these firms also need to be aware of any (future) possibilities to use elements elsewhere. The firms need to understand what elements could substitute new ones in other projects. In other words, reuse potentials of existing building elements are based on both supply and demand. Demolition contractors hence need to understand both to identify which building elements are reusable (and which ones are not). Several mechanisms to do so were found in the two circular demolition projects studied.

The demolition contractors obtained insights into the existing conditions of the buildings before and during the projects. Drawings provided a primary source of information to that end. The system builder that hired the first demolition contractor had handed over so-called disassembly drawings. This selection of construction drawings represented floor plans and cross sections of the nursing home, together with annotated disassembly instructions. One of the drawings, for example, showed how a floor was constructed with a series of separate floor slabs that were joined together. Colours were used to indicate which floor elements were planned to be reused where. The drawings also included a few generic comments, such as "disassemble plywood cladding" and "leave the angle profiles on the floor elements". These drawings appeared to resemble existing conditions quite well, given that the building was only a few years old. This was different for the duplex houses, which were much older and had had many adaptations over the years that were not properly documented.

The demolition contractor responsible in that contract, consequently, relied more on a pre-demolition audit. That audit, conducted by a third party, had resulted in an extensive material inventory: a table with pictures and associated information of characteristic building elements within the to-be-demolished houses. The report also showed where asbestos was present in the buildings, which had to be removed at the start of the project. Information from this audit was complemented with visual inspections. The wooden back doors of the duplex houses, for example, had already caught the eye of the project leader before the project commenced. These were still in a good condition and, thus, he considered them as potentially reusable. The other demolition contractor made similar assessments for many appliances in the nursing home, such as the microwaves and refrigerators.

Those assessments were complemented with relevant market information. The ethnographic researcher heard the site manager and foreman in the nursing home project frequently using the phrases "one can make money" or "one cannot make money" with certain elements. They appeared to have developed a fine understanding of the market demand for existing building elements. This was also found in the other project. The project leader there had contacts with various traders in second-hand building materials. While trying to sell the wooden back doors and certain plastic frames to one, that trader also expressed an interest in ventilation systems and fuse boxes and eventually bought those as well. There was, however, no interest in other elements, such as radiators or toilets, because those were considered "aesthetically not modern". The project leader was nonetheless confident that he could sell (most of) the wooden beams and planks that were part of the duplex houses' structures. Market demand from traders, which provided indirect sales channels, helped the demolition contractors to identify reusable elements.

Direct demand for certain building materials was also observed. A school was planned to be built with most of the modular elements of the nursing home. The system builder had informed the demolition contractor about the corresponding new location for many of the floor, column, roof, façade, and other elements. Those were pictured on construction drawings for the school as well as listed in various demolition project plans. Some other elements appeared reusable because of passersby asking for them. In both demolition projects, the ethnographic researchers regularly observed (and sometimes participated in) conversations with people passing by the demolition sites, usually living in the same neighbourhood. These people inquired about certain elements, such as large flowerpots, kitchen appliances, and timber beams from the nursing home and high-pressure laminate plates and timber beams from the duplex houses, with some people willing to pay for these and others hoping to receive them at no cost. Other elements, such as glass wool plates, were finally offered for sale on online marketplaces.

Demolition contractors thus appeared to identify reusable building elements based on their assessments of, on the one hand, the conditions of available elements and, on the other hand, the market demand for such elements. Any elements that were not identified as reusable were treated as waste, hence recycled at best.

### *Harvesting reusable elements*

Elements identified as reusable are harvested during a sequential activity. Demolition contractors instruct workers about disassembling and storing the valuable building elements. This can be seen as harvesting: just like farmers harvest crops to gather their value, demolition workers harvest the elements with reuse potential. In this way, the hidden value of buildings is unlocked through disassembling elements and storing them on site until they are distributed to their target locations. A series of practices aimed at harvesting reusable elements was observed in the two demolition projects.

Demolition workers were first instructed about the elements with reuse potential. Site management was responsible for assigning demolition workers to specific demolition activities. These men informed the workers about how to handle any elements. Sometimes they did so by referring to floor plans, exterior views, and other drawings on which reuse intentions were indicated with colour coding. More frequently, however, they provided disassembly instructions on the site. From within the nursing home itself, for example, the responsible site manager instructed two workers to use certain rolling scaffolding for "clearing the ceiling" so that roof elements could be reused. As another example, the site manager sprayed graffiti on specific metal-stud interior walls to signal within which of those walls reusable wind bracing could be found; the mini excavator operators were told to handle those walls with special care. Likewise, workers in the duplex houses project were instructed to keep certain kitchens intact after the ethnographic researcher had initiated contact with a new auction company interested in buying (and reselling) those kitchens. Many similar examples of instructions regarding how to disassemble certain elements or where to store them were found in both projects.

Building elements were, subsequently, disassembled based on those instructions, with the ease of disassembly depending on the type, accessibility, and number of connections present. The nursing home was, in contrast with the duplex houses, designed with disassembly in mind. Most of the elements had been prefabricated, and they were assembled together on site. Those elements could, consequently, be disassembled with relative ease through reversing the assembly sequence. The ethnographic researcher participated in, for instance, disassembly of the distinct façade elements. This started with one worker removing the three screws on the bottom of the element that connected it to a floor element. Another worker, who was standing on the roof, would thereafter attach the façade element to two ropes hanging from a crane hook. That worker, nicknamed the "director of the roof", subsequently removed the three remaining screws that attached the façade element to the roof so that it would be completely detached from the rest of the building. The involved crane operator moved the element to the ground, where a third worker (together with the researcher) assisted to put it in a yoke. The element could be kept intact during this procedure as its connections were reversible (unscrewing), there were few of them (six screws) and those were also accessible (from the outside).

Disassembly procedures for smaller elements followed similar principles, but required greater manual labour compared to their heavier or larger counterparts. Demolition workers, for example, used their hands to take down curtains, lift ceiling tiles, and unplug kitchen appliances. Standard tools were used for activities such as clearing ceilings or disconnecting installations. Such activities sometimes required other activities to be completed first. In the duplex houses project, for instance, the demolition workers were first observed removing gypsum ceiling plates with a crowbar before they could access and remove the reusable glass wool plates behind the ceiling. Even the bricks were manually cleaned from mortar after a crane operator had torn down the façades there. This was only economically feasible for parts that were not mixed up with other rubble, which points to the importance of skilful and thoughtful handling in disassembly activities. Likewise, an experienced excavator operator demonstrated being able to lift an entire wooden floor, including beams and planks, and to put it upside down so that other workers could manually separate the beams from the planks. These reuse-oriented disassembly activities were, for that reason, called "more strategic" by workers in the nursing home project.

All reusable elements were thereafter stored on site for some time. Site management typically instructed demolition workers where and how they needed to store any disassembled elements. The roof tiles of the duplex houses were, for example, collected on a central outside location. Small disassembled elements in the nursing home, such as air conditioning units, lighting fixtures, and toilet grab bars, were quickly wrapped into some protective plastic foil and then stacked on pallets. These elements were stored in specifically designated rooms where larger machinery was prohibited. These sets of disassembled elements nonetheless had to be moved to other (dry and safe) locations from time to time when those rooms had to be demolished as well. Larger elements, such as façades or floors, were also stored outside, on the site. The specific storage locations for such elements depended primarily on the crane operating area and the accessibility of waste containers.

Demolition contractors thus harvested the elements identified as reusable through executing specific disassembly procedures and then temporarily storing those elements on site.

### *Distribute reusable elements*

Distributing the harvested building elements completes realising of reuse potentials. Demolition contractors are usually not reusing any elements themselves. Most elements are reused somewhere else. Distribution of harvested elements – from so-called donor to target building – is, thus, an essential activity. Demolition contractors handle this through packing, transporting, and monitoring the (reverse) element flows. Variations in these practices were observed in both direct and indirect flows: whereas some elements were distributed directly to their target locations, other ones moved to an intermediary warehouse first.

The harvested building elements were prepared for further transportation through packing. For instance, labourers in the duplex houses project were instructed to put roof tiles in intermediate bulk containers. They would typically separate each row of roof tiles with wooden planks so as to prevent damage during transportation. The masonry bricks of the façades were stacked on top of each other in alternating layers and then wrapped around in plastic foil. Kitchen cabinets were, likewise, stacked and wrapped. Similar observations were also made in the nursing home. Pallets there held microwaves and other appliances, often separated with planks. Foil not only protected such smaller elements against dust, which was omnipresent during demolition works, but also helped preventing scratches. Certain larger building elements were also protected. For example, labourers affixed extra timber bars to the nursing home's façade elements before lifting them into the yokes so that protruding ventilation rosters would not damage neighbouring façade elements in the same yoke. Packing, accordingly, protects and organises harvested elements for transportation.

Some elements were moved to a new (target) building directly. The duplex houses' packed roof tiles, for example, substituted new ones in a construction project elsewhere. The demolition contractor organised transportation to that site accordingly. Likewise, many structural elements of the nursing home were planned to be reused for building a school. Those two consecutive projects were coordinated in such a way that most of the nursing home's elements could be moved to the target location without intermediate storage. As such, the demolition contractor put – in reverse construction order – different roof, column, and floor elements on trucks that drove between the two sites. Certain other elements, like radiators and glass wool insulation, were sold through online sales platforms, often to private parties. Transportation efforts for the demolition contractor were limited in such cases as those parties typically picked up those elements by themselves. The same applied to elements bought by passersby. For instance, one person admitted feeling "cheeky" to ask site management for the large flowerpots that decorated the nursing home's main entrance. The ethnographic researcher regularly observed local residents, like this one, to inquire about and, at times, buy certain elements. Transportation movements in all these cases happened between donor and target buildings directly.

Many other elements were moved to intermediary warehouses first. Both demolition projects were connected to several traders in reclaimed materials which received parts of the demolition harvest. Such firms bought, among others, kitchens, bricks, and timber – all of which they hoped to resell in the near future. Since no final target destinations for those elements were available yet, the demolition contractor facilitated transportation to the warehouses or hubs of the respective buyers instead. Those firms generally sourced elements from several projects to achieve advantages of scale. They physically traded the elements from their warehouses and also, in the case of the kitchens, through an online auction. A trader in reclaimed bricks explained that those elements first had to be tested in a laboratory so they can "put a new certificate" on them before selling them to general contractors again. The demolition contractor transported some of the larger structural elements from the nursing home back to the system builder that had produced them. That firm's warehouse made it possible to store elements with reuse potential until a suitable new project was found. The elements could also be refurbished or upgraded there. Transportation from the nursing home respectively duplex houses to any (potential) future target buildings was in these cases indirect, via an intermediary warehouse.

Monitoring actions supported the transportation. Several observations were made of labourers coding or labelling distinct elements. Notably, the nursing home's demolition contractor was instructed to label façade elements according to a scheme with handwritten numbers on the provided disassembly drawings. Those numbers matched with numbers on the construction drawings for the school that was planned to be built with the elements. Through attaching the numbering information to the façade elements, the demolition contractor could monitor that the elements would be transported to the correct locations. Although this on-site numbering was redundant for most other elements, site managers informed truck drivers about the contents and destinations of their loads. This included instructions about elements that sustained damage during the demolition works and required repairs, such as with one specific roof element that had been improperly hooked to the crane and subsequently crashed onto the floor below. Likewise, site management posted pictures and descriptive information about elements available from the duplex houses on an online marketplace. For as long as the project lasted and space permitted on-site storage, they monitored interest from potential buyers – though such interest appeared to be relatively limited.

Distribution of the reusable elements thus involved various types of packing, transporting, and monitoring actions. The demolition contractors, accordingly, organised or facilitated transport movements between the donor and target buildings either directly or indirectly to realise reuse potentials.

## Discussion

The chapter uses ethnography to present a process framework for realising reuse potentials in circular demolition projects. The process framework links together three generic activities that demolition contractors need to engage with sequentially: identifying, harvesting, and distributing reusable building elements. Using ethnographic data collected during two projects, I demonstrate that all three activities are essential for realising reuse. There are, however, local variations through which these activities are organised, depending on the type of building element. Demolition contractors identify reuse potentials based on their assessments of both the conditions of available elements and the market demand for such elements. They then harvest elements identified as reusable through following specific disassembly procedures and storing the elements on site. Distribution, finally, involves packing, transporting, and monitoring actions to move elements from a donor to a target building. This chapter's ethnographic account unpacks these three activities and links them together into a process framework that can inform future work.

The detailed ethnographically grounded insights reveal how demolition contractors can realise reuse. The process framework complements earlier research that uncovered reuse barriers. This study reveals how demolition contractors could navigate those barriers. Rather than further problematising reuse in construction, it illuminates pioneering practices performed during circular demolition projects through which reuse became possible. These insights are helpful, for example, because prior work shows that the lack of information about existing building elements represents a notorious barrier to circular practices (Mahpour, 2018; Jayasinghe et al., 2019; Wijewickrama et al., 2021). This study's ethnographic insights, instead, reveal how demolition contractors can cope with this particular barrier through recollecting such information with, among others, pre-demolition audits and visual inspections. "Immobility of buildings and huge size of building products" was identified as another barrier by Hosseini et al. (2015, p. 508); yet this study enriches these insights by showing how reversing assembly sequences, eased through modularity and prefabrication, enables distribution and, in essence, makes certain building elements mobile. The ethnographic insights presented in this chapter thus offer a fresh perspective on realising reuse potentials, moving the scientific discourse from problems to solutions.

Insights like these are much needed to promote circularity in construction. The latter has become an increasingly prominent issue among policymakers, business leaders, and scholars alike (Illankoon & Vithanage, 2023). The implementation of circularity in construction has nevertheless remained limited so far, mainly because of insufficient knowledge and attention for the end-of-life phase of buildings (Benachio et al., 2020). This study contributed with knowledge about closing material loops to that end. It conceptualises circular demolition as a process that involves three interlinked activities aimed at realising reuse. A central activity involves what Jongert et al. (2011) have started to name harvesting, the systematic disassembly of elements for reuse elsewhere. To do that, demolition contractors must nevertheless first identify reuse potentials. They need to start looking at existing building elements as resources rather than waste, which relies on an entirely different skills set (see Smeyers & Mertens, 2022). Harvested elements must also be distributed, from donor to target building. This activity essentially connects reverse and forward resource flows with each other, which requires more integration between supply chain actors (Ding et al., 2023). This study offers the empirical evidence for various possible types of distribution channels, such as intermediary wholesalers, online marketplaces, and even informal passersby. Altogether, the ethnographic insights into different mechanisms for realising reuse potentials can inspire others to adopt or incentivise similar mechanisms elsewhere. This can thus help in developing more informed measures to stimulate circularity.

Ethnographic research appears well suited to make such contributions to the field of circular construction. Several ethnographic engagements with reuse practices in other fields have already illustrated the deeply relational, situated, and cultural entanglements associated with reuse (Isenhour & Reno, 2019). The accelerated interest in reuse under circular economy agendas may also further stimulate the reimagination and revaluation of demolition waste. Andersson and Buser (2022), for instance, illustrated how demolition contractors started to rename the materials generated during demolition as products or resources instead of waste. The implications of doing so are profound and may, therefore, be understood best through reflecting on participant observations of such circularity practices. This helped me to observe surprising reuse engagement mechanisms that more traditional research methods, largely rooted in positivist traditions, could not uncover, e.g., direct sales to passersby and to auction companies. Observing such best practices is an essential first step in theory development, which is especially relevant for emergent fields like circular construction. Mhatre et al. (2021) argue that observations (which may be original, replicated, or validated) must then be categorised into frameworks or typologies before testing them with various statistical or analytical tools. Ethnography-inspired observations can, in this way, inform the quickly evolving body of knowledge on circular construction.

Applying ethnography to this relatively new field can be a rewarding experience, but it is also "hard work: physically, emotionally, and mentally exhausting" (Delamont, 2004, p. 211). Ethnography is revelatory of organisational processes, as shown here, through "intensive fieldwork, meaning-focused sensework, and 'thick' textwork" (Van Hulst et al., 2017, p. 226). Here are some reflections on these aspects. My fieldwork involved prolonged and intensive engagement with practitioners on a demolition site, a remotely familiar environment. Oswald and Dainty (2020) warned aspiring ethnographic researchers for potentially macho, masculine, or dangerous circumstances that may prevail on construction sites. I found it, in that regard, invaluable to get introduced to the site by a friendly contact person. Systematically engaging in the focal phenomenon, also during less favourable weather conditions or other circumstances, appeared important to establish and maintain trust. My sensework involved an orientation towards the socio-material complexities on site. Through actively participating in the actual demolition works, I became more and more familiar with reuse-enabling practices. Being there up close also helped me understanding the "back stage" of site work, including dimensions of meaning making that practitioners typically hope to conceal from public scrutiny (Ybema et al., 2009, pp. 6–7). For instance, I witnessed how the demolition contractor responded to a safety incident as well as how it addressed extensive oil spill as a result of equipment malfunctioning. Ethnographies based on such events may, depending on the research interests, lay bare harsh and/or hidden social realities. My textwork, finally, involved exploring and exemplifying the general through the local and the particular (Van Hulst et al., 2017). Fieldnotes and pictures are invaluable to support writing thick descriptions. In trying to bring the text alive, I especially benefitted from having written down certain striking quotes verbatim. I would also recommend sharing the ethnographic insights with key informants after completing the project to validate findings and benefit the practitioners involved. Ethnographic research experiences may, in this way, lead to direct impact and work towards higher-level transition goals.

This study does so with some restraints. It is, firstly, confined to the examination of reuse from the perspective of demolition contractors, with limited consideration for other viewpoints. Actual reuse through closed-loop material flows nevertheless involves considerable changes in the work practices of other supply chain actors, such as designers (Gorgolewski, 2008), which need to be addressed as well. Secondly, this study focuses on reuse as a preferred waste management strategy, but does not delve into the complex interplay with alternative strategies such as recycling or disposal. Future research can broaden the focus to more fully understand how demolition contractors handle combinations of waste management strategies in their projects. Finally, the insights derived from two similar project settings may not be directly transferable to other settings. The pioneering projects were selected for "the force of example" (Flyvbjerg, 2006, p. 228); future research needs to compare the exemplar insights with lessons learned in other settings to discern transferability across contexts.

## Conclusions

This chapter uses detailed ethnographically grounded understandings of circular demolition practices to present a process framework for realising reuse potentials. Participant observations conducted during two circular demolition pioneering projects in the Netherlands served as the primary source of data for this study. A process perspective was applied to the data to advance understanding how circular demolition practices unfold over time. It was found that demolition contractors engage in identifying, harvesting, and distributing reusable building elements. These three activities are organised sequentially with specific variations across different types of elements. They are oriented towards realising reuse potentials and, as such, organised when demolition contractors have an interest in doing so. Any actors striving to incentivise reuse in construction can, in that regard, benefit from the ethnographic insights that this study offers. That is, incentives need to consider the entire process – from identifying reusable elements to distributing them – and be tailored to the building elements at hand. The insights of this study can thereby offer inspiration for navigating reuse barriers. The process framework for realising reuse potentials that this study presented can, accordingly, guide transformative action in circular demolition.

## References

Andersson, R., & Buser, M. (2022). From waste to resource management? Construction and demolition waste management through the lens of institutional work. *Construction Management and Economics*, 40(6), 477–496. https://doi.org/10.1080/01446193.2022.2081989

Benachio, G. L. F., do Carmo Duarte Freitas, M., & Tavares, S. F. (2020). Circular economy in the construction industry: A systematic literature review. *Journal of Cleaner Production*, 260, 121046. https://doi.org/10.1016/j.jclepro.2020.121046

Cherian, C., Siddiqua, S., & Arnepalli, D. N. (2022). Utilization of recycled industrial solid wastes as building materials in sustainable construction. In K. R. Reddy, R. K. Pancharathi, N. G. Reddy, & S. R. Arukala (Eds.), *Advances in Sustainable Materials and Resilient Infrastructure* (pp. 61–75). Singapore: Springer. https://doi.org/10.1007/978-981-16-9744-9_4

Coelho, A., & De Brito, J. (2011). Economic analysis of conventional versus selective demolition—A case study. *Resources, Conservation and Recycling*, 55(3), 382–392. https://doi.org/10.1016/j.resconrec.2010.11.003

Dalton, T., Dorignon, L., Boehme, T., Kempton, L., Iyer-Raniga, U., Oswald, D., Amirghasemi, M., & Moore, T. (2023). *Building materials in a circular economy* (1922498696). AHURi Limited. https://www.ahuri.edu.au/research/final-reports/402

Dantata, N., Touran, A., & Wang, J. (2005). An analysis of cost and duration for deconstruction and demolition of residential buildings in Massachusetts. *Resources, Conservation and Recycling*, 44(1), 1–15. https://doi.org/10.1016/j.resconrec.2004.09.001

Delamont, S. (2004). Ethnography and participant observation. In C. Seale, G. Gobo, J. F. Gubrium, & D. Silverman (Eds.), *Qualitative Research Practice* (Vol. 217). Thousand Oaks: Sage.

Dijksma, S. A. M., & Kamp, H. G. J. (2016). *A circular economy in the Netherlands by 2050: Government-wide programme for a circular economy*. https://circulareconomy.europa.eu/platform/sites/default/files/17037circulaireeconomie_en.pdf

Ding, L., Wang, T., & Chan, P. W. (2023). Forward and reverse logistics for circular economy in construction: A systematic literature review. *Journal of Cleaner Production*, 388, 135981. https://doi.org/10.1016/j.jclepro.2023.135981

Ellen MacArthur Foundation. (2013). *Towards the circular economy: Economic and business rationale for an accelerated transition*. https://www.ellenmacarthurfoundation.org/assets/downloads/publications/Ellen-MacArthur-Foundation-Towards-the-Circular-Economy-vol.1.pdf

European Commission. (2020). *A new circular economy action plan for a cleaner and more competitive Europe* (Communication from the Commission to the European Parliament, the Council, the European Economic and Social Committee and the Committee of the Regions, Issue.

Flyvbjerg, B. (2006). Five misunderstandings about case-study research. *Qualitative Inquiry*, 12(2), 219–245. https://doi.org/10.1177/1077800405284363

Gálvez-Martos, J. L., Styles, D., Schoenberger, H., & Zeschmar-Lahl, B. (2018). Construction and demolition waste best management practice in Europe. *Resources, Conservation and Recycling*, 136, 166–178. https://doi.org/10.1016/j.resconrec.2018.04.016

Ghaffar, S. H., Burman, M., & Braimah, N. (2020). Pathways to circular construction: An integrated management of construction and demolition waste for resource recovery. *Journal of Cleaner Production*, 244, 118710. https://doi.org/10.1016/j.jclepro.2019.118710

Ghisellini, P., Ripa, M., & Ulgiati, S. (2018). Exploring environmental and economic costs and benefits of a circular economy approach to the construction and demolition sector. A literature review. *Journal of Cleaner Production*, 178, 618–643. https://doi.org/10.1016/j.jclepro.2017.11.207

Ginga, C. P., Ongpeng, J. M. C., & Daly, M. K. M. (2020). Circular economy on construction and demolition waste: A literature review on material recovery and production. *Materials*, 13(13), 2970. https://doi.org/10.3390/ma13132970

Giorgi, S., Lavagna, M., Wang, K., Osmani, M., Liu, G., & Campioli, A. (2022). Drivers and barriers towards circular economy in the building sector: Stakeholder interviews and analysis of five European countries policies and practices. *Journal of Cleaner Production*, 336, 130395. https://doi.org/10.1016/j.jclepro.2022.130395

Gorgolewski, M. (2008). Designing with reused building components: Some challenges. *Building Research & Information*, 36(2), 175–188. https://doi.org/10.1080/09613210701559499

Grosse, H. (2019). An insider's point of view: Autoethnography in the construction industry. *Construction Management and Economics*, 37(9), 481–498.

Hosseini, M. R., Rameezdeen, R., Chileshe, N., & Lehmann, S. (2015). Reverse logistics in the construction industry. *Waste Management & Research*, 33(6), 499–514. https://doi.org/10.1177/0734242X15584842

Hulsbeek, L., & Van den Berg, M. (2022, 5–7 September 2022). *Evaluating reuse conditions during a circular demolition project: An ethnographic study*. 38th ARCOM Conference, Glasgow.

Illankoon, C., & Vithanage, S. C. (2023). Closing the loop in the construction industry: A systematic literature review on the development of circular economy. *Journal of Building Engineering*, 76, 107362. https://doi.org/10.1016/j.jobe.2023.107362

Ingold, T. (2014). That's enough about ethnography! *Hau: Journal of Ethnographic Theory*, 4(1), 383–395. https://doi.org/10.14318/hau4.1.021

Isenhour, C., & Reno, J. (2019). On materiality and meaning: Ethnographic engagements with reuse, repair & care. *Worldwide Waste: Journal of Interdisciplinary Studies*, 2(1), 1. https://doi.org/10.5334/wwwj.27

Jayasinghe, R. S., Chileshe, N., & Rameezdeen, R. (2019). Information-based quality management in reverse logistics supply chain. *Benchmarking: An International Journal*, 26(7), 2146–2187. https://doi.org/10.1108/BIJ-08-2018-0238

Joensuu, T., Edelman, H., & Saari, A. (2020). Circular economy practices in the built environment. *Journal of Cleaner Production*, 276, 124215. https://doi.org/10.1016/j.jclepro.2020.124215

Jongert, J., Nelson, N., & Goosens, F. (2011). Recyclicity: A toolbox for resource-based design. *Architectural Design*, 81(6), 54–61. https://doi.org/10.1002/ad.1320

Kanters, J. (2020). Circular building design: An analysis of barriers and drivers for a circular building sector. *Buildings*, 10(4), 77. https://www.mdpi.com/2075-5309/10/4/77

Kirchherr, J., Reike, D., & Hekkert, M. (2017). Conceptualizing the circular economy: An analysis of 114 definitions. *Resources, Conservation and Recycling*, 127, 221–232. https://doi.org/10.1016/j.resconrec.2017.09.005

Knoth, K., Fufa, S. M., & Seilskjær, E. (2022). Barriers, success factors, and perspectives for the reuse of construction products in Norway. *Journal of Cleaner Production*, 337, 130494. https://doi.org/10.1016/j.jclepro.2022.130494

Korhonen, J., Honkasalo, A., & Seppälä, J. (2018). Circular economy: The concept and its limitations. *Ecological Economics*, 143(Supplement C), 37–46. http://www.sciencedirect.com/science/article/pii/S0921800916300325

Langley, A., Smallman, C., Tsoukas, H., & Van de Ven, A. H. (2013). Process studies of change in organization and management: Unveiling temporality, activity, and flow. *Academy of Management Journal*, 56(1), 1–13. https://doi.org/10.5465/amj.2013.4001

Lansink, A. (2017). *Challenging Changes: Connecting Waste Hierarchy and Circular Economy*. Nijmegen: DPN Rikken.

Leigh, N. G., & Patterson, L. M. (2006). Deconstructing to redevelop: A sustainable alternative to mechanical demolition: The economics of density development finance and pro formas. *Journal of the American Planning Association*, 72(2), 217–225. https://doi.org/10.1080/01944360608976740

Lynch, N. (2022). Unbuilding the city: Deconstruction and the circular economy in Vancouver. *Environment and Planning A: Economy and Space*, 54(8), 1586–1603. https://doi.org/10.1177/0308518X221116891

Mahpour, A. (2018). Prioritizing barriers to adopt circular economy in construction and demolition waste management. *Resources, Conservation and Recycling*, 134, 216–227. https://doi.org/10.1016/j.resconrec.2018.01.026

Mhatre, P., Gedam, V., Unnikrishnan, S., & Verma, S. (2021). Circular economy in built environment – Literature review and theory development. *Journal of Building Engineering*, 35, 101995. https://doi.org/10.1016/j.jobe.2020.101995

Munaro, M. R., & Tavares, S. F. (2023). A review on barriers, drivers, and stakeholders towards the circular economy: The construction sector perspective. *Cleaner and Responsible Consumption*, 8, 100107. https://doi.org/10.1016/j.clrc.2023.100107

Musante, K., & DeWalt, B. R. (2010). *Participant Observation: A Guide for Fieldworkers*. Rowman Altamira.

Olde Scholtenhuis, L. L., Hartmann, T., & Dorée, A. G. (2016). 4D CAD based method for supporting coordination of urban subsurface utility projects. *Automation in Construction*, 62(Supplement C), 66–77. https://doi.org/10.1016/j.autcon.2015.10.013

Oswald, D., & Dainty, A. (2020). Ethnographic research in the construction industry: A critical review. *Journal of Construction Engineering and Management*, 146(10). https://doi.org/10.1061/(ASCE)CO.1943-7862.0001917

Oswald, D., Sherratt, F., & Smith, S. (2014). Handling the hawthorne effect: The challenges surrounding a participant observer. *Review of Social Studies*, 1(1), 53–73. https://www.rossjournal.co.uk/wp-content/uploads/2016/12/RoSS-Vol1-No1-Oswald-et-al-53-73.pdf

Park, J., & Tucker, R. (2017). Overcoming barriers to the reuse of construction waste material in Australia: A review of the literature. *International Journal of Construction Management*, 17(3), 228–237. https://doi.org/10.1080/15623599.2016.1192248
Phelps, A. F., & Horman, M. J. (2009). Ethnographic theory-building research in construction. *Journal of Construction Engineering and Management*, 136(1), 58–65. https://doi.org/10.1061/(ASCE)CO.1943-7862.0000104
Pink, S., Tutt, D., Dainty, A., & Gibb, A. (2010). Ethnographic methodologies for construction research: Knowing, practice and interventions. *Building Research & Information*, 38(6), 647–659. https://doi.org/10.1080/09613218.2010.512193
Pink, S., Tutt, D., & Dainty, A. (2012). *Ethnographic Research in the Construction Industry*. Abingdon: Routledge.
Pomponi, F., & Moncaster, A. (2017). Circular economy for the built environment: A research framework. *Journal of Cleaner Production*, 143, 710–718. https://doi.org/10.1016/j.jclepro.2016.12.055
Shah, A. (2017). Ethnography? Participant observation, a potentially revolutionary praxis. *Hau: Journal of Ethnographic Theory*, 7(1), 45–59. https://doi.org/10.14318/hau7.1.008
Smeyers, T., & Mertens, M. (2022). *Reuse toolkit: The reclamation audit – A guide to creating an inventory before demolition of potentially reusable construction products*. https://vb.nweurope.eu/media/19516/fcrbe-inventory-guide-en.zip
Spradley, J. P. (1980). *Participant Observation*. New York: Holt, Rinehart and Wineston.
Van den Berg, M. (2024). Digital technology use cases for deconstruction and reverse logistics. In C. De Wolf, S. Çetin, & N. M. P. Bocken (Eds.), *A Circular Built Environment in the Digital Age* (pp. 197–212). Springer International Publishing. https://doi.org/10.1007/978-3-031-39675-5_11
Van den Berg, M., Voordijk, H., & Adriaanse, A. (2020). Recovering building elements for reuse (or not) – Ethnographic insights into selective demolition practices. *Journal of Cleaner Production*, 256, 120332. https://doi.org/10.1016/j.jclepro.2020.120332
Van Hulst, M., Ybema, S., & Yanow, D. (2017). Ethnography and organizational processes. In A. Langley, & H. Tsoukas (Eds.), *The Sage Handbook of Process Organization Studies* (pp. 223–236). Thousand Oaks: Sage.
Wiarda, M., Coenen, T. B. J., & Doorn, N. (2023). Operationalizing contested problem-solution spaces: The case of Dutch circular construction. *Environmental Innovation and Societal Transitions*, 48, 100752. https://doi.org/10.1016/j.eist.2023.100752
Wijewickrama, M. K. C. S., Chileshe, N., Rameezdeen, R., & Ochoa, J. J. (2021). Information sharing in reverse logistics supply chain of demolition waste: A systematic literature review. *Journal of Cleaner Production*, 280(124359). https://doi.org/10.1016/j.jclepro.2020.124359
Wu, Z., Yu, A. T. W., Shen, L., & Liu, G. (2014). Quantifying construction and demolition waste: An analytical review. *Waste Management*, 34(9), 1683–1692. https://doi.org/10.1016/j.wasman.2014.05.010
Ybema, S., Yanow, D., Wels, H., & Kamsteeg, F. (2009). Studying everyday organizational life. In S. Ybema, D. Yanow, H. Wels, & F. Kamsteeg (Eds.), *Organizational Ethnography: Studying the Complexities of Everyday Life* (pp. 1–20). London: Sage.
Yin, R. K. (2014). *Case Study Research: Design and Methods* (5th ed.). Thousand Oaks: Sage.
Yu, Y., Junjan, V., Yazan, D. M., & Iacob, M.-E. (2022). A systematic literature review on circular economy implementation in the construction industry: A policy-making perspective. *Resources, Conservation and Recycling*, 183, 106359. https://doi.org/10.1016/j.resconrec.2022.106359
Zhang, C., Hu, M., Di Maio, F., Sprecher, B., Yang, X., & Tukker, A. (2022). An overview of the waste hierarchy framework for analyzing the circularity in construction and demolition waste management in Europe. *Science of the Total Environment*, 803, 149892. https://doi.org/10.1016/j.scitotenv.2021.149892
Zhang, C., Hu, M., Yang, X., Miranda-Xicotencatl, B., Sprecher, B., Di Maio, F., Zhong, X., & Tukker, A. (2020). Upgrading construction and demolition waste management from downcycling to recycling in the Netherlands. *Journal of Cleaner Production*, 266, 121718. https://doi.org/10.1016/j.jclepro.2020.121718

# 21

# AN AUTOETHNOGRAPHY TO REVEAL THE BITTER TRUTH

## I'm the problem

*Henning Grosse*

*IU International University Campus Hamburg, Erfurt, Germany*

*Gloucestershire Business School, University of Gloucestershire, Cheltenham and Gloucester, United Kingdom*

**Key takeaways**

- When pursuing autoethnography the practitioner-researcher must anticipate that
- They need to be patient, accept moments of doubt.
- They must be critical about themselves to an uncomfortable level and beyond.
- They will need to be courageous as they find out unfavourable aspects out about the way they think and act and what implications their actions have on others.
- They will inevitably change throughout the process of researching based on the gained insights.
- The learning and changing can be challenging because it exposes failures and vulnerabilities but also learns one to act better or more appropriately in future.

### Introduction

One of the most discussed issues in ethnography is the role of the researcher. In ethnography, the researcher as a "tool" strongly influences the outcome of the ethnographic exploration. Even more in autoethnography, the researcher often reports on their own experiences and conveys feelings and emotions (e.g., Ronai, 1995; Ellis, 2019). Autoethnographers observe their environment, hence the self and the observation of oneself play a central role in autoethnography. Therefore, researching autoethnographically inevitably affects the autoethnographer. Subsequently, the researcher transforms.

The autoethnographer's transformation during the research process is not a side effect but a central feature of autoethnography. It is "inherent in the autoethnographic method itself" (Hernandez et al., 2022, p. xv). This chapter illustrates this process and discusses the upsides and downsides of this transformation. This autoethnography, a form of reflexive individualistic action research, has not been discussed in construction management. Therefore, my awareness of and learning about gender inequality in construction management serves as an example of

DOI: 10.1201/9781003379584-25

this rare type of research. I explore what transformation means for the researcher and the practitioner doing research. I showcase what one might expect when embarking on an autoethnographic journey.

Practitioners using autoethnography may produce highly contextual insights into the industry. Although the recent trend is promising, in construction management research, only a few have embarked on an autoethnographic journey (e.g. Thiel, 2005; Kanjanabootra & Corbitt, 2016; Mlecnik, 2016; Jones, 2019; Whaley, 2020; Crascall, 2021; McCarthy, 2021; Encheva, 2022, 2023). Autoethnographers are still rare in construction management. I will demonstrate some of the advantages of autoethnography and emphasise some difficulties. Not to argue against the method but to manage expectations. I highly recommend using it, but not under false assumptions. For me, it is first and foremost the reflexive engagement with one's role in the everyday environment, which is challenging and rewarding, as I will show.

The remainder of this chapter briefly explains what distinguishes autoethnographic research from (non-auto-) ethnography. Subsequently, I give some personal background and explain how I use the method. After that introduction, I present vignettes and reflections to highlight some aspects of autoethnographic writing but emphasise the transformational nature of the research method. Finally, reflecting on my learning about gender inequalities, I emphasise a need for a critical stance towards oneself. The critical stance incorporates and produces the main challenges but also the most significant benefits of such a transformation process.

## Autoethnography

Since others in this book have sufficiently elaborated on ethnography, I highlight what distinguishes autoethnography. Simply put, autoethnographers stay on their home turf, whereas ethnographers research elsewhere and usually leave the field afterwards. Ethnographers observe a research field, i.e., a cultural setting they do not naturally belong to. In contrast, I predominantly write fieldnotes about my experiences in my business. Although the distinctions are somewhat fluid, autoethnographers generally research in their own backyard (Wolcott, 1999). Both auto- and non-autoethnography need to reflect on their relation to the field of research. However, being native to the setting is different. I am a natural part of what I research and, therefore, the subject and object of research simultaneously. As I go along and research, I learn and change. I cannot avoid the influences of my research and teaching on what I do in business and vice versa. This learning requires a reflexive engagement with myself to cast the research as credible.

Research at home – in my business – is a form of opportunistic research (Riemer, 1977). Since 1999, I run my own construction business in the wider Berlin area in Germany. Today, I employ 25 staff most of them bricklayers. My company builds a range of things from concrete and bricks; it might be a 7-storey residential building in Berlin, a detached house in Berlin's suburbs or a garage. As the owner of a construction company, I began researching autoethnographically in 2012/2013 for my PhD and continue to this day. Against that backdrop, I started teaching teamwork and project management at the polytechnical school of Berliner Hochschüle für Technik (BHT) in 2017.

Calling it opportunistic research is almost a euphemism: I couldn't go anywhere else to do my research. My company is my breadwinner. So, I have chosen to do ethnography, but the auto-prefix was not a choice but a necessity. For predominantly practical reason, I could not leave my business behind to do research elsewhere. But this chapter shows that my business is a valuable source of ethnographic material.

After finishing my PhD in 2018, I continued fieldnote-writing. I often sit down and write in the evenings – like many other ethnographers – but I write about my business and what I find worthwhile to note of the day's events. I seek to connect the notes to earlier events – some even historic – and to what I read in the literature. After a while, meaning days, weeks or even months, I identify interesting-looking patterns or narratives and explore them further. That might be how I encounter questions about sustainability in practice (Grosse, 2022), how I experience claims management on site (Grosse, 2023) or anything else that provides an intriguing perspective. Again, I try to connect what I find to the literature. I start a hermeneutic process of reading, observing, writing and reflecting.

One needs a good amount of patience. I often write and find nothing exciting, but suddenly, I spot something I had overlooked. One cannot push this epiphany one has to wait. As Bukowski (1994) says, "Don't try". One has to wait for it to come. If I am lucky, I produce something worthwhile to show others. But one needs faith in the method, as Alec Grant (University of Bolton, UK) wrote in a personal conversation.

Writing something for publication is one aspect of autoethnography; another aspect is writing to learn personally. Ellis (1999, p. 677), similarly to Herr and Anderson (2015), describes autoethnography "as action research for the individual". In this sense, I use it to learn about my business, role and myself. While also drawing on literature, I understand the construction industry differently from before I embarked on my autoethnographic journey. Sharing these understandings, I seek to contribute to scholarly debates. It is the hermeneutic process and frequent connection to the literature that separates the auto-ethnography from personal and professional reflection. My goal is not only to learn but to develop descriptions of meaning and as such contribute to academic debates.

### *Reflecting on multiple and interacting roles*

In this chapter, I reflect on three roles: practitioner (business owner), university lecturer and academic researcher. At length, one may discuss the ensuing identity struggles (Sveningsson & Alvesson, 2003), but this should not be the main focus here. I want to show how the different roles inform each other. A key difference with other types of ethnography is that autoethnographer often test their interpretation and theories in their daily lives. Hence, it inevitably becomes some form of action research. One cannot unlearn the knowledge acquired in another role, and one does certainly act upon that knowledge.

Apart from thinking about one's identity, which is unavoidable in autoethnography, each role contributes to another. However, for me, *writing* autoethnography is critical to study the process of meaning construction more deeply. Only through writing as a method of inquiry (Richardson & St. Pierre, 2005) can I form a deeper understanding of each role and ignite a transformative process. Autoethnographers often share their insights and, subsequently, create social activism (Grant & Zeeman, 2012).

I sit down to write about the day's events. I learn about what is going on: I describe, reflect on it, see first patterns, connect it to prior events and what I read, and subsequently understand what's going on more profoundly. Most of it I can do on my own. But as I will demonstrate later, reading other's autethnographic accounts, such as Kamelia's and Chrissi's work (McCarthy, 2021; Encheva, 2023), helped me understand their perspective better and reflect on my own views and actions. I would not have written this piece without reading their work in the first place.

I do have three different roles, practitioner, lecturer and academic researcher, which advance my learning. I write about my practice and the events that I encounter in the industry. I use academic literature to understand my experiences and connect my writing personal insights to the broader debate in academia. I present my findings to a research audience and receive feedback. Inevitably, if I am asked whether I have thought about a problem in this or that way at a conference, I do not just think as a researcher about it. I always carry that thought into practice. It does inform my practical doing. Similarly, my students' questions are a valuable opportunity to reflect on my understanding. In turn, the new understanding has consequences for both roles – academic and practitioner.

I use autoethnography in a more personal and self-reflexive form. This is different from what I observe other construction management researchers do. They often use autoethnography in a less personal fashion. They observe what's around them and employ a somewhat analytical (Anderson, 2006) use of the method. Yet often, construction management researchers shy away from exploring their own role. I lament the lack of more personal, evocative–reflexive applications of autoethnography.

A strength of the method is that it can access the feelings, emotions and thoughts of the autoethnographer more easily than other approaches. Hernandez et al. (2022, p. 106) note that "[e]motions speak and offer clues to meaning-making sometimes even more profoundly than logic. [there is] the art of learning to listen and be attuned to their emotions". Listening and tuning in may provide exciting insights hidden about the construction industry that other methods fail to unveil. Usually, this listening and tuning happen behind closed doors when I write my fieldnotes and reflections. What bothers me during this sense making process (in the broadest sense) often hints at some underlying, undiscovered and ill-understood pattern. Often, small and seemingly unimportant events reveal their importance only later (Rilke, 1929). Through writing about events that bother me, I start to form interpretations and theories (Richardson & St. Pierre, 2005) and may put them to the test the next time I encounter similar events (Herr & Anderson, 2015). In so doing, I form a hermeneutic circle, which often provides a deeper understanding but may also lead to more confusion and self-scrutiny.

### *Facing criticism*

One may argue that the focus on the self is uninteresting (Delamont, 2009) and not the researcher, but "others remain infinitely more interesting and sociologically significant" (Atkinson, 2006, p. 403). Some argue autoethnography is self-indulgent/-absorbed, narcissistic or naval-gazing (Mykhalovskiy, 1996; Sparkes, 2002; Madison, 2006; Allen Collinson, 2013). Some even argue that autoethnography isn't proper research. This argument is more prominent in engineering departments: after reading an autoethnographic piece, a colleague in my engineering department at the university wondered, "This is an interesting, well-written and accessible story, but this isn't serious research, is it?" Others experience similar fates (Mykhalovskiy, 1996; Sparkes, 2002) in academic fields more familiar with qualitative studies.

To address these criticisms, autoethnographers need to write about things they are passionate about but in a way that is accessible and makes the reader think. Self-criticism and reflexivity are essential tools (Pillow, 2003). Both are required for a personal transformation and to render it rigorous research. Since autoethnography cannot be planned in advance and usually is messy and full of surprising twists and turns it needs some guiding principle. Critically and reflexively questioning one's actions, observations and interpretations is key to make a systematic inquiry. This questioning inadvertently is an uncomfortable process and reveals unfavourable aspects of oneself as reflexivity often does (Pillow, 2003). However, it leads to a deeper understanding of oneself and one's environment. Therefore, it can highlight vistas of culture through the self, otherwise difficult to access but worthwhile to read.

## Learning about gender inequality

I write about my learning about gender inequality in construction management. It started in 2016 when I was working on my PhD, but it took me some time. So, the last observations are more recent as of 2022. As I said earlier, it is not straightforward but a journey with twists and turns.

I will not dwell on numbers, the share of women in the construction workforce or anything alike, but I will reflect on my recognition of women's experiences in construction management. Below, I depict the change in my perception of the problem, seeking to make a case for more insider accounts from both sides – male and female – to understand gender inequality, its causes and its consequences better. At the same time, I seek to highlight autoethnography's transformative nature. I show how writing notes and reflecting on the topic helped deepen my understanding.

### *Becoming aware*

From my very early days on sites, I encountered some women working in construction management who developed a "thick skin" while on building sites and around. Some talked about not being taken seriously by male colleagues, being belittled or dealt with as if they were not competent or naïve. I heard stories of uphill battles against dominating men. And I saw women having grown hard and disillusioned and frustrated. However, reading Chrissi's experiences was different from what I had heard:

> A different reality
>
> In 2106, I got introduced to Chrissi McCarthy. She was writing her PhD on equality in the British construction industry at the time. After a brief email exchange about autoethnography, Chrissi shared some autobiographical notes on her experiences as a construction manager. One of them burned itself into my memory:
>
> 'It was after 6pm. Only my site manager and I remained on-site as I finished up the as-built drawings, not being entirely sure that everything was as precise as I would have liked it to be. My site manager came into the room to enquire how long I needed and as I told him I was finishing off, he replied casually "I'd like to fuck you over that drawing board, good and hard"' (McCarthy, 2021, p. 4). Reading it, I felt my jaw dropping. It isn't easy to describe what I thought. The ingrained power dynamics in this and the following incidents were even more disturbing. I knew building site banter; I knew all the talk about 'just jokes', but how severe the implications are for women on construction sites wasn't that clear.

I knew construction sites were a tough place for women to work. Still, I had not imagined that to happen. Sure, I was naïve to believe so. I had not encountered, nor had I had a conversation with women from the industry intimate enough to tell me about such experiences as Chrissi's.

However, the severity of Chrissi's writing made me aware that there was much more hidden than I had known before.

This autobiographical note stayed with me over time. It made me more sensitive to how women are treated differently in the construction industry. However, apart from that, I did not explore the matter further research-wise. But then an opportunity loomed:

> Would you supervise?
>
> When teaching project management on a sustainable building design course at the BHT Berlin, I regularly hand out group activities to my students and provide them feedback on their hand-ins. During the coronavirus pandemic, I met a group of my students in our weekly MS-Teams meeting. We discussed their work, and I gave them feedback. Somehow, we came to talk about gender issues in the construction industry, and I may have mentioned Chrissi's work. Since the group comprised three females and two male students, we had a lively discussion. Kamelia – one student – talked about her experiences working part-time as a construction manager while studying. After a while, she asked whether I would supervise a master's thesis on gender diversity. First, as she later said, she asked more jokingly, but I had a genuine interest. I said I sure would. Since reading Chrissi's notes, I was primed. I knew I had to support her project.
>
> It was not just an opportunity to support and supervise her research project. I was keen to do it. It was a topic in dire need of investigation. Over time it turned out to become some qualitative research – interviews with female practitioners were one of the first ideas. Additionally, some of Kamelia's personal experiences on the building site should underpin her findings too.

There was the chance to pursue another form of "opportunistic research" (Riemer, 1977). Or was it some calling? One could argue that how I was *thrown* into the situation made it impossible to act differently. To Heidegger (1927), "throwness" is the sum of the baggage one carries to situations one encounters. It is "my rootedness in a culture, my already established preferences, skills, habits, and so on" (Hoffman, 1993, p. 208).

Thrownness is more about the "who I am" as a victim of my history than arriving at a new situation. As an ethnographer, but maybe especially as autoethnographers, we approach an experience always with our history, values, norms and predispositions. As autoethnographers, we hold vested interests in the situations we explore since it is our life we write about. We are never free of intentions in our research efforts. This thrownness guides what we are able to see and what we may fail to recognise. Having read Chrissi's fieldnotes (McCarthy, 2021) was an essential part of that thrownness. It was almost a predicted course of action.

However, the possibility of critically and reflexively questioning our observations and interpretations remains open to the autoethnographer. Heidegger recognises the limiting influence thrownness poses for the individual. Yet there is still a potential to free oneself – to some degree – from these limits. Heidegger calls that being authentic – to engage with the situation critically and reflexively – one finds oneself in. Hence, the possibility of being authentic rests in thrownness, yet one has to choose to live authentically (Large, 2008). So, not supervising this thesis wouldn't have been an option for me as the opportunity presented to support a student exploring a topic I felt strongly about but had not found the chance to explore myself.

But, thrownness is not a stable entity. New experiences, especially the reflective engagement with our experiences, allow us to fulfil our opportunities. As Heidegger (1927) says, we can become authentic if we make an effort. But we can also fall into this situation. Heidegger's falling (1927) is a form of losing oneself in mundane activities without much thought. Not falling but taking the opportunities requires one to challenge oneself.

The point is that, on one hand, autoethnographers experience thrownness while researching. It is unavoidable. On the other hand, an awareness of thrownness and how it works opens the opportunity to see things differently. Only when accepting that I'm *thrown* I can engage in an hermeneutic-reflexive exploration of myself, thoughts and actions. Knowing about thrownness makes authenticity (Heidegger, 1927) possible. We understand our *being* – what we personally are (Gemeinigkeit) – from the inside – while we are what we are (Heidegger, 1927). We cannot step outside of our selves but have to understand our selves from the inside.

In that sense, Kamelia's approach and my reaction bring together three different aspects I introduced earlier: I felt strongly about the subject of her proposed thesis, which relates to "the art of learning to listen and be attuned to their emotions" Hernandez et al. (2022, p. 106) write about. It shows how my thrownness is the basis for the emotions I am supposed to listen to. But without reflexively engaging with "what's going on?" and without the uncomfortable (Pillow, 2003) process of unforgivingly reflexively questing oneself, one fails to lift the potential (Heidegger, 1927) that is offered. Thrownness is the point of departure, but to reflexively learn about oneself, fulfil one's potential, and become authentic, is the route Heidegger offers. It is a hermeneutic approach to a deeper, although never complete, understanding.

I have to admit that I never researched anything I had no personal interest in. For me, there is no relevance in only theory-driven research. It's always about my personal issues – topics contributing to my business or of interest to me otherwise. I want to extend Riemer's (1977) opportunistic research further. Not only should the field of research be easily accessible, but it also should the subject matter to the researcher. I would argue that intrinsic motivation and some passion of the researcher to understand a subject are incredibly beneficial. This is the case for me, but I also see it in others like Kamelia (Encheva, 2022) or Chrissi (McCarthy, 2021).

Supervising this master thesis was similar, and I was interested in the subject. I knew I had sufficient methodological knowledge to supervise a qualitative study but also substantial gaps regarding gender inequality. This was my chance to gain insights I would have had difficulties accessing otherwise.

The challenge in supervising this research was that I knew I would learn about my male colleagues and myself things I would not appreciate. I would learn unfavourable truths seen from a woman's perspective. I knew the topic was sensitive. After a while, it became an autoethnographic study, which posed more hurdles in the engineering department of the BHT and challenges for me as a male supervisor. I could easily hurt my student's feelings with the wrong question or a clumsy remark, especially as a male, powerful supervisor talking to a junior female student writing about vulnerabilities and sensitive issues. Still, or even more so, I had to do that project.

Similar patterns

During the early stages of her research, Kamelia shared some of her fieldnotes. They showed parallels to Chrissi's work. These notes dealt with other places, people, and events big and small but the same patterns: power dynamics. Yet they often were minor comments that signalled not being respected, taken seriously, belittled and excluded. The examples range from not being provided information to not being asked whether to join for a beer after work or go to the Christmas market.

When we discussed these notes, she often said they appeared unimportant and might not interest anybody. However, they tell a story that is too seldom voiced in the construction industry. It is one of the suppressed narratives which Grant and Zeeman (2012) mention in other contexts. Listening to these stories, however, slowly but incessantly changed my view on the subject. It reminded me of "the little things that hardly anyone sees", which unexpectedly become significant (Rilke, 1929, p. 34). That there are micro-obstacles for women in the industry is often unrecognised, ignored or silenced. These little things were not present to me either. In some sense, these conversations brought across more than a rational explanation but an emotional or ethical (Levinas, 1974) understanding. It was more like the other in Levinas' work: "development is neither purely descriptive nor purely deductive. It proceeds with the infinite insistence of waves on a beach: return and repetition, always" (Derrida, 1967, p. 398). I repeatedly needed to listen to Kamelia to feel and understand their importance. The little stories brought home to me what male privileges I enjoy. I only see them because Kamelia told me how it is to miss them.

### *Learning from Kamelia*

It didn't take long for more severe incidents to surface. Kamelia talked about working with 25 male colleagues in a building site office. A new female colleague was expected to start working the next day, but office space was limited, and one colleague wondered why there was no available desk for her. Kamelia wrote:

> Just a joke
>
> "She could sit on my lap at first," another answered. As if that was not enough of an absurd joke, a third colleague suggested, "She could go directly under the table," as this would have made it easier for her to "start" (performing "oral sex" was the implication of his hand movements).
>
> My jaw dropped, [...] also because it had escalated so quickly. It was as if each of the participants was trying to say something to one-up and impress the other "dudes".
>
> As soon as the whole theatre happened, I shared my disappointment by stating: "This was a sexist joke." What I met in return was "It's just a joke!" – a very, very common dude argument used over and over again, as if it is a reasonable argument I must accept and respect ...
>
> *(Encheva, 2022, p. 10)*

Not that the sexist joke itself was insulting enough. The notion of "just a joke" made it worse. Hearing this argument "over and over again" exhausts the victim even more. Not even "a tough skin and good banter will see you through" (McCarthy, 2021, p. 3); the tough skin wears out over time.

I recall situations when I heard similar male banter and the same "it's just a joke" justifications. Listening to the receiving edge of such "jokes" changed my understanding. These jokes are massive manifestations of power. They are relegating women to subaltern places only because of their gender. It is not too difficult to see other groups on building sites subject to similar jokes and how they might feel and think about them.

> Denial of gender inequality
>
> Some weeks later, we met on a rainy dark afternoon in a café around the corner from my place. Kamelia told me about an argument with a colleague roughly her age who denied the

existence of gender inequality. She was astonished and shocked by the denial. But then she also told me he is otherwise a nice guy. After talking to his girlfriend, he began to change his mind, which gave Kamelia some comfort.

How is it possible to deny the existence of inequalities from gender? I wondered.

This young lad had a blind spot. Kamelia unveiled it, and he reflected. In the end, he could see a bit better. Yet there are indeed many blind spots that Kamelia's colleague has and does not know about. But the same holds for anybody else, me included, of course. Equally, I often heard the "just joke" justification and used it too. I cannot imagine any particular occasion, but I am sure I used it. It feels uncomfortable to think of oneself in similar terms. To arrive at this understanding, I benefitted from working with Kamelia. One must not forget "to be reflexive demands both an other and some self-conscious awareness of the process of self-scrutiny" (Chiseri-Strater, 1996, p. 130). Here, Kamelia was the other I needed.

Others are pushing me to think and scrutinise myself. They push me to explore my *thrownness* and to become *authentic* (Heidegger, 1927). I usually put these thoughts on paper. These thoughts are a feature of autoethnographic writing that I appreciate and loathe simultaneously. Once I start writing about a topic – be it a fieldnote or a piece to publish – I come across things I do not like about myself. Often, I find out that I could have acted more appropriately and dealt with the situation better. That is not very surprising, but oneself does not appear in the most favourable light. That's the part I don't like. I think it is a challenge to be hard on oneself. It raises self-doubt and lowers self-esteem. That is incredibly challenging in an environment like the construction industry – marked by masculinity, loud and emotional conflicts – where showing weaknesses and shortcomings is often exploited. It certainly is of value to be reflexive and doubt oneself, but one has to be careful about showing vulnerabilities and weaknesses.

However, after writing reflexively about an event, I have at least a slight idea of how to act better and respond more appropriately. I enjoy the action research facet of autoethnography. I like the learning that takes place.

Therefore, after reading Kamelia's experiences and thinking and writing about them, I am much more reluctant to use the "just a joke" phrase. The engagement with Kamelia and her writing demonstrated again that once I investigate, I find an unknown world hidden behind a veil of ignorance. I observe differently. That is how "[r]eflexivity can promote deep professional learning and bring sustainable change" (Feucht et al., 2017, p. 234). I arrive at new situations with a new *thrownness*. Now, things catch my attention that I would not have seen before.

This is, for example, how women and men solve conflicts. Men tend to act more aggressively and competitively (Galea et al., 2021). Kamelia's experiences support this claim. Her boss suggested she should "become louder" and "show more teeth". She, however, would have preferred resolving the disputes by communicating about differences, not by using "raised voices and emotional discussions" (Encheva, 2022, p. 44). Adopting an aggressive and loud approach to conflicts adversely affects women's and men's well-being. But women's well-being deteriorates unequally more (Galea et al., 2021).

### *Translating into practice*

This experience in academia as a supervisor and as an academic interested in the topic has also impacted me in my role as a construction professional. Considering what I heard about women's experiences and what they mean to them, I experience construction sites and their surroundings differently. I could take advantage of the understanding gained in my supervisor role. In essence, the supervisor experiences laid the foundations for the following fieldnote, which derives from my role as construction company owner.

It depicts how, in a male-dominated environment, conflicts are dealt with.

Rapid escalation

Nadine,[1] the architect in charge, invited my site manager Mark and me, George, the subcontractor for the windows, and maybe one of George's staff members to a meeting at 2 p.m. in the site office. Nadine planned and managed the construction site for the architect's practice. She is a skinny woman in her thirties, enthusiastic about her work and knows almost everything concerning 'her construction site'.

George is a tall guy in his mid-fifties with blond hair, a deep voice and a masculine appearance. Often, he acts a bit bossy. Well, he is the boss of his own company. But then he makes his employees look, let's call it, inferior. I had my own bunch of conflicts with him, so I'm probably not free of prejudices here. Anyway, I had to work with George on that large building site. He was employed to fit the windows and the cladding after my company finished the concrete structure of the house.

I met Mark when I arrived at the building site. We chatted a bit. We did know what the meeting would deal with but did not expect anything hugely problematic. George's staff member – Simon – arrived and joined our chat; it turned already ten. Although George had not arrived, I suggested going in. We went to the meeting room, and Nadine greeted us friendly. From then on, the conversation went relatively smoothly. We discussed the thermal insulation and what kind of water stopper we should use. I took Simon's notepad and drew a section to explain what we would do immediately and later do once the cladding was finished. We also talked about our crane having been removed from the building site. I express some anger for not being able to use it any longer, mainly because the new crane of the subcontractor isn't installed and, hence, not in use. It was a calm and productive meeting so far. After more than an hour, we moved to the site where a particular detail had to be explored. When explained there, it was immediately evident.

Finally, George arrived. He was (passive) aggressive from the very first moment. He said without having it in hand that one detail was wrong. Nadine was irritated because she knew the details and couldn't see what should be wrong with it. She asked clarifying questions, and George struggled to answer. Although he made bold claims that the detail would be wrong, he seemed unfamiliar with the situation we were discussing. He didn't exactly know what he was talking about but was still very assertive.

Then, the discussion turned to another topic. I can't recall the details, but it concerned the scaffolding and space to deposit materials. Although George said he could bare loud environments, he got louder and louder while talking. Finally, he started shouting at Mark -my site manager.

Then George also began yelling at me. I was already unnerved because he had left us waiting. I told him, "Listen, you're 1:15h late and didn't even apologise." Maybe I shouldn't have done that since it spilt oil into the fire. He replied, accusing me of finishing my work on site one year late, although he didn't know my contract at all. He came as close as 25–30 cm to me. I felt the need to push him gently but decisively back. To which he reacted with a "Don't touch me!" I stepped aside and stopped talking to him.

Then the atmosphere relaxed a bit afterwards. We demarked an area which we would empty for him. But it was still a bit of haggling about it. In the end, we agreed on something.

Aside from the group, I talked to Nadine. She was upset about the situation but sought to keep her temper. 'You know, I can't bear that. I am unwilling to work this way.' she told me, referring to the loud and aggressive argument.

Kamelia, Chrissi and others described similar patterns: hypermasculinity, aggressiveness and emotional rows. In particular, the last paragraph shows how alienated Nadine became during the aggressive exchange. Although she was not the direct target of the aggression, I sensed a massive unease on her side. In contrast, the male participants merely shrugged and went on with their business, supporting the observations of Galea et al. (2021).

First, we were content-driven, friendly atmosphere despite differences in the way forward. We discussed friendly and politely and resolved some differences. It all changed when George entered. Hence, not joining his way of disputing may have made a difference. Yet, I failed to do so. Maybe I could have sidestepped George's accusations and relaxed the atmosphere earlier. Somehow, I didn't act reflectively. It's often an effort to step away from previously learned but inappropriate reflexes. I fell into this situation (Heidegger, 1927).

Writing about oneself should "convey an acceptable-to-me-for-the-moment portrayal" (Ronai, 1995, p. 398). In other words, "the act of writing about oneself places an ethical duty on oneself to responsibly, responsively, appropriately and sympathetically live according to the self that is constructed" (Grant, 2010, p. 378). This look in the mirror will inevitably show some dark sides of oneself. That is quite tough and makes writing autoethnography an utterly self-reflective endeavour.

Reflecting on these events is making me itch; it's uncomfortable. I see a version of myself which I don't like. But I know if I am not alert, I will fall into the same pattern next time. This is not an excuse, but it shows that changing behavioural patterns is not done by recognising them but is an ongoing challenge to establish a new tactful habit. There is potential, but one needs to make an effort to lift it. However, given the attitude towards conflict and tough masculinity in the industry, I wonder what sort of professional embarks on this journey. As one male student said, acknowledging gender inequality and speaking up is sometimes seen as a weakness by other males. When, from the outset, it seems evident that the researcher-managers themselves will turn out to be causing one of the significant gender equality problems, then the researcher needs a sufficient portion of willingness and courage to expose vulnerabilities and own failures to start. Autoethnography's ethical demand often turns on the self. In my case, I embrace a stoic attitude about others (Aurelius, 1992), which also means that I seek to look at what I can do to resolve situations and what coping strategies are available to me, rather than planning another. This also means that I hold myself responsible all the time and problematise my own actions, thinking and behaviour. It's probably best said in Dostoyevsky's words (2009, pp. 376–377): "we are each responsible to all for all [...] In truth, perhaps, I am more than all others responsible for all[...]". Maybe the Dostoyevsky quote signs the biggest challenge of autoethnography: doing it properly, the autoethnographer first and foremost recognises their own responsibility to act. There is no chance to escape; there is no way to be ignorant. It becomes an ethical duty.

Autoethnography's approach to problematise contradicts the problem-solving attitude of practitioners (Antonacopoulou, 2010), which is very common among construction professionals. Because it favours problematising over solution-finding, it requires a change of mindset from the practitioner doing autoethnography. It makes thinking necessary and relates to Rilke's thought. It is more of an approach to life than a research method. One has first to "love the questions" and then to "live the questions" (Rilke, 1929, p. 33) even though they are very uncomfortable.

It's easy to complain about George, but I have to ask myself how I could have made a difference. I could have made the situation better for all participants if I sidestepped George's aggression. I could have taken a second to reflect instead of thoughtlessly replying. It would even be seen positively by Nadine. When I managed to stay calmer, feedback on other disputes shows that staying calm is seen as a strength rather than a weakness. So, the advice from Kamelia's boss to be louder does, at least in these situations, not help. The next time we – Nadine, Mark, George, I and a few others – met in a larger round, I retreated to listening and only intervened when necessary.

> Raised voices
>
> We met for one of these crisis talks. Very early, we discussed faulty window-opening sizes. For a while, I was in the line of fire because George held my company responsible for building the window openings not according to plan. He had already fitted the windows in, and now he claimed they were – because of the faulty openings – wrongly fixed, too.
>
> "But shouldn't you have checked the measures before?" Nadine asked George, somewhat suggesting. Then, in an aggressive tone, George replied that she had told him in one of the earlier meetings that the architect's office had checked them; hence, it was her accountability, too, but not his.
>
> Although she talked calmly, I could sense the anger in her voice when she replied that it was not the case. However, George insisted he was right, putting his louder voice and masculine appearance behind his words.
>
> Nadine did not further elaborate on the matter. Again, she appeared disgusted by the nature and tone of the discussion – it felt like resignation having to fight a loud voice. I knew Nadine was right; she always has been. She knew her building site better than anybody.

Two things are here: do I act similarly in other situations, and how could I have mitigated George's behaviour? I could see myself using aggressive tactics and raising my voice; hence, there are situations in which I do not act much better. So, I am not in a position to judge about George. Like Ellis (2019, p. 51), "I do not wish to whine, cast blame, complain, attack, point a finger at anyone, or claim to have the Truth. I don't want to make others feel bad". I could blame others for their failures, but that's easy but cheap. Hence, as Ellis (2019) did, it is necessary to depict the situation in which events occur but to be critical of oneself for the most part. Again, I am more responsible than others (Dostoyevsky, 2009), particularly when I am putting my thoughts on paper.

During the meeting, I didn't know how to mitigate the tone or move the exchange towards a content-driven agenda rather than an emotionally aggressive one. I felt a responsibility to act upon oit. One might call it an ethical duty. But, of course, I was not in charge of the meeting, nor did I have the central role. Still, I felt I should have done something but lacked the power to do what would have been necessary. Still, I was not alone in that feeling.

> Comforting
>
> Talking about this meeting, a few days later, Karl, another participant, told me about his frustration with George's rude tone and aggressive behaviour, particularly against Nadine. Of course, it did not touch him as much as it frustrated and upset Nadine. He recognised Nadine

seemed to wrestle more with the aggression, personal insults, and emotional discussions. He sought to comfort her by giving positive feedback and listening to her. It was his way to brighten her mood and try to help.

Karl, I and some others as well recognised the inappropriate behaviour of George. However, none of us effectively intervened. The recognition is a good start, and Karl's attempt to comfort Nadine afterwards, too. However, it is not enough to address and mitigate gender inequality effectively.

For this situation, a crucial question remains: what is the best response to behaviours like George's? It appears as if Nadine would have needed some support. However, what kind of support would Nadine have welcomed? I don't know. There is a chance that supporting Nadine could amount to benevolent discrimination. Yet refusing to act, fearing discriminating benevolently against women, isn't an option either. It needs to be balanced and tactful.

Acknowledgement and raising the topic sensibly in discussions may help to find a way to respond as appropriately as possible. As a male construction professional, I see these discussions differently than my female colleagues. That's owed a great deal to gender differences and my male privileges. However, each person has different needs because of our fundamental otherness (Levinas, 1961). Hence, finding the best way to act is each time an individual search. As a student's account showed, for others, aggressiveness and a loud voice can signal something different:

Turn the tables

One of the feedback sessions drew to a close at the BHT Berlin. I sat with three female students and began discussing gender issues in construction project management. One student talked about men raising their voices and starting to shout. "You know", she said, "it's not nice, but when they get louder, I know I touched their weak spot. So that's how they react, and I just have to stick to the point, and I'll find out how to get them. I know I already had won."

This student frames loud responses differently and again shows that a one-fits-all conceptualisation of loud voice and resulting response isn't possible. In this student's case, my envisioned "support" would have to look very different. Its very individualistic nature is the problem of autoethnography. As with ethnography, it deals with individual cases that never provide statistically significant evidence but demonstrate insights that are otherwise difficult to see. Still, readers can relate to these cases; as I discussed ideas for this chapter with some friends and students, they resonated especially but not exclusively with females. It felt like the female students welcomed the effort to discuss gender inequality, while male students showed a bit of defensiveness or unease. For what reason, I cannot say. In any case, it made them think and discuss. That is what autoethnography should do: It "should stretch and trouble the reader" (Alec Grant, 2023, personal exchange).

But not only does the reader feel this stretching and troubling – the autoethnographer needs to experience it, too. Writing autoethnography always produces unanswered questions. In private conversations, I am sometimes accused of not taking a position but remaining vague about my judgements. Not to jump too fast to conclusions is a consequence of doing autoethnography. I begin to understand but often find more questions than answers. Questions raised in autoethnography cannot be answered immediately – if at all. I think, given my own experience, one needs time to digest. One has to wait for the answers; they cannot be given immediately. One has to "love the questions", and someday, one may "grow into the answers" (Rilke, 1929).

## Conclusion

Autoethnography draws researchers' attention to the little things that become important over time (Rilke, 1929). This chapter shows based on the example of studied gender inequality by graduate student Kamelia that these subtleties can reveal a lot also about the autoethnographic researcher. Therefore, if one embarks on an autoethnographic journey, *patience*, *doubt*, *courage* and *self-criticism* are part of the process.

The researcher needs some patience to draw one's attention to the little things that become important over time (Rilke, 1929). This chapter shows, based on the example of studied gender inequality, that these subtleties can reveal a lot about the autoethnographic researcher. Answers to questions cannot be given immediately but need time to take shape (Rilke, 1929). Autoethnography helps one to understand different perspectives and meanings (e.g. the joke or raised voices) while both meanings and perspectives shift. Awareness of this constantly shifting meaning sometimes confuses me but also lets me act more tactfully – "a kind of situated practical knowledge" (Van Manen, 1995, p. 42). One becomes hesitant to jump to conclusions.

To trigger the process of self-reflection and change, doubt is necessary. I still wonder what a woman in front of me needs and expects. The existential otherness must prevent me from submitting the other to my categories (Levinas, 1974). Hence, I will never be able to "really" understand the other. However, that should excuse my inaction, but there is difficulty in applying male theories to women's realities (Cunliffe, 2022). For women, the appropriate response might differ from mine. What I think is right and how I act should be questioned.

One ought to be courageous to learn uncomfortable things about oneself – about one's own failures, mistakes and vulnerabilities, as I did above regarding my ignorance and insensitivity.

It is vital to engage in self-criticism to the level of severe discomfort in order to learn about oneself profoundly. There is no shortcut to engaging with one's own thoughts and actions reflexively. Autoethnography makes you look differently; it makes you question your assumptions; it makes you more cautious about jumping to immature conclusions, and with this healthy insecurity, you tend to act more tactfully.

My example showed that I underwent an ongoing transformation, which according to Hernandez et al. (2022) is inevitable. I didn't choose to change; it happened to me. At the same time, the process forced me to act authentically. It is an effort to overcome one's thrownness (Heidegger, 1927) and to understand events differently, which needs learning and adaptation. This kind of transformation should be part of every reflexive autoethnography.

The example in this chapter shows that my autoethnographic worked led me to be aware of my responsibility to do something about gender inequality. To learning about gender inequality puts an ethical duty on the researcher. I understood, there is always something I should or could have done. Despite this transformation, I also realised that the case of gender inequality remains an uphill battle against a deeply ingrained culture which is not changed immediately and easily. This still makes it difficult for me to see how I can – through my different roles – (re-)act appropriately and tactfully in various situations.

## Note

1 The names in this and the following vignettes are changed.

## References


Allen Collinson, J. (2013). Autoethnography as engagement of self/other, self/culture, self/politics, and selves/futures. In S. L. Holman Jones, T. E. Adams, & C. Ellis (Eds.), *Handbook of Autoethnography* (pp. 281–299). Oakland: Left Coast Press.

Anderson, L. (2006). Analytic autoethnography. *Journal of Contemporary Ethnography*, 35(4), 373–395. https://doi.org/10.1177/0891241605280449

Antonacopoulou, E. P. (2010). Beyond co-production: Practice-relevant scholarship as a foundation for delivering impact through powerful ideas. *Public Money & Management*, 30(4), 219–226. https://doi.org/10.1080/09540962.2010.492182

Atkinson, P. (2006). Rescuing autoethnography. *Journal of Contemporary Ethnography*, 35(4), 400–404. https://doi.org/10.1177/0891241606286980

Aurelius, M. (1992). *Meditation* (A. S. L. Farquharson, Trans.). New York: Alfred A. Knopf.

Bukowski, C. (1994). *Selected Letters* (2nd ed.). Boston: Black Sparrow Press.

Chiseri-Strater, E. (1996). Turning in upon ourselves: Positionality, subjectivity, and reflexivity in case study and ethnographic research. In P. Mortensen, & G. E. Kirsch (Eds.), *Ethics and Representation in Qualitative Studies of Literacy* (pp. 115–133). Champaign: National Council of Teachers of English.

Crascall, L. (2021). *Re-Imagining Built Environment Education: Investigating The Pracademic Through Autoethnography* [Anglia Ruskin University]. Newcastle, UK.

Cunliffe, A. L. (2022). Must I grow a pair of balls to theorize about theory in organization and management studies? *Organization Theory*, 3(3), 1–28. https://doi.org/10.1177/26317877221109277

Delamont, S. (2009). The only honest thing: Autoethnography, reflexivity and small crises in fieldwork. *Ethnography and Education*, 4(1), 51–63. https://doi.org/10.1080/17457820802703507

Derrida, J. (1967). *Writing and Difference* (A. Bass, Trans.). Routledge. (L'écriture el la différence [Translation published 1978]).

Dostoyevsky, F. (2009). *The Brothers Karamazov* (C. Garnett, Trans.). The Lowell Press.

Ellis, C. (1999). Heartful autoethnography. *Qualitative Health Research*, 9(5), 669–683. https://doi.org/10.1177/104973299129122153

Ellis, C. (2019). Failing to communicate in a communication department. In J. Brower, & W. B. Myers (Eds.), *Critical Administration in Higher Education: Negotiating Political Commitment and Managerial Practice* (pp. 49–76). Lanham: Lexington Books.

Encheva, K. (2022). *Gender inequalities in the German construction industry: Interpretations and recommendations of a female insider* [Master Thesis]. Berliner Hochschule für Technik.

Encheva, K. (2023). Gender Inequality in the German Construction Industry: Interpretations and Recommendations of a Female Insider. Proceedings of the 39th Annual ARCOM Conference, 4–6 September 2023, Leeds, UK.

Feucht, F. C., Lunn Brownlee, J., & Schraw, G. (2017). Moving beyond reflection: Reflexivity and epistemic cognition in teaching and teacher education. *Educational Psychologist*, 52(4), 234–241. https://doi.org/10.1080/00461520.2017.1350180

Galea, N., Powell, A., Salignac, F., Chappell, L., & Loosemore, M. (2021). When following the rules is bad for wellbeing: The effects of gendered rules in the Australian construction industry. *Work, Employment and Society*, 36(1), 119–138. https://doi.org/10.1177/0950017020978914

Grant, A. J. (2010). Writing the reflexive self: An autoethnography of alcoholism and the impact of psychotherapy culture. *Journal of Psychiatric and Mental Health Nursing*, 17(7), 577–582. https://doi.org/10.1111/j.1365-2850.2010.01566.x

Grant, A. J. (2023). Personal exchange in May 2023.

Grant, A. J., & Zeeman, L. (2012). Whose story is it? An autoethnography concerning narrative identity. *The Qualitative Report*, 17(36), 1–12. https://doi.org/10.46743/2160-3715/2012.1735

Grosse, H. (2022). Exploring Sustainability and Identities from an Investor's Perspective: An Autoethnography. Proceedings of the 38th Annual ARCOM Conference, 5–7 September 2022, Glasgow, UK.

Grosse, H. (2023). Thrown into a construction dispute: the personal toll of aggressive claims management. Proceedings of the 39th Annual ARCOM Conference, 4–6 September 2023, Leeds, UK.

Heidegger, M. (1927). *Sein und Zeit*. Tübingen: Max Niemeyer Verlag.

Hernandez, K.-A. C., Chang, H., & Bilgen, W. A. (2022). *Transformative Autoethnography for Practitioners: Change Processes and Practices for Individuals and Groups*. Saratoga Springs: Meyers Education Press.

Herr, K., & Anderson, G. L. (2015). *The Action Research Dissertation: A Guide for Students and Faculty* (2nd ed.). Thousand Oaks: Sage.

Hoffman, P. (1993). Death, time, history: Division II of being and time. In C. Guignon (Ed.), *The Cambridge Companion to Heidegger* (pp. 195–214). Cambridge University Press. https://doi.org/10.1017/CCOL0521385709.008

Jones, K. (2019). *A Framework for Assessing Interventions to Promote the Implementation of Material Innovations on Construction Projects* [University College London]. London, UK.

Kanjanabootra, S., & Corbitt, B. (2016). Reproducing knowledge in construction expertise: A reflexive theory, critical approach [Article]. *Construction Management & Economics*, *34*(7/8), 561–577. https://doi.org/10.1080/01446193.2016.1151064

Large, W. (2008). *Heidegger's Being and Time*. Bloomington: Indiana University Press.

Levinas, E. (1961). *Totality and Infinity* (A. Lingis, Trans.). Duquesne University Press. (Totalité et Infini: Essai sur l'extériorté [Translation published 1969])

Levinas, E. (1974). *Otherwise than Being, or, Beyond Essence* (A. Lingis, Trans.). Duquense University Press. (Autrement qu'être ou au-delà de l'essence [Translation published 1998]).

Madison, D. S. (2006). The dialogic performative in critical ethnography [Article*]. Text & Performance Quarterly*, 26(4), 320–324. https://doi.org/10.1080/10462930600828675

McCarthy, C. (2021). *Towards behavioural equality: the impact of employee perception of justice on employee attitudes towards company equality approaches in large UK main contractor companies* [Loughborough University]. Loughborough, UK.

Mlecnik, E. (2016). Activating the adoption of innovation: Lessons from a passive house network. *Built Environment Project and Asset Management*, 6(2), 205–217. https://doi.org/10.1108/bepam-08-2014-0036

Mykhalovskiy, E. (1996). Reconsidering table talk: Critical thoughts on the relationship between sociology, autobiography and self-indulgence [journal article]. *Qualitative Sociology*, 19(1), 131–151. https://doi.org/10.1007/bf02393251

Pillow, W. S. (2003). Confession, catharsis, or cure? Rethinking the uses of reflexivity as methodological power in qualitative research. *International Journal of Qualitative Studies in Education*, 16(2), 175–196. https://doi.org/10.1080/0951839032000060635

Richardson, L., & St. Pierre, E. A. (2005). Writing: A method of inquiry. In N. K. Denzin & Y. S. Lincoln (Eds.), *The SAGE Handbook of Qualitative Research* (3rd ed., pp. 959–978). Thousand Oaks: Sage.

Riemer, J. W. (1977). Varieties of opportunistic research. *Urban Life*, 5(4), 467–477. https://doi.org/10.1177/089124167700500405

Rilke, R. M. (1929). *Letters to a Young Poet* (M. D. Herter Norton, Trans.; F. X. Kappus, Ed. Revised ed.). W. W. Norton & Company. (Briefe an einen jungen Dichter [Translation published 1954])

Ronai, C. R. (1995). Multiple reflections of child sex abuse: An argument for a layered account. *Journal of Contemporary Ethnography*, 23(4), 395–426. https://doi.org/10.1177/089124195023004001

Sparkes, A. C. (2002). Autoethnography: Self-indulgence or something more? In A. P. Bochner, & C. Ellis (Eds.), *Ethnographically Speaking: Autoethnography, Literature, and Aesthetics* (pp. 209–232). Lanham: Altamira Press.

Sveningsson, S., & Alvesson, M. (2003). Managing managerial identities: Organizational fragmentation, discourse and identity struggle. *Human Relations*, 56(10), 1163–1193. https://doi.org/10.1177/00187267035610001

Thiel, D. J. (2005). *Builders: The Social Organisation of a Construction* Site (Publication Number U206485) University of London: London School of Economics and Political Science). London.

Van Manen, M. (1995). On the epistemology of reflective practice. *Teachers and Teaching: Theory and Practice*, *1*(1), 33–50. https://doi.org/10.1080/1354060950010104

Whaley, A. (2020). *Conflicts and Claims Culture in Construction: An ethnographic investigation into conflict over claims in the GCC construction industry* (University of Salford). Salford.

Wolcott, H. F. (1999). *Ethnography: A Way of Seeing*. Lanham: AltaMira Press.

# 22

# CONCLUSION

*Léon olde Scholtenhuis and David Oswald*

*University of Twente, Enschede, The Netherlands*

*RMIT University, Melbourne, Australia*

The construction industry is a people industry. As Sir John Egan once reported in the United Kingdom: 'much of construction does not yet recognise that its people are its greatest asset and treat them as such' (Egan, 1998, p. 14). To understand the industry, to realise its potential, and to make change, we should still remember to focus on the people of the construction industry. It is the people that make up construction organisations, the practices they follow and the sites they create. They make the workplaces of the construction industry unique places that dynamically change with time. Within local physical contexts and institutional arrangements these people continually shape and reshape situated, local work cultures.

The great construction management research debate of the mid-late 1990s stressed this point again when Seymour and Rooke (1995; Seymour et al. 1997) argued that the 'objects of study' in construction management research should be people. People attribute meaning, both to what they do and to the presence of the researcher. Their meaning systems are socially constructed and, thus, reflected in the realities of construction management practices. In the social sciences we use interpretative methods to study peoples' perceptions, their sensemaking and meaning systems.

This argument, for a greater role for social sciences in construction, still has a place today. It was Seymour and Rooke, who stated almost three decades ago, that: 'if researchers are to have a role in changing the culture of the industry, then the culture of research must change also' (Seymour and Rooke, 1995). While we have witnessed a slow change through a growth in qualitative and interpretative methods, they still account for only a proportion of research undertaken within construction management (Sherratt and Leicht, 2020). Traditionally, this field has been dominated by thinking that follows what Chan (2020) refers to as engineering ontology: studies were involved with finding factors for improvements in project performance through positivist methods and report about this through statistical tests, and criteria such as objectivity and generalisability. These approaches will not deeply explore the micro practices in everyday construction life and, thus, leave many questions unanswered about peoples' detailed practices and their beliefs. The answers are the 'hidden gems' that are waiting to be exposed. However, construction research has not been tempted to do so often through use of ethnography.

Thus, we call for construction researchers to embrace ethnography – in all its forms. The construction management research community has diversified from the early 'classic' ethnographies – which typically involve many months or years of being embedded within a cultural

DOI: 10.1201/9781003379584-26

setting, participating, observing and writing fieldnotes. This diversification of ethnographic methods (to be embraced) contain those discussed and reflected upon within this book. These include short-term or rapid, long-term or classic, online and digital, multi-sited, inter-organisational, being combined with action research, or are autoethnographies.

This methodological variety supports the argument of Pink et al. (2013, p. 11), who stated that: 'there is no single way to do ethnography in the construction industry, and moreover, no single "right" or "better" way'. Through showcasing the variety of ethnographic methodologies within our book, we build on this movement towards embracing ethnographic diversification within construction and demonstrate how ethnographic research designs are being rethought and implemented to still achieve the richness of knowledge that follows from 'classic' anthropological approaches (but in less time).

We have presented how ethnography is being applied in innovative, diverse and flexible ways within the construction industry. These applications adapt to the reality that 'classic' ethnographies are often not viable, since extended periods of time in one location may not be possible for the researcher, or because of the temporal and dynamic nature of construction where there is movement of people and places.

These recent innovative and flexible approaches have encouraged immersive and intense encounters to reach data that had previously been almost left untouched. For example, Galea (Chapter 4) adopted twinned ethnographic approaches that reflect on fieldwork from researchers with different social identities to further expose the details of the gendered nature of construction. Other innovative approaches such as video ethnography show how complementary data can be collected for later reflection and analysis (see Brett, Chapter 10), and to create intense and immersive participatory encounters (see Oswald and Lingard, Chapter 8). Innovation also is possible with respect to researched themes, as Pink (Chapter 9) shows. She used ethnography to explore potential futures for construction robotics by studying activities, materialities and identity of an innovative construction robotics firm.

While we argue that diversification should be 'embraced', that does not mean 'accepted' unconditionally. When deciding about ethnographic methods, researchers need to think about the philosophical underpinnings of their work (see Voordijk, Chapter 7) and ethics, when designing and executing their research projects (see Hayes and Maslen, Chapter 6). Further, new forms of ethnographic research should still include hallmarks of ethnography, such as immersion, reflexivity and observations. The previous chapters have exhibited these hallmarks and draw important conclusions to advance the innovative use of ethnography in construction research.

This concluding chapter addresses two final questions that guide the advancement of ethnographic research in the future. We continue addressing these in the two main sections entitled: *why embrace ethnography?* and *what to consider when doing ethnography?*

## Why embrace ethnography?

Over the past decade, influences from sociology and management research directed attention to analysis beyond the 'what' (i.e., the identification critical factors) to why and how things in practice emerge and what impact this has (Chan and Räisänen, 2009). These questions were central to the first book on ethnography in the construction industry by Pink et al. (2013) and form the basis for all chapters in *Embracing Ethnography*. Below, we propose five reasons of why ethnography should be more widely embraced by the construction research community to investigate the essential why and how questions.

*Ethnography provides in-depth understanding into the local culture of construction work.* The topic of culture in construction tailors to ethnography given it is a research approach that is rooted in anthropology. The background, context, practices, values and behaviours that are required for an understanding of culture can be explored through ethnographic approaches. The insights that can be gathered from the Western society context chapter (see Porter, Chapter 2), where work is masculine and male-dominated, compared to an Eastern society context (see Swider, Chapter 3), where almost all work is informal and female participation is much higher, begin to indicate what complexities can be captured. These Eastern and Western cultural differences exemplify clearly how contextual elements influence construction culture within a particular setting.

*Ethnography is conducted close to action and within the natural setting of the construction industry.* This provides opportunity to deeply understand the complexities associated with an industry that is dynamic, fragmented, multi-sited, has multiple contractors, and has physical site changes as work progresses. This is not limited to the 'natural setting' of operational processes taking place at construction sites. As Chapter 13 (Martinius and Volker) shows, ethnographic work can cover various organisational layers and places. It exemplifies that multi-site ethnography can help grasp the – sometimes contradicting – realities at project offices, construction sites and home offices in inter-organisational settings of utility construction. Further, Chapter 12 (Yunyan Jia) expands this notion and suggests that multiple fieldsites could be studied within the scope of one ethnographic project and that these sites not only include multiple places, but also different hierarchical levels, communication channels and moments in time.

*Ethnography provides insights on topical construction management challenges that have been left mostly untouched due to the prevalence of traditional positivist approaches.* This is partly because ethnographic approaches provide immersive encounters, which help access types of detailed data that traditional methods could not gather. For example, gender inequality within construction is a topical challenge that can be deeply studied through ethnography. In Chapters 4 (Galea), 11 (Oswald, Smith and Sherratt ), and 21 (Grosse), we show how the male-dominated culture influences inequality through, for instance, aggressive and sexist behaviour and commentary. Chapters 6 and 21 show particularly how positionality and identity of the researchers themselves can influence the extent to which they are exposed to this behaviour, as well as the challenges with women gaining acceptance and becoming insiders within the field.

*Ethnography identifies where change is needed to improve practice, proposes how changes can be made, and explores the impact of implemented change.* In particular, when combined with action research, ethnographic approaches offer an opportunity for pragmatic and practical improvement in construction processes. By combining action research with ethnographic work, researchers gain a detailed insight in the lifeworlds of professionals and iteratively improve these through intervention. We show that ethnographic action research can culminate in information systems that structure coordination between project teams and collaboration workshops (Chapter 14, olde Scholtenhuis) but also in strategic improvements towards productivity challenges (Chapter 18, Raiden and Manu). Other forms of ethnography can also trigger change. For example, in Chapter 15, Ongodia and Hartmann showed how their use of ethnographic video interviews provided detailed insight in the workflows of project engineers. The method helped them explore how historical knowledge could be re-used for improvements in early design stages of a mega-project. Further examples include autoethnographies that triggered an activist identity to make change regarding gender inequality (Chapter 21, Grosse) or reflection to strive for changes in built environment education curricula (Chapter 17, Crascall and Sherratt).

*Ethnography builds theory by challenging existing assumptions, while also making practical contributions to knowledge.* Critical theory can be built through ethnography by, for example, through exploring practices of health and safety (Chapter 8, by Oswald and Lingard). This led to findings that contrasted 'with other theories that emphasize the development of worksite safety through the enforcement of universal rules with little context and flexibility' (Garza, 2021, p. 1). Further, they even found that, while frequent H&S communication in practice and research is usually assumed to be positive for H&S, some construction workgroups explained they had no H&S issues and that they did not send H&S messages frequently (see Chapter 19). This again contrasted theory by suggesting that infrequent (rather than frequent) H&S communication was good, as there were no problems to report.

## What to consider when doing ethnography?

With ethnography gradually becoming more mainstreamed in construction research, there are still plenty of considerations to ensure that the quality of the research approach is critically contemplated. Even an experienced expert ethnographer such as Pink (Chapter 9) explains that her methodological skills are under continuous development and evolve depending on the contexts she is in.

Below, five considerations are presented that can equip construction researchers with input to further the development of their ethnographic approaches. These considerations relate to philosophy and ethics in research design, researcher-researched relationships, data gathering and writing, the researchers' role and continuous reflection.

### *Contemplate research philosophy and ethics while planning ethnographic encounters*

Ethnographic work involves following the opportunities that occur at the fieldsite and being dependable on what key informants wish to share. Despite this spontaneous nature of the approach, the methodology still requires a systematic dealing with data collection, sensemaking, analysis and reporting. Philosophical and ethical considerations lie at the back of this process and help plan immersive encounters, and decide on data gathering techniques.

Chapter 7 (Voordijk) addresses three philosophies that are relevant to ethnographic work in the construction context. Voordijk explains that ethnography can be rooted in phenomenology, hermeneutics and critical theory. Influenced by their perspectives, researchers could be focusing on describing the details of lifeworlds through phenomenological lenses and narratives, interpreting human engagement and their sensemaking processes on fieldsites through hermeneutics, or adopting a critical stance to challenge the taken for granted beliefs, ideologies and discourses. Determining where your research fits philosophically will help make consistent choices in data collection, analysis, and reporting later.

Another key consideration in the research design is ethics. Hayes and Maslen (Chapter 6) suggest that researchers should not just see ethics as a bureaucratic or administrative process that involved filling forms and obtaining approval from universities' ethics committees. Instead, it is a planned process that involves approval, implementation and continuous reflection – especially if unexpected issues emerge. Hayes and Maslen argue that anticipating ethical issues is still particularly important when planning ethnographic research. This includes thinking about how to minimise harm, preserve privacy of the researched, respect their autonomy, honour their own commitments, and be fair. Following ethical principles is often also straightforward but should be seen as a continuous process throughout that not ends at the project planning stage.

Minimising harm is one of the foremost ethical concerns in research. Construction researchers should avoid that participants become exposed to physical risks of, for example, being hit by heavy machinery or falling from heights. Hayes and Maslen stress that the type of qualitative research that ethnographers conduct is often not harmful. Researchers do, however, need to think about how they cope with health issues and their role as researcher. Do they, for example, intervene when they observe unsafe practices onsite? Or do they stay in their 'shadowing' role? These questions can be thought of already during research design stages.

Ethics also includes privacy of the researched, as is discussed in Chapter 16 (Osburn, Snider, Dossick). When the COVID-pandemic started, they moved their study to an online ethnography, this necessitated them to think about how they cope with ethics (e.g., intentions or research, participant autonomy, consent), when being part of a digital collaboration platform where they could access several types of visual meeting data (e.g., recorded meetings) and project documents. Another moment where ethics need to be considered is with innovative methods such as videotaping. Videos become more appealing and accessible ways to collect data, but they also may complicate the dealing with anonymity, consent and privacy of the participants (see Chapters 5 and 9). It may, for example, be challenging to obtain consent from all participants that are videotaped.

On balance, all ethnographic methods pursue research where high-quality data gathering is prioritised through rich encounters. Ethics need to be dealt with responsibly but flexibly in this process. Ethics always need to be observed, but sometimes the administrative means to justify that researchers have 'conducted their work ethically' is not practical. Keeping the ethics process as non-intrusive as possible will then benefit the research. Chapter 10 (Brett) proposes approaches such as 'ethics on the fly', where ethics and participant information sheets were not shared at the outset of a study, but while the researcher engages with informants and as the moment was opportune, these documents were presented to them.

### *Anticipate variation in depth and rapport with people in the field*

Ethnography focuses on the informants, participants, gatekeepers and other types of stakeholders. These words represent categories of the researched people that feed construction ethnographers the narratives to build theories. While the words may sometimes be used interchangeably, the involvement and rapport building with each of these people may be different (Chapter 11, Oswald, Smith and Sherratt). Gatekeepers can have short encounters with the ethnographer. These could be people from senior management that only give permission to the site, and managers that introduce researchers to speak to crew on construction sites (e.g., Chapters 12, 13, and 20). Participants and informants, on the other hand, can be people with whom the researcher spends considerable time while having long conversations, ethnographic interviews and other collegial interactions.

Researcher-researched relationships are essential to following the social trail to knowledge. Valuable relationships should be built through what Chapter 12 (Yunyan Jia) refers to as empathy. She explains that the researcher continuously positions their empathic self in diverse and changing situations to get most out of them. This involves feeling how certain responses, questions and requests are appropriate in given situations. It is essential to create legitimacy first, before furthering such interactions by the researcher.

In general, researchers need to decide where to focus on when at the fieldsite and follow those participants they wish to engage. Building rapport with them is mostly essential, as the deep interactions with them produce the data needed for analysis and reflection. While spending time to establish rapport is needed, it is also valuable to realise that these participants typically have busy roles in a production-pressured industry. Critically, the researchers need to be considerate about their impact in these forms of ethnography, since they would not normally want to overly intrude with the participant's work. Intrusion can damage the relationship with the participant and even block access to valuable data.

Establishing rapport can be more of a challenge in shorter-term or rapid forms of ethnography, where there is limited time to answer particular research questions. Thus, in short-term ethnography researchers may need to make pragmatic decisions about their presence and access to data. Brett (Chapter 10) illustrates this and shows how asking targeted questions to participants (relevant to their research study aims) can still be a suited data collection technique in such a situation.

While being mostly at the background of the investigated fieldsite, researchers themselves sometimes also need to be responsive to actions and expressions from participants. So, while the researched may feel vulnerable to being exposed and looked after, the researchers also need to have mental resilience during their encounters with people in the field. Galea (Chapter 4), Oswald (Chapter 11), olde Scholtenhuis (Chapter 12) and Grosse (Chapter 21) show, for example, that researchers' identities associated with students, apprentices and woman can invoke confrontational or even intimidating reactions from participants. These reactions may be exactly what the researcher investigates and wants to describe at length, but they could also be moments that require a response from a researcher that wants to bring about change to unwanted practices. They would need to decide how their role allows them to respond to those situations: shall they intervene with inappropriate behaviour and risk being obstructed from further rich data collection, or shall refrain from an activist role and exactly describe the event at length without disrupting the natural process? These considerations are important to continually address and reflect about.

### *Gather data through multiple methods, and write continuously and extensively*

Embracing the various forms of ethnography allows researchers to choose from a wide range of data gathering methods, including fieldnote taking, photography, videotaping, making online screenshots, accessing physical documents and conducting interviews. There is no customary practice or rule that prescribes which method needs to be used when. The situational nature of the research process requires the researcher to continually determine about what means of gathering data is effective. This not only means that the method should be appropriate for the researched subject, but also that it needs to assemble the data in ways that are rich enough to later consult for analysis and reporting.

Throughout this book, various examples – such as fieldnotes and picture (Chapter 20, Van den Berg) – are given that can absorb rich context even from the outset. A variety of data gathering techniques is encouraged to ensure that patterns can be confirmed and that key details that cannot all be memorised are stored. Brett (Chapter 10) illustrates such a technique by showing how video can supplement traditional fieldnote taking. Especially in his short-term ethnographic work, where time was limited to be onsite, video allowed him to revisit his short, intense encounters and experiences. Other means of data collection, such as participatory observations and interview approaches, can also produce powerful encounters. At these stages of the data collection period, rapport is usually established with participants and the researcher is familiar with the context the interviewees will be discussing (Chapter 19, Lingard and Oswald). We also give examples of where participant observations are combined with action research activities such as designing focus groups, workshops and technology evaluation sessions (Chapters 14, by olde Scholtenhuis, and 18, by Raiden and Manu).

In many ethnographic approaches, multiple methods could be combined for data gathering. By collecting data through different methods, triangulation (Hartmann et al., 2009) can be achieved. This is important and stressed by Osburn et al. (Chapter 16) for the context of online ethnographies, where data online collection appeared insufficient to capture informal dynamics of project teamwork.

Beyond data collection, supplementary fieldnotes and reflexive accounts should be produced after researchers have spent time periods at the fieldsite. These pieces are an integral part of the research process itself. Martinius and Volker (Chapter 13) provide examples of such data in the form of vignettes, while Crascall and Sherratt (Chapter 17) and Grosse (Chapter 21) show reflections on their experiences and how these writings have become a strategic part of their research.

Within the field of construction management, and within this book, the writing process, presentation and reporting of ethnographic accounts are varied. With the methodology being applied by researchers with anthropological, management and engineering backgrounds, the writing styles differ across the various forms of ethnographic research. For instance, scholars using ethnography to identify conditions and rules to support site decision-making (Chapter 20, Van den Berg), use it in the context of management practice (Chapter 2, Porter; and Chapter 17, Crascall and Sherratt), combine it with action research or system development (Chapter 14, olde Scholtenhuis, and Chapter 15, Ongodia and Hartmann) and communicate using unambiguous language and brief formulations. This differs from the narratives and reflections that are formulated in other chapters that use 'anthropology-inspired' writing styles (e.g., Chapters 9 and 13 by, respectively, Pink and Martinius and Volker).

While we need to embrace this variety within the multidisciplinary field of construction research, it always remains important to remind scientific authors that ethnographic writing is not *only* an opportunity to share rich insights and highlight main contributions, but that the writing needs to demonstrate methodological rigour. Irrespective of the writing styles and structure, it is hence essential for ethnographic research papers to show that rapport was established and that self-reflexivity was deep and continuous throughout the research.

### *Be aware of what roles to take while in the field*

Ethnographers need to carefully consider their role in the field and to be self-reflexive throughout the research. They should be mindful of the ways in which their position within the natural setting influences the ongoing processes, but also account for their own backgrounds (positionality) and how that shapes their biases.

Whether the adopted role either helps establishing rapport, or inhibits building relations, can depend on how participants' view your positioning within the setting. Oswald, Smith and Sherratt (Chapter 11) reported success in establishing rapport when creating more structure around the researcher's role that fit in with the natural setting. This included: writing fieldnotes on his laptop in the office setting to fit in with the other workers, going on structured site walks with the safety team, and to have the same work hours and schedules as the construction professionals. While that led to cancelling all academic commitments on fieldwork days, it reduced the participants expressing that he was a lazy part-timer, who just chatted and observed and would come and go as he pleased.

Critical consideration on the ethnographers' role and continuous reflectivity throughout the fieldwork offers opportunity to collect deeper and more meaningful data. For instance, the twinning approaches for immersive encounters used by Galea combined reflections of ethnographer with different gender identities, revealing insights in the gendered nature of construction rules and practices (Chapter 4). Further, differentiation between intense, immersive and a minimally intrusive role taken in the ethnographic action research of olde Scholtenhuis (Chapter 14) shows the importance of the researchers' own awareness of how one shapes a rigorous research process. Next, Grosse's work (Chapter 21) demonstrates that autoethnography highlights changes in the self-perception of the researcher, and by this changing the natural setting, altering his responses to undesirable situations in the field.

### *Reflect and acting with respect to place, time and people*

Construction takes place in projects, at material production plants, engineering offices, client offices and on the roads where stakeholders travel between these different places. Place is thus prominent in construction ethnography. Places or sites, as (Chapter 14) Jia refers to in the context of multi-site ethnography, can also mean more than just physical location. It can include the different levels within the organisational hierarchy, different channels of communication, different points in time on the project and various subcultures within an organisation. The interfaces between these levels could all be visited and compared in ethnographic work.

Other than place, the elements of time and people also shape natural settings in construction. When a construction project passes from design to engineering, planning, execution, operation and demolition phases; clients and designers leave the stage while contractors, material suppliers, operators and supervisors gradually gain prominence. Throughout time, ethnographic encounters cope with a lack of participants willing to share (as rapport has not yet been built), recurrent insider and outsider dynamics within the natural setting, and a continuous effort to secure access to these changing fieldsites. In Chapter 13 the researcher had to negotiate access multiple times and often reflected on characteristics of a multi-site field to decide where she would gather data. It meant that repeated observations and reflections with people at places 'where the action took place' were needed to develop patterns and discriminate between different logics.

Time also plays a vital role in ethnographic action research. In this approach, action research interventions must be proposed timely to achieve intended outcomes (Chapter 14). For example, collaboration workshops on projects should align with actions of participants to contribute to the processes at the fieldsite and build practical knowledge with practitioners (Chapter 18).

Place and time may suddenly be disrupted because of crises. In case of the global COVID-19 pandemic, this not only changed research methods by moving online to gather data, but also the very work culture project team altered themselves rapidly. These changes were not reported upfront and required strong improvisation by researchers. Chapter 16 shows, for example, that online ethnography during the crisis complicates elements of place. Spotting valuable project activities, following the social trail of information, establishing rapport with people and accessing informal data through hallway conversations, for example, became limited when work could be executed solely on digital collaboration and information exchange platforms.

## Closing remarks

Ethnographic research within construction continues to grow from a few sporadic studies pre-millennium to a broader range of methods and application areas. This book demonstrates what is possible through ethnography and serves as an in-depth complementary addition, to the short methodological sections of the few ethnographic articles that are published in contemporary construction journals. Through the examples in this book, we allowed greater focus on the ethnographic researchers' journey, and we hope to have approached a point where ethnography is no longer disregarded through a positivistic lens as risky, unstructured or lacking rigour.

This book discusses, reflects and demonstrates many distinct types of ethnographies ranging from short-term or rapid, long-term or classic, online or digital, multi-sited, and inter-organisational to autoethnography, ethnographic interviews and ethnographic-action research. Construction researchers should embrace these different forms that can all lead to new routes of knowledge. We call to embrace new ethnographic research designs, which do not only innovate with respect to the type of ethnography, but also regarding positionality (e.g., twinned researchers), place (e.g., inter-organisational) and immersion (e.g., intense encounters through participatory video).

In embracing the different types of ethnography, researchers coming from engineering, management and social science fields can learn from one another. Engineering researchers may be inspired by the variety and possible depth that new ethnographic approaches provide when exploring a local culture in the context of intervention and management change research. Alternatively, anthropologists may benefit from the pragmatism and change-making roles that emerge when combining ethnography with engaged techniques for practical research.

Construction researchers aspiring to use ethnography may initially feel resistance because their lack of experience with the method, or their lack of time to conduct long-term classic ethnography. This book shows, hopefully, that a variety and flexibility of methods can help also aspiring ethnographers to start studying construction. This may start basically with some form of close observations of participants to collect data, a subsequent reflection on the position of the researchers, and a reporting through rich descriptions of the researchers' encounters. The variation of additional methods to further use is wide and allow for intense interactions, short encounters, invasive and less invasive processes, and gathers data in a multitude of ways.

While the debate of the mid-late 1990s on interpretivist research has faded, we argue that the attention should now shift to one of how to assess the quality of ethnography, rather than of defending the credibility of ethnographic approaches. The methodology offers complementary nuances, insights and subtleties to the body of construction knowledge. It is flexible, innovative, diverse, and should be judged through its own assessment criteria – not a positivist framework. This means that focus should re-shift to the assessment of ethnographic work in all its various forms. We propose this includes not only include assessment of work *by ethnographers*, but also how *construction journals* and *funding organisations* respond to it. Directions may be needed for editors, reviewers and funders to help assess quality of the various possible ethnographic approaches covered in this book.

Lastly, we notice that ethnography may not be equally widespread yet across the globe. This provides an unexplored territory that can be contrasted with current work. For example, in the

Eastern context, ethnography has been applied to a lesser extent, while the characteristics of construction in this setting differ from the West, for example, in terms of the formality of its practices. This represents greater opportunity as some contexts and cultures have had little or no construction ethnographies undertaken and published in mainstream construction journals. There may be other future directions worth considering such as innovative ethnographic encounters that help to investigate the evolving present. This has demonstrated to lead to new, more realistic and even optimistic visions of phenomena such as Construction 5.0 (Chapter 9). This role of ethnography to re-envision new futures and develop alternative theoretical perspectives to worldviews are promising and worthwhile embracing.

## References

Chan, P.W., 2020. Revisiting basics: Theoretically-grounded interesting research that addresses challenges that matter. *Construction management and economics* 38(1), 1–10.

Chan, P.W., Räisänen, C., 2009. *Informality and emergence in construction*. Taylor & Francis, Milton Park.

Egan, J., 1998. *Rethinking construction*. Department of Trade and Industry, London.

Garza, dela J.M., 2021. 2020 Best paper award. *Journal of construction, engineering and management* 147, 6, 01221001–1

Hartmann, T., Fischer, M., Haymaker, J., 2009. Implementing information systems with project teams using ethnographic–action research. *Advanced engineering informatics* 23, 57–67.

Pink, S., Tutt, D., Dainty, A., 2013. *Ethnographic research in the construction industry*. Routledge, London and New York.

Seymour, D., Crook, D., Rooke, J., 1997. The role of theory in construction management: A call for debate. *Construction management & economics* 15, 117–119.

Seymour, D., Rooke, J., 1995. The culture of the industry and the culture of research. *Construction management and economics* 13, 511–523.

Sherratt, F., Leicht, R., 2020. Unpacking ontological perspectives in CEM research: Everything is biased. *Journal of construction engineering and management* 146, 04019101.

# INDEX

FSC
www.fsc.org
MIX
Papier | Fördert gute Waldnutzung
FSC® C083411

Kolibri 360 GmbH
Ferdinand-Jühlke-Straße 7
99095 Erfurt, Deutschland
produktsicherheit@kolibri360.de